THE NEIGHBORHOOD PROJECT

Also by David Sloan Wilson

Evolution for Everyone
Darwin's Cathedral
Unto Others
The Literary Animal

THE NEIGHBORHOOD PROJECT

Using Evolution to Improve My City, One Block at a Time

David Sloan Wilson

LITTLE, BROWN AND COMPANY
NEW YORK BOSTON LONDON

Little, Brown and Company
Hachette Book Group
237 Park Avenue, New York, NY 10017
www.hachettebookgroup.com

First Edition: August 2011

Little, Brown and Company is a division of Hachette Book Group, Inc. The Little, Brown name and logo are trademarks of Hachette Book Group, Inc.

Library of Congress Cataloging-in-Publication Data
Wilson, David Sloan.
The neighborhood project : using evolution to improve my city, one block at a time / David Sloan Wilson. — 1st ed.
p. cm.
Includes bibliographical references and index.
ISBN 978-0-316-03767-9
1. Cities and towns — Growth. 2. Civic improvement. I. Title.
HT371.W55 2011
307.76 — dc22 2011002752

10 9 8 7 6 5 4 3 2 1

RRD-C

Printed in the United States of America

For TCV

Contents

THE NEIGHBORHOOD PROJECT

INTRODUCTION

The Listener

ISAAC BASHEVIS SINGER, the famous novelist, began his life as the son of a rabbi. Not a famous rabbi, just one of many who held court in the crowded Jewish neighborhoods of Warsaw before the cataclysmic events of World War II. Holding court was not as royal as it sounds. It meant consulting with anyone who might tread up the stairs and knock on the door of the rabbi's humble apartment. Young Isaac listened as the visitors poured out their troubles and his father advised them on the basis of a tradition that had been perpetuated in this way for two thousand years.

In one story from Singer's memoir *In My Father's Court,* a woman bursts into their apartment with the astounding news that two geese continue to shriek even after they have been slaughtered and their organs removed. She demonstrates by hurling one carcass against the other, and indeed, an unearthly sound emerges from their hollow bodies. The rabbi, who is inclined toward mysticism, proclaims it a miracle and proof of the Creator. But the rabbi's wife is a rationalist. "Slaughtered geese don't shriek," she insists. She solves the mystery by plunging her hands into the geese and emerging with their windpipes, which the woman had neglected to remove with the rest of the organs. I will let Singer, the master storyteller, take it from here.

> Father's face turned white, calm, a little disappointed. He knew what had happened here: logic, cold logic, was again

> tearing down faith, mocking it, holding it up to ridicule and scorn.
>
> "Now, if you please, take one goose and hurl it against the other!" commanded my mother.
>
> Everything hung in the balance. If the geese shrieked, Mother would have lost all: her rationalist's daring, her skepticism which she had inherited from her intellectual father. And I? Although I was afraid, I prayed inwardly that the geese *would* shriek, shriek so loud that people in the street would hear and come running.
>
> But alas, the geese were silent, silent as only two dead geese without windpipes can be.

Already, as a little boy hiding behind his mother's skirt, Singer was experiencing the tension between rationalism and faith. Events were to obliterate his entire neighborhood and hurl him across the Atlantic, where he wrote as perceptively about the immigrant experience in America as about life in the old country. As a novelist, Singer used the medium of literature, rather than religion, to listen and reflect on the human condition.

I am a scientist who began my life as the son of a novelist. My father, Sloan Wilson, was certainly known in his own place and time, with novels such as *The Man in the Gray Flannel Suit* and *A Summer Place* that were said to define his generation. His experience was very different from Singer's—born in America, fighting in World War II, joining the corporate army after the war, divorce and remarriage, alcoholism and recovery—but he greatly admired Singer. As a boy, I did not experience a parade of visitors consulting my father for advice, but I did experience a torrent of words from my father as he tried to make sense of the world around him. At times, I felt like a trout, flexing rhythmically to keep my place in the rushing stream of his words, listening, listening.

Now I am grown and a scientist, rather than a novelist or a religious sage. Science is the flowering of rationalism, the intellectual tradition that enabled Singer's mother to declare that dead geese don't shriek. Her insistence that all aspects of the world can be explained on the basis of natural processes, without any kind of divine intervention, has proven itself again and again, transforming our world at an ever-increasing pace. Already, my father's America of the 1950s seems as remote as Singer's prewar Poland or the South Sea Islands described by the first European explorers. Not all of the changes are benign. It often seems that we are hurling toward a new cataclysm of one sort or another. Some blame the triumph of reason over faith, as if there is too much science and not enough religion. Others blame the triumph of faith over reason, as if religion is the main culprit. The whole world seems to be engaged in a standoff similar to that of the rabbi and his wife, arguing over the shrieking geese on either side of their kitchen table.

I see things differently. For me, science is a medium for listening and reflecting on the human condition, much like religion and literature. What does it mean to listen? It means acquiring information from the outside world. What does it mean to reflect? It means processing that information to achieve some kind of desirable outcome, such as surviving another day, or, more expansively, making the world a better place. Science can expand our capacities for listening and reflecting. Indeed, unless we use these expanded capacities, there is no hope of solving the problems of modern existence.

It takes an evolutionist to appreciate the feebleness of our natural listening and reflecting abilities. The outside world contains too much information for humans or any other creature to process. Every species on earth is adapted to perceive only the information that matters for survival and reproduction and to be deaf to everything else. For bats, that means listening to the echoes of sound pulses that bounce off flying insects at night. For elephants, it means listening to low-frequency sounds that can be heard for miles. For us, it means listening to a certain portion of the sound spectrum, seeing a certain portion of the light spectrum, detecting only a tiny fraction of chemicals with our senses of taste and smell, and so on. We would never know about such things as electrical and magnetic fields or the existence of outer space if we didn't expand our sensory abilities with the tools of technology and science.

As for listening, so also for reflecting. Every species is adapted to process information only in ways that contribute to survival and reproduction. For bats, this means using the echoes to fly toward and capture their prey. For elephants, it means using information about other elephants within a radius of several miles to guide their movements and other behaviors. But reflection — information processing — is designed to benefit the individual organism, perhaps its family or some other social unit, never the beastly equivalent of the human yearning for "peace on earth, goodwill toward men."

We are unique among all species in our ability to gather and reflect on information, but that does not place us outside the orbit of evolution. *Human diversity is like biological diversity,* because both are the outcomes of evolutionary processes. We are the product of evolution at a variety of timescales. First, there is the timescale of genetic evolution, which is usually regarded as slow but at times can be quite fast. Then there is the timescale of cultural evolution, which is usually regarded as fast but at times can be quite slow. Finally, there is the timescale of psychological processes, which operate over the course of a human lifetime or even within a fraction of a second. When you make a decision, for example, it is often the result of neuronal processes that count as Darwinian, of which you are totally unaware. The tools of science are required to listen and reflect on our inner space — what happens inside our heads — no less than gazing into outer space.

Cultural and psychological evolution differ from genetic evolution in their details, but once we take the differences into account, we can explain human diversity in the same way as biological diversity. I call this the *evolutionary paradigm.* It enables all aspects of humanity to be approached in the same way that biologists approach the rest of life.

The evolutionary paradigm has profound implications for understanding the human condition, the province of religion and literature. Indeed, religion and literature are among the phenomena that need to be understood. We are currently witnessing a coalescence of knowledge about our species comparable to the coalescence of knowledge in the biological sciences during the twentieth century. I am lucky to be part of this movement as a practicing scientist and also to step into my father's shoes by writing about it for the general public, considering themes that have been pondered by religious sages, philosophers, and storytellers throughout the ages.

In this book, I try to play marriage counselor by resolving the conflict that divided young Isaac's mystical father and rationalist mother. I show that logic need not be cold and need not tear down faith. Almost everything that we value associated with faith can be understood scientifically in a way that makes it stronger, not weaker. Science and evolutionary theory can clarify what it means to have a soul in addition to a body. Bodies and souls can transcend the skins of single individuals, making us part of something larger than ourselves. A city can have a body and a soul, for example. The whole earth can become a single body with a soul, although that would be a tall order.

And then, if evolutionary theory can be used to *understand* the human condition, it can also be used to *improve* it. When young Isaac heard a knock on the door of his family's apartment, it was usually not someone who wanted to ponder the big questions but someone who urgently wanted to solve the problems of everyday life. A great religious tradition such as Judaism functions in both capacities. Evolutionary science will not fully prove itself until it, too, can deliver practical answers to the urgent questions that confront us in everyday life. This book is therefore dedicated first and foremost to making a difference in the real world at all scales, from single individuals, to neighborhoods, to cities, to the planet. Our problems have arguably never been greater. Fortunately, the tools of science and evolutionary theory enable us to listen and reflect as never before.

CHAPTER 1

Evolution, Cities, and the World

IT'S EARLY MARCH in Binghamton, New York, and I'm lined up with the rest of the citizenry to watch the Saint Patrick's Day parade. We hold our parade before the real Saint Patrick's Day to avoid competing with the City—New York City—for the marching bands and Irish music groups that will be playing in the bars far into the night. Binghamton is also a city, New York State's sixth-largest, but it's minuscule compared with the City that gets capitalized and needs no qualifiers.

New York is a vast state, and most of it is rural, as I must inform people who ask where I'm from and confuse the state with the City. Binghamton is located 150 miles northwest of the City in Broome County, close to the Pennsylvania border and south of the Finger Lakes, in an area called the Southern Tier. Its population is just less than 50,000—down from a high of 80,000 in the 1970s—and fewer than 200,000 people inhabit the entire county. I live on the edge of the city and can be biking along roads amid fields, forests, and dairy farms within a few miles from my home.

The weather in early March could be anything, but today we are blessed with a fine promise of spring. The sky is cloudless, and the sun warms the air just enough for the braver celebrants to show some skin. I'm not a parade person and wasn't planning to come. I'm the nerdy professor type and was planning to work in the morning and cross-country ski in the afternoon to take advantage of the disappearing snow. Then I got a call from Dan O'Brien, one of my graduate students, who said that I really, really, must come to the parade. If I didn't meet him and my other grads at the Court Street Bridge before the start of the parade at noon, my name would be mud.

As soon as I arrived at the bridge, I realized that Dan was right. A sizable fraction of the county must be here, judging from the throng stretching along Court Street as far as I can see. The long winter is breaking its hold, and these folks are ready to party. Even though I'm not a party animal, it feels great to mingle with this vast ribbon of humanity festooned in green. Binghamton is more culturally diverse than you might think for a city in Upstate New York. The ancestors of the people here could come from literally anyplace on earth, but today we are all Irish.

Dan has a legitimate claim to being Irish on his father's side, but his mother is Italian. According to him, whenever these cultures intermarry, an Irish man marries an Italian woman, because no Italian man could endure a lifetime of Irish food. Today he looks quintessentially Irish in a tweed cap and winks as he hands me an iced-tea bottle for a nip of whiskey. My newest grad student, Ian MacDonald, is of Scottish descent by way of Canada, but that's close enough today. A large man who reminds me of Hagrid in the *Harry Potter* books, if only by virtue of his size and remnant of a brogue, I barely recognize him when he shows up, because he's wearing a neon-green wig and a pair of *Simpsons* shorts over a pair of New York Yankees pajama bottoms. "I love this town!" he exclaims, fitting right in even though he's only been here a few weeks.

Eventually, my other grads arrive, and I snap a photo of them with their arms around one another. They look no different from the other revelers. Their parents and grandparents plied the same trades as those of the others. But they are young scientists, and we are studying the city of Binghamton in a way that no city has ever been studied before: from an evolutionary perspective.

EVOLUTION AND CITIES? I admit that the two words are seldom placed next to each other. For most people, evolution is about the vast stretch of time since the origin of life. It's about fossils, dinosaurs, and human origins. Cities didn't exist until roughly 5000 years ago, a nanosecond on the evolutionary timescale. How can evolution tell us anything about cities?

This reasoning confuses evolution with *genetic* evolution. Moreover, it assumes that genetic evolution must be a slow process requiring hundreds and thousands of generations. As soon as we think about evolution more broadly to include cultural and psychological processes, in addition to rapid genetic evolution that can easily take place within 5000 years, the connection with cities begins to make more sense. After all, evolution is fundamentally about change. Perhaps the accelerating pace of human change reflects evolution in warp drive rather than a mysterious suspension of evolutionary processes.

Change is not necessarily for the better. Just as it is wrong to equate evolution with "slow," it is also wrong to equate it with "progress." Evolution doesn't make everything nice. It results in the full spectrum of outcomes that we associate with good and evil, thriving and decay. The kind of change that we associate with progress *can* emerge as a robust product of evolution but only under certain environmental conditions. With the right conditions, the world becomes a better place. With the wrong conditions, evolution takes us where we don't want to go. That is why we must learn to become wise managers of evolutionary processes.

Cities represent both the best and the worst of modern human existence. They pulse with activity in the arts, learning, and commerce. They are centers of tolerance, where people from all cultures and persuasions can celebrate instead of fighting over their differences. People flock to cities to realize their dreams for a better life. Yet many become trapped in poverty, crime, despair, and prejudice from which there appears to be no escape.

The festive and inclusive atmosphere of the Saint Patrick's Day parade reflected the best of Binghamton. To glimpse the worst, let's take a walk down Court Street on an average day. The older buildings date back to the 1800s and have ornate Victorian charm, but Binghamton reached the peak of its prosperity in the mid-1900s. Now it suffers from the ills that afflict so many cities in America and around the world. Malls clog the arterial highways, and the downtown businesses struggle to survive. People walking purposefully in business suits mingle in equal proportion with those who have nothing to do but wander the streets. A city, like a human body, decays when it is not maintained, and signs of decay are everywhere. People seem to take their cue from disorderly surroundings to behave in a disorderly manner themselves. Even the newly constructed buildings look ugly and ephemeral compared with the older buildings. Aren't we supposed to be the most powerful and affluent nation on earth? Then why does it seem that the inhabitants of Binghamton are squatting on the ruins of an earlier civilization?

I'm studying my city of Binghamton not as an aloof scientist but because I think my expertise can make a difference. Just as we need the physical sciences to build the physical structure of the city, we need evolutionary science to understand and manage the life that takes place within the city. I'm here to provide the science with the help of a few friends, including my posse of grad students. If we can improve the quality of life in Binghamton, then we can become a model for cities everywhere, even big ones like the City. The proportion of the earth's human population that resides in cities recently passed the halfway mark. If evolutionary science can make cities work better, we want to know about it.

WHEN I WAS A GRADUATE student in the 1970s, I never imagined that I would be studying a city. I expected to be an ecologist specializing in the study of zooplankton, tiny creatures that inhabit the open water of lakes and oceans. Even this was not a considered decision but was more like the trajectory of a ball in a pinball machine. From an early age, I loved and admired my famous novelist father but also wanted to escape his shadow. I announced that I would be a scientist, something that he would respect but couldn't really understand. I didn't understand it either and vaguely imagined becoming a brain surgeon or finding a cure for cancer. What I really loved was spending time outdoors fishing, exploring, and daydreaming.

In college, I discovered that being a scientist didn't necessarily require donning a white coat and spending all my time in a laboratory. I could be an ecologist, spending lots of time outside doing what I already loved. An independent-study project on zooplankton in an ecology course caught the eye of my professor, who allowed me to work in his laboratory, which enabled me to work more on zooplankton, which made it a logical career choice for graduate school. If you could rewind the tape of my life and replay it (a famous thought experiment that the late evolutionist Stephen Jay Gould posed for the history of life on earth), I could have become many things, just as that silver ball in the pinball machine never travels the same path twice.

The more I learned about evolution in graduate school, the more I realized that I did not need to restrict my attention to zooplankton. In the last paragraph of *The Origin of Species,* Charles Darwin contemplated the many species inhabiting a tangled bank, from worms in the soil, to the tapestry of plants visited by insects, to the birds and mammals, so different from one another and yet produced by the same laws acting around us. That vision became the modern science of evolutionary biology. When I asked questions about zooplankton, such as how they eat algae or why they migrate up and down the water column, I made predictions about how they might be expected to behave as a product of evolution, based on their history as a species and the environmental forces influencing their survival and reproduction. I then performed experiments to see if real zooplankton confirmed or refuted my predictions. I wasn't always right—nobody bats 1000, in science any more than in baseball—but alternating between prediction and test, prediction and test, enabled me to discover how real zooplankton behave, better than any other method.

The meaning of Darwin's tangled-bank passage was that I could employ the same methods to study any aspect of any species. After all, how zooplankton eat algae and why they migrate up and down the water column are completely different questions, even if they are asked about the same species. If I could use the same methods to answer those two questions, I could also ask how dragonflies eat their prey, why some birds migrate south for the winter while others remain up north, or an infinity of other questions. Darwin's theory of evolution transformed biology, the study of life, from many disparate subjects into a single subject, which accounts for its profound significance. Instead of becoming a zooplankton ecologist, I therefore became an all-purpose evolutionist as a graduate student and have been studying the entire tangled bank ever since—from microbes to humans.

THE INCLUSION OF HUMANS in Darwin's tangled bank was controversial in the 1970s. It was obvious from the start that evolutionary theory would profoundly alter our conception of ourselves, yet the theory became confined to the biological sciences and avoided most human-related subjects for most of the twentieth century. This apartheid was maintained by people who fully accepted Darwin's theory for the rest of life—including our physical bodies and our basic drives for things such as food and sex—but who argued that our rich behavioral and cultural diversity had a life of its own about which evolution had little to say.

The publication of Edward O. Wilson's *Sociobiology* in 1975, the year I obtained my PhD, provides a vivid demonstration of this apartheid. The whole point of *Sociobiology* was to show how there could be a single science of social behavior, based on evolutionary theory, that included everything from ants to primates. *Sociobiology* was celebrated as a triumph except for the final chapter on human social behavior, which was vigorously attacked by people who were not going to allow a breach in the wall.

I was excited by the prospect of including humans in Darwin's tangled bank, even before the publication of *Sociobiology*. I was trying to escape my father's shadow by becoming a scientist, but I was still drawn to the novelist's quest to understand the human condition. My dad described himself as a "middlebrow" writer, without the intellectual pretensions of a highbrow such as Norman Mailer but with more in mind than a lowbrow who's only trying to score a commercial success. I grew up with him pacing the floor, puffing away on his pipe, snapping his fingers to keep rhythm with his thoughts, and drinking too much. Picture me as a boy of fourteen, pressed against the back of my chair in a cocktail lounge, as if pinned by a great wind, as my dad used me as a sounding board to rehearse the themes of his current book. I can attest that he had much more in mind than commercial success. He was trying, above all, to make sense of his own life in a way that would resonate with the lives of his readers.

The moment I realized that evolutionary theory provided a way for me to view the human condition through a scientific lens, the storyteller in me awoke. I knew little about the complicated history that caused evolution to become taboo in academic disciplines such as sociology and cultural anthropology. I only knew that the most enduring themes of human life and literature, such as the eternal conflict between benefiting oneself and serving one's group, were also fundamental evolutionary themes and that insights for the birds and bees might well apply to humans. I regarded the final chapter of *Sociobiology* as uncontroversial and was surprised when it created such a ruckus.

The wall separating the study of humanity from the study of the rest of life did start to crumble in the aftermath of *Sociobiology*. I feel lucky to have been present during this period of intellectual history, when entire disciplines began to be rethought from an evolutionary perspective. Consider that terms such as *evolutionary psychology* and *evolutionary anthropology* weren't even coined until the late 1980s. The study of religion from an evolutionary perspective, a special interest of mine, didn't gather steam until the early 2000s. The first formulations weren't necessarily the right ones. If you disagree with some of the conclusions that initially seemed to emerge from the nascent field of evolutionary psychology, for example, you might be surprised to discover how much I agree with you. Nobody bats 1000, and the game of science is still very much in progress.

BY THE EARLY 2000S, I was comfortably established as a professor at Binghamton University, one of the four major research universities in the State University of New York (SUNY) system. Life was great. My wife, Anne B. Clark, was also my colleague, with an office just a few doors down from my own. Our two kids, Katie and Tamar, who spent their early childhood following us like baby ducklings as we did our work, increasingly were exploring life on their own. Our graduate students and faculty colleagues were also our friends. I traveled widely giving seminars and had a worldwide network of associates. The entire tangled bank was my stage. I danced from topic to topic, species to species, as if disciplinary boundaries didn't exist—not because I'm so special but because anyone who learns the basic principles of evolutionary theory is welcome to join the dance.

Yet something was missing. My rich intellectual life was not yet reflected in higher education. The vast majority of college professors remained within their disciplinary boxes, and the vast majority of college students learned about evolution strictly as a biological subject. The academic world can be disturbingly conservative for people who regard themselves as free thinkers. Decades might be required for the intellectual ferment that I was experiencing to be reflected in textbooks, formal courses, and the thinking of the average college professor.

What would it be like for evolutionary theory to be reflected in the entire culture of a university? The basic principles would be made available to all students early in their academic careers. Humans would be described as part of the tangled bank from the very beginning. Then students would deepen their evolutionary knowledge throughout their academic careers, in every human-related department in addition to the biological sciences. In essence, they would function as my worldwide network of colleagues and I were already functioning. Everyone would dance across disciplinary boundaries, and the university would become a single intellectual community. This has always been the ideal of a liberal-arts education, but it had failed to materialize. Knowledge about humanity remains fragmented into a vast archipelago of academic tribes, each absorbed by its narrow concerns, trading a bit here and there, but mostly oblivious or even hostile to one another. Ironically, the same theory that was excluded for the study of humanity for so long can fulfill the ideal of a liberal-arts education.

This grand vision gave birth to EvoS (for *Evo*lutionary Studies and pronounced as one word), a campuswide evolutionary studies program that I started at Binghamton University in 2003. It also marked the first time that I tried to build something larger and more enduring than myself, a program that could survive and replicate after its builder was gone.

As you can tell from this description, I thought of EvoS as a product of cultural evolution and myself as a manager of evolutionary processes. But I didn't try to implement EvoS as a grand vision. Evolution is typically an incremental process. My edifice would not be constructed all at once from a blueprint like a skyscraper, but grain by grain like a termite mound. Evolution is often described as a tinkerer that builds new structures from preexisting parts. The "parts" of my new EvoS program were the professors at Binghamton University who, like me, were already employing the evolutionary perspective in their teaching and research. Some came from our biology department, of course, although you'd be surprised how many biology professors don't think much about evolution, especially those who study life at the molecular level. Several came from our anthropology department, but other members of the same department were aghast at the idea—they didn't want the apartheid to end in the first place. Anthropologists are famously unable to dance with one another, much less melding with other disciplines. Then there was a sprinkling of faculty from other departments such as psychology, economics, and philosophy, who were beginning to speak the common language of evolutionary theory but had little company among their own colleagues.

EvoS brought these faculty members together and made their courses available to students as a package. Any student could enroll in parallel with his or her major, resulting in a certificate in evolutionary studies along with the degree. They needed to take a certain number of courses with evolutionary content, chosen from the menu of courses offered by the program. In this way, students could tailor their evolutionary training to their particular interests. To engage students early in their academic careers, I turned my upper-level course on evolution and human behavior, which I had been teaching to small numbers of juniors and seniors for many years, into a large introductory lecture course without any prerequisites, which I proudly titled "Evolution for Everyone."

My dean gave me a few thousand dollars to organize a campus-wide EvoS seminar series, which I stretched by cost-sharing with departments that had their own seminar series. Now my EvoS colleagues at Binghamton and I could provide a stage for our worldwide network of colleagues. A talk on the evolution of disease organisms (cohosted with the biology department) might be followed by a talk on the deep structure of the arts (cohosted with the art history department), substance abuse (cohosted with the psychology department), the moral emotions (cohosted with the philosophy department), or the species diversity of warblers (back to the biology department). The EvoS seminar series enabled all professors and graduate students to witness and consider joining the interdisciplinary dance.

I also created a way for undergraduate students to join the dance along with graduate students and professors. A little-known fact about university life is that undergraduate education and graduate education are almost totally segregated from each other. Graduate education is centered on original research, while undergraduate education is centered on formal coursework. A few undergraduate students are lucky enough to become involved in original research, but the vast majority interact with faculty and graduate students only as professors and teaching assistants in their formal courses.

EvoS changed that through the simple device of an undergraduate "current topics" course built around the seminar series. The course involved reading an article by each speaker from the primary literature (written for other scientists, not undergraduate students), writing a commentary on the article due before each seminar, attending each seminar, and attending a social event following each seminar that combined food, drink, and a continuing discussion with the speaker. Undergraduate students enrolled in the EvoS program repeat this experience twenty times as part of the requirement for earning the certificate, providing a vivid demonstration of how many subjects are being approached from an evolutionary perspective at the level of cutting-edge scientific research.

In grain-by-grain fashion, EvoS was built from these humble parts. It did not transform the culture of the university all at once, but it comes close to providing the grand vision of the university as a single intellectual community for those who enter the program. I had a blast teaching my "Evolution for Everyone" course, and my students loved thinking about themselves as part of Darwin's tangled bank. They weren't threatened by the material, even though we boldly considered taboo topics such as such as religion, race, and sex. They didn't need to be science geeks to absorb the basic principles of evolution. This is not just my opinion but something that I have rigorously documented. Courses can be studied scientifically, along with any other observable phenomenon, and I can say with authority that my "Evolution for Everyone" course succeeded with every type of student who chose to enroll — religious and nonreligious, liberal and conservative, freshman and senior, humanities major and science major. Glimpsing the full scope of evolutionary theory is like reading a great novel — nearly everyone resonates.

EVOS CREATED A LEARNING ENVIRONMENT intriguingly similar to a martial-arts dojo that I was visiting with my daughter Tamar. All ages and skill levels were on the dojo floor, from middle-aged white belts such as myself to teenage black belts. We paired up for each exercise, the person with the higher skill level becoming the teacher and the person with the lower skill level becoming the student. Teaching is also a powerful form of learning, as every teacher knows. The sensei, a legendary martial-arts expert named Hidy Ochiai, walked the floor, observing the pairs and occasionally making suggestions. When he demonstrated a kata, it was pure poetry. Bowing and other rituals created a culture of respect that kept everything moving at a brisk pace, even for a bunch of unruly Americans. Individuals could advance at their own pace, and the belt colors provided a tangible, publicly recognized reward to work toward.

As a participant, I was amazed at how motivated I became and how eager I was to abide by the rules. When I earned my yellow belt, I goofed by standing on the wrong side of the sensei during the brief recognition ceremony. Everyone was nice about it and kindly showed me the right way, but inside I burned with shame, and I never repeated the mistake again. What made me so attentive to the rules of right and wrong conduct? The dojo was a lean, mean learning machine, an island of cultural order in a sea of disorder just outside the door. It existed as a tradition that survived, replicated, and adapted to local conditions, perpetuating a complex body of knowledge that would otherwise be lost.

EvoS shared key elements with the dojo, even if it lacked many of its rituals. The mingling of undergraduate students, graduate students, and professors was like the martial-arts students with their different belt colors sharing the same floor. During the discussions that followed the EvoS seminars, I noticed that students new to the program were hesitant to ask questions but listened intently to questions asked by more experienced students. After all, these questions were being posed to a world authority—the speaker—in the presence of their professors. A question was like performing a kata in the presence of the sensei and other black belts. If the question was regarded as perceptive, then the student who asked it was functioning at a very high level indeed. Gradually, the newer students gained the confidence to speak and basked in the public recognition of having their questions treated respectfully. In this fashion, students advanced at their own pace, and some achieved very high degrees of sophistication. Speakers often expressed amazement that undergraduate students attended their seminar and asked such great questions.

Just as I could join the dojo as an out-of-shape adult and learn along with much younger novices, professors at Binghamton who did not receive evolutionary training during their own higher education—which meant most faculty in human-related departments—could become involved and start building their skills by attending the seminars and interacting with other EvoS faculty participants. In this fashion, the number of faculty participants swelled to more than sixty, representing nearly every human-related department on campus. Once they started interacting, they also started collaborating on research projects that transcended disciplinary boundaries. EvoS became such a good incubator for research that it was eventually designated an Institute for Advanced Studies at Binghamton University, in addition to an undergraduate teaching program, and was provided a budget for facilitating new collaborative research.

Like an organism, once EvoS began flourishing at Binghamton, it started to replicate. The second program was initiated by Glenn Geher, a psychologist, and Jennifer Waldo, a cell biologist, at SUNY New Paltz, a four-year college north of the City that is especially strong in the arts. In just a few years, it became one of the most popular programs on campus. Together we received funding from the National Science Foundation to create a nationwide consortium, with more than forty programs at various stages of development as of this writing. EvoS is even becoming international, with a program that was inaugurated at the University of Lisbon in 2011. Each program is built on the template provided by EvoS Binghamton but can also be freely modified to fit local conditions. With many programs, we will be able to compare the different ways in which we operate and select the best practices as a form of managed cultural evolution.

EvoS gave me a taste for building something more enduring than myself. You'd think that it would exhaust my time and energy, but time and energy can be created, not just expended, when you work with other people. In chemistry, a catalyst is a substance, usually used in small amounts relative to the reactants, that modifies and increases the rate of a reaction without being consumed in the process. My programmatic efforts were catalytic in this sense, causing the evolutionary paradigm to spread at my university, the nation, and even the entire planet much faster than it would otherwise. Hoarding my time and energy for myself held little appeal once I appreciated the power of acting as a catalyst.

NOW THAT I WAS FUNCTIONING in catalytic mode, I began to contemplate the idea of studying my city of Binghamton. Evolution is fundamentally about the relationship between organisms and their environment. That's why field studies, which follow the lives of animals as they go about their daily lives, are such an important component of evolutionary research. One of the most beautifully written books on evolution, Jonathan Weiner's *The Beak of the Finch,* describes Peter and Rosemary Grant's field study of the finches on the Galapagos Islands, called Darwin's finches because of the role they played in helping him formulate his theory. Millions of television viewers have enjoyed episodes of *Meerkat Manor,* which follows the lives of those strange skinny mammals in Africa, based on fieldwork led by Tim Clutton-Brock at Cambridge University in England.

Field studies have two objectives that seem contradictory. First, the whole point is to study organisms in relation to their natural environment without intervening. Second, the organisms must be studied with precision, which requires intervening. Individuals must be caught and measured from stem to stern to know what traits they possess and how they differ from one another. They must be marked if they do not already have natural markings that enable them to be identified. Sometimes they must be outfitted with radio transmitters so that they can be reliably located. They must be observed and filmed to obtain a permanent record of their behaviors. Their children must be counted to measure how they fared in the Darwinian contest. While all of this poking and prying is going on, they're supposed to act naturally!

Remarkably, most species do act naturally after they have acclimated to the people and paraphernalia of a field study. If you have watched *Meerkat Manor,* you know that Flower and other members of the Whiskers Clan hop onto the scales to be weighed for a small food reward, climb onto the heads and shoulders of the scientists for a better look around, and otherwise treat the scientists as an unthreatening part of their environment. Some species have even moved into our cities, such as pigeons long ago and deer and crows more recently, where they go about their daily lives in our midst, unconcerned by our presence unless we pose a threat.

I imagined Binghamton as a field site for studying people in modern everyday life from an evolutionary perspective. It would be business as usual for an evolutionist but radically different from the way people are usually studied. Most psychological research is conducted on college students without any reference to their everyday lives. Sociologists and anthropologists are accustomed to studying people in their everyday lives but not from an evolutionary perspective. Armies of scientists join the battle against problems such as poverty, crime, and substance abuse, but each problem tends to be studied in isolation — the very opposite of the tangled-bank approach.

I also had personal reasons for starting the Binghamton Neighborhood Project, or BNP, as it came to be called. Binghamton had been my hometown for more than twenty years, but I was not engaged with it. I hadn't joined the PTA, attended council meetings, given blood, or served turkey to the homeless on Thanksgiving. My heart was in the right place, and I was happy to write checks for good causes, but I reserved my precious time and effort for my work and family. It was enough, I told myself, to raise two healthy kids and work at changing the way the planet thinks about evolution.

My lack of civic virtue was ironic, given that virtue was one of my main interests as an evolutionist. I was best known for reviving a theory called group selection, which explains how altruism and other traits that are "for the good of the group" can evolve, despite being vulnerable to exploitation by more self-serving individuals within groups. My professional career was all about how groups and communities can evolve into adaptive units, but in my own community, I was a slacker! For me, the BNP was a way to walk the walk, not just talk the talk. If evolutionary theory could be used to *understand* the human condition, then it could also be used to *improve* it. This was a personal epiphany for me. I wouldn't be satisfied until I had used my scientific expertise to make my city of Binghamton a better place.

The Binghamton Neighborhood Project can become a model for cities everywhere. I call it a whole-university/whole-city approach to community-based research. EvoS provides the "whole university" part in the form of a network of professors from all academic disciplines and an eager cadre of students who speak the same evolutionary language. The "whole city" part is a network of partners representing all sectors of the city — the mayor's office, the police department, the public school system, the housing authority, the health and social services and environment departments, the neighborhood associations, even the churches. Unlike most community-based research, which can be first-rate but is almost invariably restricted to addressing a single problem, the BNP is organized to tackle any and all problems as part of the same tangled bank. It's a grand vision, but it can be built grain by grain, like a termite colony, from existing parts.

That is how I ended up studying a city, which I could never have foreseen as a graduate student expecting to study zooplankton. And then, when I thought that my plate was full with EvoS and the BNP and wasn't looking for additional projects, the pinball machine of life sent me hurling in another direction in the form of an e-mail from a man named Jerry Lieberman, president of the Humanists of Florida Association. Jerry wanted to start a humanist think tank and was persuaded by my book *Evolution for Everyone* that it should feature the evolutionary paradigm. He had the experience to create a think tank. Might I be interested in a partnership?

As you have probably guessed by now, I have a hard time saying no to an exciting scientific prospect. I knew nothing about think tanks, and Jerry was a total stranger, but his offer had the same appeal as the BNP: using evolutionary theory to improve the human condition. A think tank could do for the world of public-policy formulation what EvoS was doing for the world of higher education. I could bring evolutionary expertise to bear on any policy issue, and the recommendations could be implemented anywhere on the planet, not just in my hometown. Another think tank called the Discovery Institute had millions of dollars at its disposal to spread misinformation about evolution. I'd give anything to have resources comparable with the Discovery Institute to show what the evolutionary paradigm can really do! I wrote Jerry a welcoming reply and the Evolution Institute was born. I would be studying not only evolution and cities but evolution, cities, and the world.

I am convinced that evolutionary science provides an essential tool kit for making the world a better place at all scales, from individuals seeking to thrive, to nurturing neighborhoods, to nations that responsibly manage their affairs at a worldwide scale. I am confident that the evolutionary paradigm will eventually become the accepted view for all things human in addition to the rest of life. The question is how fast it will happen. This book is written as a catalyst to accelerate the change because the problems that threaten modern human existence will not wait.

AS THE SON OF a novelist who became a scientist, I rankle at the way science is understood by the general public, especially when it is portrayed as a sacred body of knowledge presided over by a priestly caste. Knowledge *is* sacred, or at least it should be, and science *does* have authority, but only thanks to a collection of practices, loosely called the scientific method, that holds people accountable for their factual statements about the world. As soon as scientists start acting as oracles and what they say is accepted on faith, the scientific method has been violated and what is sacred about it has been profaned. So I aim to describe science as it is actually practiced by fallible people trying to uphold an ideal, similar to religious believers trying to uphold the ideals of their faiths. I'm even comfortable calling science a religion that worships factual reality as its god.

The more we focus on the creative side of science, the less it resembles science at all. Albert Einstein famously said that if we knew what we were doing, it wouldn't be called research. The initial stages of a new scientific insight are often more like a Woody Allen movie than a deliberative process, and I'm not afraid to admit it. The deliberative part comes later and has a drama that is seldom shown to the general public. A good experiment is like a work of art or a well-executed chess game. Often it requires a heroic amount of effort, like building the Great Wall of China. It also requires accountability: the way the ideals of science hold people accountable for what they say, so thoroughly that it becomes second nature for them, is something to admire and, above all, to emulate in other walks of life.

As for scientists, they're just everyday people who wandered into a certain line of work. When I decided to become a scientist and started to become proficient at skills such as algebra, my entire family regarded it as a miracle. They were so mystified by math that it boggled their minds how I could solve a linear equation. Since I came from a nonscientific background, I have always been curious about how my colleagues became scientists and gratified to learn that they, too, wandered into this line of work from every which way, like that silver ball in the pinball machine of life. Studs Terkel did the world a great service by telling the stories of everyday people in books such as *Hard Times* and *Working*. I aim to provide the same service by telling the stories of scientists along with their work—arguably the best way of conveying the true nature of science as it happens.

* * *

MY GRADUATE STUDENTS AND I might have looked like the other revelers standing by the Court Street Bridge, waiting for the parade on that early March day, but we were seeing our city in a different way: As deeply continuous with the other creatures in our midst, such as the crows flying overhead and even the insects skating on the river surface below the bridge. As people bearing genes that arose in the far distant past, in Africa or anyplace else on the planet that our ancestors inhabited on their way to this spot. As people bearing the cultures that arose hundreds and thousands of years ago, including the religions that are visible from the church spires and golden domes gracing our skyline. As people bearing the experience of their own lifetimes on their road to riches or ruin. As a city with a charm and vitality of its own but with so much future potential. All of this as part of Darwin's tangled bank, produced by the same laws acting around us.

It might seem that a little city in Upstate New York can't bear the weight of such great expectations, but not if you are the son of a novelist. Some of the greatest stories are enacted on the tiniest stages. Literature is transcendent because timeless and placeless themes are manifested in particular times and places. So it is for the city of Binghamton, New York.

CHAPTER 2

My City

THERE IS NOTHING like a bicycle for making me feel like a boy. As I coast down a hill on a sparkling June morning, I could be thirteen and thrilling to my first day of summer. In truth, I am fifty-eight, with wind rushing through thinning hair, but if you think that a boy looks forward to summer, try becoming a professor. Classes are over, and I have ten glorious weeks to indulge in scientific play.

Today I have decided to take a bicycle tour of Binghamton with new eyes and ears. It has been my hometown for more than twenty years, but there is a sense in which I don't know it and haven't even seen it. I only attended to aspects of the city that bore on my personal life, and everything else was invisible to me. Like so many professors, I spent most of my time absorbed by my work and remained aloof from community affairs. I was pondering the big questions. Who had time for the "Local" section of the *Press & Sun Bulletin*?

All that changed when I had my epiphany about using evolutionary science to improve the quality of everyday life. Suddenly, the real test was whether I could make a difference in my own community. I could talk the talk, but could I walk the walk?

Today, on my bicycle, I want to listen and reflect on my city as a scientist and evolutionist, not as an inhabitant. Binghamton is located at the confluence of two rivers, the Susquehanna and the Chenango, both named for Native American tribes that once occupied their banks. The confluence itself is an inspiring sight, showcased by a lovely park that is the first destination of my tour. The broad Susquehanna is on my left, about 100 meters wide here on its way to the Chesapeake Bay. The Chenango is to my right, about 50 meters wide, and joins the Susquehanna in a broad expanse that disappears around the bend. The banks are mostly forested, concealing the highway that runs along one side and the residential neighborhoods along the other. The surrounding hills are not majestic but have an allure of their own. They are steep, therefore undeveloped, and surround the river valley in a warm embrace. Thanks to the forested riverbanks and surrounding hills, the view of the confluence is surprisingly wild for being in the center of a city. Remove two bridges, a few buildings, an electrical line, and some radio towers on the hilltops, and a Native American standing on this spot 1000 years ago would be greeted with much the same view.

Native Americans were standing on this spot as early as 7000 years ago. How can I estimate this number, as provisional as it might be? I can't directly see it, as I can the width of the rivers. I only know thanks to archeologists who have been painstakingly digging up artifacts and piecing together the story of our past. To the best of their knowledge, the entire North American continent was uninhabited by people until a few traveled from Siberia into Alaska about 12,000 years ago. Their descendants spread south and east, eventually reaching this area. At first, they lived as hunters and gatherers entirely off the land. The confluence must have been a choice spot for them, teeming with fish, waterfowl, and game along the banks. Then they learned to grow crops or were replaced by agriculturalists from elsewhere — the first residents of what became my city.

When the Puritans landed on Plymouth Rock in 1620, this river valley was already densely populated by Indians who lived in villages enclosed by sturdy wooden walls and surrounded by fields of corn, beans, and squash, the famous "three sisters" that most of us learned about in school. They even governed themselves in a federation called the Haudenosaunee and known by the Europeans as the Six Nations. Oren Lyons, a living chief of the Onondaga nation, put it this way:

The history of North America began a long time ago. For us it began thousands upon thousands of years ago, longer than our white brothers care to acknowledge or admit. During those times, indigenous people resided here and prospered. They understood life and the laws of nature and they lived, by and large, in peace.

When Benjamin Franklin and his compatriots were trying to create a federation of their own in 1751, he wrote ironically about the Haudenosaunee:

> It would be a very strange thing, if six nations of ignorant savages should be capable of forming a scheme for such a union, and be able to execute it in such a manner, as that it has subsisted Ages, and appears indissoluble, and yet that a like union should be impracticable for ten or a dozen English Colonies, to whom it is more necessary, and must be more advantageous; and who cannot be supposed to want an equal understanding of their interests.

I don't know about you, but I feel good that our nation, now the most powerful on earth, can afford to spend a little bit of its money recording its history as carefully and as far back as possible. Every time there is a construction project in New York State, it is required by law that a team of archeologists first investigate the site for important historical and archeological remains. Just such an excavation took place on the east side of the Chenango River, a stone's throw from where I am sitting at the confluence, in preparation for Binghamton University's new downtown building. The main campus, several miles to the west, on the other side of the Susquehanna, includes a public archaeology facility for just this kind of work. Kevin Sheridan, a former student of mine, works for the facility and told me what happened.

First, a grid was mapped for the entire site with the points spaced 15 meters apart. Then a hole was dug at each point by a two-person team using shovel and trowel, as deep as necessary to reach soil that existed before human habitation. By a river, that can be pretty deep because of soil deposited by floods. The soil was carefully examined for artifacts, and more closely spaced holes were dug at locations that appeared promising. All of this was to locate the best sites for the real excavation, which consisted of removing and screening the soil in 1-meter squares at 15-centimeter depth intervals. Most of us have seen photographs of such archeological digs, and I am amazed at the patience that must be required to do such hard and careful work.

Even the soil provides clues to our past. The Indians inserted poles into the ground to support their long houses. When the poles eventually rotted, they created soil with a different color from the surrounding soil, clearly visible as round discs in the excavation. Thanks to the archeologists, the outlines of the long houses standing centuries ago on the banks of the Chenango River can be measured with the same precision as the width of the river itself.

THE DEPARTURE OF THE INDIANS from the current site of Binghamton can be determined more accurately than their arrival. In 1779, as soon as the Americans had achieved victory over the British, George Washington directed a quarter of the Colonial Army to destroy the crops and villages of the Indians on what was then the edge of the frontier. The Six Nations had found it impossible to remain neutral during the Revolutionary War. They were eagerly courted by both the Americans and the British, and some had chosen the wrong side. Now it was time for payback and to ensure the safety of the frontier for all time. It is chilling to read Washington's directive, given the ethnic conflicts raging elsewhere in the world today:

> You will not by any means listen to any overture of peace before the total ruinment of their settlements is effected.... Our future security will be their inability to injure us...and in the terror with which the severity of the chastisement they receive will inspire them.

The orders were carried out by Major General John Sullivan and General James Clinton. Sullivan moved up the Susquehanna from the south, starting from the settlement of Wyoming, Pennsylvania, which had been the site of a major massacre at the hands of the British and Indians. The survivors of that massacre wrote sorrowfully of how women and children were forced to make their way back east through a wilderness that they named "Shades of Death." Clinton moved down from Otsego Lake, the headwater of the Susquehanna River, in an audacious military maneuver. He built a dam across the outlet of the lake, waited for the water level to rise, then broke the dam, riding the flood waters down the Susquehanna with his boats.

The colonial troops carried out Washington's orders, burning villages, destroying crops, and girdling the fruit trees standing in orchards. A poignant letter from a soldier to his sweetheart was recovered upon his death:

> Yesterday we attacked an Indian village called Shemung several miles distant and ruthlessly destroyed all their habitations and their grain. Poor savages! I don't imagine they are worthy of much pity but nevertheless I do pity them from my in most soul I do. I really felt guilty as I applied the torch to the huts that were Homes of (at least) Content until we ravagers came spreading desolation everywhere. Oh this cruel business! I may not deny the necessity of Retaliation this severe but... my mind's eye did not picture any such scenes as these that evening when I came to you and proudly told I had listed for the war.

Mary Jemison, a Scotswoman who had been captured by Indians as a child and elected to stay with them, wrote from their perspective:

> We all returned; but what were our feelings when we found that there was not a mouthful of any kind of sustenance left, not even enough to keep a child one day from perishing with hunger.

The Americans spoke about retaliation, but as the years marched by, a different motive began to be more openly acknowledged.

William Tecumseh Sherman, the Civil War general who knew a thing or two about scorched-earth tactics, said this during a speech celebrating the 100th anniversary of the Sullivan-Clinton campaign:

> I know it is common, and too common a practice to accuse General Sullivan of having destroyed peach trees and cornfields, and all that nonsense. He had to do it, and he did do it.... Why does the Almighty strike down the tree with lightning? Why does He bring forth the thunderstorm? To purify the air, so that the summer time may come, and the harvest and the fruits. And so with war. When all things ought to be peaceful, war comes and purifies the atmosphere... we are better for it; you are better for it; we are all better for it. Wherever men raise up their hands to oppose this great advancing tide of civilization, they must be swept aside, peaceably if possible, forcibly if we must.

A. C. Flick, the official historian of New York State, spoke with less bluster during the 150th anniversary of the campaign:

> Washington and other leaders saw that independence with a mere fringe of land along the sea coast would scarcely be worth the cost of the struggle [without access] to the... potential wealth of the fertile regions of the interior of the continent.... Hence in the Sullivan-Clinton Expedition an inland empire was at stake for which Washington was playing and not merely the punishment of dusky foes on our border.

In the years following the campaign, colonial pioneers began filtering into the area. The Indians had also returned, and, amazingly, the rules of engagement had changed. The pioneers negotiated with the Indians for their land. The Indians kept an area for themselves with the understanding that they could also roam more widely to hunt, fish, and gather. Accounts spoke admiringly of the skill with which the Indians could throw spears and hit distant fish swimming upstream. Indian and pioneer children played on the riverbanks, and young pioneer women accompanied squaws to gather whortleberries. The first pioneers also spoke of the love they felt for one another, as if they were a single family, in contrast to an indifference that would set in when the area became more thickly settled.

Ultimately, the local Indians were swindled out of their remaining patch of land by a rascal known to history only by his last name: Patterson. Thinking that they were trading bearskins for a rifle, the chief and his son signed a piece of paper that was the deed to their land. Patterson sold the deed to the Massachusetts Company, which bought and sold blocks of land, and retired with his profits to Ohio, where he boasted about his exploits, thinking that he was safely out of reach. He was wrong. The Indians had a social network of their own that extended more widely than most of the pioneers imagined. The chief's son, Abraham Antonio, had himself been to Ohio previously to join the fight against the pioneers there. He tracked Patterson down and murdered him and his family in retaliation, but the deed to the land could not be recovered.

IN COLONIAL DAYS, large blocks of land were awarded to prominent men to develop as they pleased. The pioneers were squatters who could be removed in principle, but most owners of patents were eager to see their land developed. Few people back then regarded wilderness with the reverence that we do today. William Bingham, a wealthy Philadelphia banker who helped to finance the Revolution, was awarded the patent to this area. He never actually visited but worked through agents to help map out the streets and lots of a village between the two rivers in 1800. The first buildings were a storehouse, a grain house, a blacksmith shop, a tavern, a drug store, and a courthouse with two log jails and a residence for the jailor. By 1810, a blacksmith named Atwell incongruously ran a dancing school at night, attended not by women but by the wealthiest men, so anxious were they to acquire the trappings of civilization. The first newspaper was published in 1811, and by 1818, there were two papers espousing different political views. In the 1830s, a canal was dug along the Chenango River to connect Binghamton to the Erie Canal. It was 95 miles long, 46 feet wide, and 5 feet deep; it had 116 locks — and was accomplished by manual labor in three years. The Erie Canal, of course, was an even greater engineering feat and turned cities such as Buffalo, Rochester, and Syracuse into boomtowns. The Chenango Canal enabled Binghamton to share the prosperity by shipping millions of feet of lumber and other goods to market.

The arrival of the railroad in the 1840s brought more prosperity. Stately homes lined Riverside Drive, and sturdy buildings in the Victorian style were erected in the center of the city. Many of them still grace the skyline, intermixed with more recently constructed buildings that are boxy and ugly by comparison. On my bicycle, I can travel in just a few minutes to the major churches that were built in the mid-1800s, such as the Episcopal church on Main Street or the Catholic church on Leroy Street. I marvel at their size and durability, as if I were an Egyptian peasant staring at the pyramids. The ceiling of the Episcopal church is made of ornately carved wood and stands forty feet high, supported by wooden columns. Photographs of the era show a city bustling with activity. Boosters called it the "Parlor City," as if it offered the same kind of high society as New York, Philadelphia, and Boston.

The wonderful stories about Binghamton's early days are known largely because a schoolteacher named John B. Wilkinson took the trouble to collect them and write them down in a book published in 1840. Today his tradition is carried on by the Broome County Historical Society, which has an impressive facility on the second floor of the County Library Building on Court Street. In addition to dedicated amateur historians, the county has just enough money to employ two professional historians to keep track of our local past. Before, I would have regarded their interest as quaint and provincial, but now I am beginning to think of them as keepers of a flame of information that I will need to consult as I begin to think about my city in a new way.

BINGHAMTON'S ERA OF PROSPERITY might have ended with the nineteenth century except for two men, George F. Johnson and John B. Watson. Johnson founded the Endicott-Johnson shoe company, which for a period was the largest manufacturer of shoes in the world.

Watson built a small company that made time clocks into the International Business Machine Corporation (IBM), initiating the mechanization of business and the age of computers. Both took a paternal interest in their employees that stands in stark contrast to the Enrons of today yet failed to serve their interests over the long term.

Johnson was a shoemaker with humble roots. He came to Binghamton in 1882 to work in a failing shoe company owned by a Boston financier named Henry B. Endicott, and he ended up borrowing money from Endicott and buying a half-interest in the business. He then proceeded to recruit immigrants from Europe, including southern and eastern Europe, by the thousands. At its peak in 1917, his payroll included more than 20,000, and two new towns had been erected on the Susquehanna downstream from Binghamton—Johnson City and Endicott—to hold the factories and houses. In addition to the familiar spires of Catholic and Protestant churches near the center of the city built during the mid-nineteenth century, a few more minutes on my bicycle can take me to the golden domes of several Eastern Orthodox churches that rise from what was the periphery of the city during the early twentieth century.

Unlike the cruel factory owner in a Dickens novel or the sweatshops of today, Johnson took a passionate interest in his employees. He called it "Industrial Democracy," but it was more paternalistic than democratic. He offered affordable housing in neighborhoods that rose away from the river toward the steep hills. He provided health care and made sure that the hospitals were sufficiently staffed with doctors. Every baby received a ten-dollar gold piece and a bank account with another ten dollars as an opening deposit. He voluntarily shortened the workday from nine and a half to eight hours without any loss of pay, prompting a parade to be organized in his honor. The factory walls were adorned with wholesome proverbs such as "LIVE AND LET LIVE" and "FOR THE BENEFIT OF ALL—SAVE!" Workers were encouraged to publish their criticisms in the company magazine. He created parks and other recreational facilities, including the world's largest aboveground swimming pool for its time, which could accommodate 2000 bathers under the big block letters "COME ON IN—THE WATER'S FINE!" As a boy, Johnson had been so poor that he couldn't even afford to ride a carousel, so he built carousels throughout the city that could be ridden without charge. Thanks to an endowment provided in his will, the carousels are maintained, and children continue to ride without charge to this day. One of them is located across the street from my house, and I garden to its merry calliope music. Johnson was so popular among his employees that they erected stone arches over Main Street, entering and leaving Johnson City, with the words "Home of the Square Deal" carved along the top. Efforts to unionize the Endicott-Johnson shoe company failed because Johnson had already given his employees everything they could imagine asking for.

Not everyone was happy with such an influx of immigrants. Back then, it was Germans rather than Muslims who were supposedly out to get us. Congress passed the Espionage and Sedition Acts during World War I, which made our current curtailment of civil liberties seem tame by comparison. A private organization called the American Protective League operated as a vigilante force with the approval of the attorney general and had more than 250,000 members in 600 cities. In Binghamton, one German-American attracted suspicion by taking notes at public meetings, looking behind him when walking down the streets, and receiving money and foreign books from Philadelphia. Secret agents were called in to investigate, but the poor fellow turned out to be a religious scholar with a brother in Philadelphia who took notes at meetings to practice stenography.

The xenophobia that was cultivated during the war easily shifted from Germans to immigrants after the war. Back then, if you were called "black," you might be from Ireland, Italy, or Lithuania, as well as from Africa. The Ku Klux Klan, which originated in the South, reinvented itself as a movement with nationwide appeal. New York City was already too ethnically diverse for the Klan to obtain a foothold, but they received a warmer reception in the upstate cities, where the first colonists from northern Europe resented the newcomers—and Binghamton became the Klan's state headquarters. Major Emmitt D. Smith, with the absurd title of King Kleagle of New York, settled in Binghamton in 1923. By most accounts, he was an unsavory character, who made a healthy profit and romanced his two secretaries behind his pious and patriotic façade. Fortunately, his reign lasted only a few years and ended with a clever, if dirty, political trick. In the 1928 mayoral race, one of the candidates was backed by the Klan. On the night before the election, crosses could be seen burning on the hills around the city. The alarmed citizens voted decisively against the candidate backed by the Klan—but those crosses were erected and set ablaze by the campaign manager for the *other* candidate.

Watson followed in the paternalistic footsteps of Johnson in his development of IBM. Johnson convinced Watson that he could stay in the area and that the business would come to him. A huge complex of buildings was built in Endicott, downstream from Johnson City on the Susquehanna. Engineers were recruited by the thousands, as the shoemakers were a few decades before. Just as Johnson painted wholesome proverbs on his factory walls, Watson made a mantra out of the word *THINK*. Before IBM became known for computers, it was filling an insatiable demand around the world for time clocks, adding machines, scales, automatic payroll machines, and other devices that made businesses run efficiently. I still remember when my father purchased an IBM selectric typewriter, with its ball that magically leaped up to strike the page through the ribbon, making manual typewriters seem unbelievably boring by comparison.

IBM wasn't the only local company to have a global impact. Edwin Link, the son of an organ manufacturer, became interested in aviation and invented the first flight simulator in 1929, using pumps and bellows from his father's factory. Ansco was a major manufacturer of cameras and film and the chief competitor of the Eastman Kodak Company in Rochester, located westward on the banks of the Erie Canal. Even though Upstate New York was a cultural backwater compared with the major Northeastern cities, it had a way of producing world-changing technological innovations, such as copying machines (by the Xerox Corporation in Rochester) and fiber optics (by the Corning Glass Company in Corning).

As for technology, so also for religion. Upstate New York was called the "Burnt Over District" in the 1800s because of the religious fervor that spread through it like a wildfire. In the 1820s, as Binghamton was growing from a village into a city, a farmer named Joseph

Smith was receiving the word of God in nearby Palmyra, New York. Less than two centuries later, his Church of Jesus Christ of Latter-day Saints has spread around the world and is growing at a rate comparable to Christianity during the first millennium. What is it about cultural backwaters that makes them so innovative, for technology and religion alike?

Alas, Binghamton's prosperity was not to last. For all of his benevolence, Johnson was slow to modernize his factories and could do nothing about the economic forces that caused the shoe industry to march west and eventually overseas. IBM followed. The workers who had placed their trust in Johnson and Watson found themselves defenseless against less enlightened leaders and harder economic times. Today, when IBM is mentioned, it is mostly in connection with a toxic plume that contaminates the ground water of sections of Endicott. The remains of Endicott-Johnson's physical plant include the arches, still standing, and the abandoned factories. I have been told that the primary ethos of the Tri-Cities area (the collective name for Binghamton, Johnson City, and Endicott) is the sense of having been betrayed.

TODAY BINGHAMTON SUFFERS THE SAME afflictions as so many American cities. Republican and Democratic mayors come and go, each bravely trying to revive the city in his or her own particular way. Houses and storefronts stand empty and are laughably cheap compared with more prosperous cities. The children disperse like dandelion seeds, my own included. More immigrants arrive, not because we are the city of opportunity, as in the past, but because they are escaping even more desperate circumstances elsewhere in the world. I was astonished when I learned from Peggy Wozniak, our school superintendent, that Binghamton's high school has recent immigrants from so many places that they speak eighteen different languages as their first language. In addition to those arriving from foreign shores, we can add a sizable number of African-Americans emigrating from the City to seek a better life. Just as technological and religious innovations from this region have spread over the planet, the people I pass on my bicycle could truly have come from anyplace on the planet, in their own lifetime or within only a few generations.

I never knew. Even though I have lived here for more than twenty years, I was so focused on my own narrow concerns that all of these wonderful facts passed me by. No doubt, I encountered them from time to time. The Victorian buildings and golden-domed churches are impossible to ignore. I dimly remember hearing that we were once the headquarters for the Ku Klux Klan, but that fact was worth only a single "Really!" as it went in one ear and out the other. It didn't stay inside my head because it didn't mean anything to matters of importance in my daily life.

That has now changed. Now that my intellectual life and my everyday life have been thrown together, I can almost feel the connections taking place inside my head. Like a Shakespearean play, the length and breadth of human nature are being enacted in front of me on a local stage. Virtually every problem that we might want to solve and every asset that we might want to nurture on a planetary scale has made its appearance at the confluence of these two rivers: from ethnic cleansing to peaceable relations, from cooperation so effortless that everyone feels like family to indifference and exploitation, from innovation to stagnation, from economic prosperity to decline, from corporate responsibility to abandonment. No boy was ever more excited on his bicycle than I, as I pedaled through the neighborhoods of Binghamton on my new voyage of discovery, visiting places and noticing things for the first time—the old buildings, the fading signs, the dry canal bed. These things are now caught inside my head by an ever-expanding web of meaning.

All very interesting, you might be thinking. The history of your hometown is probably just as colorful—but what's evolution got to do with it?

CHAPTER 3

The Parable of the Strider

IT IS AUGUST, and I am 20 feet high in a tree house on some property that we own about a half-hour drive from Binghamton. This is my version of Darwin's beloved sand walk, the path around his property at Down House that he circled to observe and reflect on nature. I come here whenever possible to ramble the old field that slopes down to a swamp, to sit by the pond ringed with cattails or to jump in on a hot day, and to walk the banks of the stream that meanders through the dark hemlock forest.

Anne and I owned the property for years, merely for the pleasure of walking on it. We toyed with the idea of building a cabin but had neither the time nor the skills. Then, by luck, one of Anne's graduate students, Michelle Berger, had a partner (now her husband) named Kevin Bach who was a master builder. He always wanted to build a tree house and found us easy to convince. What a tree house! It is actually a cabin built on a platform between a triangle of hemlock trees, with a wood stove, a sleeping loft, and a balcony on each end. The trees rise through holes in the balconies. Thanks to Kevin's skill, the tree house is still standing after thirteen years, and the trees have grown to the point that we had to enlarge the holes to accommodate their enlarged trunks.

There is something inspiring about being up here, almost like entering a cathedral. It is an old forest, and the trees are present at all stages of their life cycle, from tiny saplings, to tall giants, to rotting logs blending back into the soil. When our kids were younger, we visited a special rotting log that we called "the couch" because it was covered so thickly with moss that you could sink into it, as if it was your living-room couch. From the balcony of the tree house where I now sit, the sun filters through the canopy as through a stained-glass window. The stream meanders through the trees in front of me. My ears, which on most days are battered by the sounds of city life, are soothed by a light wind rustling through the leaves and a medley of bird songs, from the flutelike tones of the hermit thrush to the distant raucous call of a crow. On windier days, the trees sway and the tree house creaks like a ship upon the sea. It's hard to imagine that I am only 25 miles away from my city of Binghamton.

IF WE WANT TO CONSULT nature to think about ourselves, we needn't travel to Africa's forests to study our closest primate relatives. We can stay right here. Darwin wrote that all species inhabiting the tangled bank are produced by the same laws acting around us. If we are the product of the same laws, then every creature taken in by my gaze as I sit on the balcony of my tree house can tell us something profound about human nature. There are many that I could choose, but I will focus on an insect that actually does what Jesus was reputed to do: it walks on water.

Down the steep steps of the tree house, across the forest floor dappled by sunlight, to the stream, which almost always draws me to its banks on my walks. It exudes peace with its quiet pools and musical rapids. On every pool, there are ripples expanding in concentric circles, as if it were raining. They are not caused by raindrops, however, but by a marvelous insect called the water strider and known to scientists by the Latin name of *Aquarius remigis*.

Striders glide over the surface of the water with the ease and grace of ice skaters. They are predators, feeding largely on other insects that don't normally live on the water surface and fall in by accident. Those insects don't have the grace of the striders. They *float,* because any light tiny object floats on the surface of water, but they don't *stride,* because that ability requires special adaptations.

The body of a strider is actually held above the surface of the water by its front and back legs, while the middle pair of legs act like oars. This can be seen even from the shore. Because the front and back legs bear the weight of the body, they create dimples on the water surface without breaking through. Each dimple acts like a lens that reflects a tiny point of sunlight and casts a disc-shaped shadow on the stream bottom below. The middle legs rest more lightly on the water, creating smaller dimples and shadows. Even though the strider itself can be difficult to see on the water surface, the diamond points of light and shadows are highly conspicuous. With every stroke of the oars, ripples spread in concentric circles and the points and shadows move in a visual display as mesmerizing as any kinetic sculpture.

Why don't the weight-bearing legs of the strider break through the water surface? Water molecules are attracted to one another, which is why they form little droplets when suspended in air. They are also attracted to or repelled by other substances. If you place a drop of water on your kitchen counter, it spreads into a thin film because the water molecules are more attracted to the counter surface than to one another. Place the same drop on a piece of felt or waxed paper, and it forms into a little globe, because the water molecules are repelled by these surfaces. Water-repelling surfaces are poetically called *hydrophobic,* which literally means "fear of water."

Striders have thousands of tiny hairs on their feet that are coated with a waxy substance, like felt and waxed paper combined. How can I state this fact with such confidence? I can't see the tiny hairs with my naked eye, as I can the points of light and disc-shaped shadows gliding around the pool. Fortunately, people have been inventing and perfecting ways to extend our vision for centuries. I have two microscopes in my laboratory that are descended from the ones invented by Antoni van Leeuwenhoek in the 1600s. One is called a compound microscope and can magnify objects as much as 1000 times, but they must be thin enough for light to shine through, and the microscope can only focus on one thin layer at a time. The other is called a stereo microscope and can magnify three-dimensional objects up to 200 times by shining light on them rather than through them. I love them both for their beautifully engineered parts, much as photographers love their cameras. They are not up to the task of viewing the hairs on a water strider's leg, however, which are too small for the stereo microscope and too opaque for the compound microscope.

A different kind of microscope, so large that it requires its own room, resides on the ground floor of the building where I work. A scanning electron microscope (SEM), it relies on electrons, rather than light waves, to view tiny objects. First, the object must be prepared by gluing it onto a little metal platform called a stub, putting it into a vacuum chamber, and blasting a piece of gold foil with argon gas. This creates a fog of gold molecules that adhere to the surface of the object itself. The gold-plated object is placed in another vacuum chamber, where it is scanned with a tiny beam of electrons. After a lot of information processing, the object appears on a video screen with knobs for rotating it and for zooming in and out.

Whenever I sit at the console of an SEM, rotating an object such as an insect, I feel like Captain Kirk circling a planet on the starship *Enterprise.* Zooming in is like beaming down onto the planet. Thanks to this technological marvel, we can see the hairs on the foot of a strider as clearly as the needles on a hemlock tree. It is amazing to see such exquisite patterns at such a small spatial scale—they would be invisible to us without the tools of science.

Water striders' feet were not designed for their aesthetic appeal but to support the striders without breaking through the water surface. They do this so well that scientists call them *super*-hydrophobic. They are so water-repellent that you could stack fifteen water striders on top of one another, like Dr. Seuss's Yertle the Turtle, and the feet of the bottom strider *still* wouldn't break through the surface.

Water striders' feet are awesome in the perfection of their design. They are so well suited for the task of repelling water that scientists study them in part to discover how to make more water-repellent surfaces for human use. In one article published in 2008, a group of scientists figured out how to take a mold of water strider feet, much as Boy Scouts take plaster molds of animal tracks, in a way that preserves the fine structure of the hairs. The molds were super-hydrophobic, just like the feet themselves. The scientists determined that the hairs contact the water surface at an angle of 164.7 degrees, which can now be emulated by man-made hairy surfaces. In the future, you might need to thank water striders' feet and the scientists who study them for your new high-tech rain jacket.

HOW DID WATER STRIDERS' FEET become so well designed for their task? Before Darwin, there was only one conceivable answer: a supernatural being with both the ability and the desire to create organisms adapted to their environments. The philosopher William Paley famously made this argument in the 1700s by asking his readers to imagine finding a watch while taking a walk. The watch has beautifully crafted parts that interact in just the right way to accomplish the task of keeping time. It is different from a nonliving natural object such as a stone or even a crystal, whose intricate parts serve no purpose. The watch, unlike the stone or the crystal, was clearly designed by someone or something. If this is true for a man-made object such as a watch, then surely a designer is required to explain the existence of living organisms so perfectly designed that they put watches to shame. Darwin grew up with Paley's argument from design and regarded it as a definitive argument for the existence of God, just like everyone else at the time—until he formulated his theory of natural selection.

For the first time, a plausible alternative to Paley's argument from design could be based on the laws acting around us, as Darwin put it. That didn't make Darwin right, but it did set the stage for a scientific contest—two theories that make different predictions about the world, which can be compared on the basis of evidence.

Today we are told again and again by defenders of creationism that the contest is still undecided, that Paley's argument is still a contender, and that creationists are being unfairly excluded from the scientific playing field. It is therefore instructive to turn back the clock to Darwin's day, when creationists were in the majority and evolutionists were the embattled minority. Even then, the weight of evidence for evolution was so great that creationism was muscled off the scientific playing field. The victory came not from a single decisive experiment—that's not how science works for the big questions—but from accumulating evidence from many sources. A scientific theory itself is a bit like a watch. Humans must construct its interrelated parts to explain phenomena as diverse as the geographical distribution of species, their anatomical similarities, the fossil record, and those exquisite adaptations that demand an explanation. Creationism was in trouble long before Darwin, starting with the study of astronomy and geology in preceding centuries. The heavenly bodies and geological features of the earth could be explained much better on the basis of unchanging physical laws operating over millions of years than by biblical accounts of heaven and earth. Perhaps there is a God who set the natural processes in motion, but the idea of *a God who intervenes in natural processes* was gradually abandoned, even by scientists who remained religiously devout. William Whewell, a celebrated scientist and theologian at Cambridge University when Darwin was a student there, put it this way in his *Bridgewater Treatise* in 1833:

> But with regard to the material world, we can at least go so far as this—we can perceive that events are brought about not by insulated interpositions of Divine power, exerted in each particular case, but by the establishment of general laws.

Darwin quoted this passage on the first page of the first edition of *Origin of Species* to remind his readers that he was merely taking Whewell's advice to the limit.

We can use creatures that walk on water to appreciate the kind of evidence that made Darwin's theory so compelling from the start, even before the mountains of additional evidence that we have today. *Aquarius remigis* is not alone. Around the world and throughout time, the surface of water has been loaded with dead and dying insects waiting to be eaten. The water surface is no ordinary habitat, however. Specialized adaptations are required to walk on it, which is why most of the insects that fall in are entrapped. A God who wanted to design a class of organisms to walk on water presumably would have done so and stocked them throughout the earth. Evolution, by contrast, is a more haphazard process. Whenever there is water but no water walkers, an empty niche is waiting to be filled—usually by creatures that live at the water's edge and fall in sufficiently often that they have evolved adaptations to stride quickly to shore. Striding to escape the water was probably a first step toward striding as a full-time occupation. Many different kinds of creatures live by the water's edge. Which particular creature invades the empty niche would be largely a matter of chance. It might be a spider, a beetle, a bug (scientists define bugs as a particular kind of insect rather than a catchword for all insects).

Once a particular creature becomes adapted to walk on water, it trounces the competition, the way established companies exclude start-ups in economic competition. Additional species will arise from the first species to make the transition, not from new origination events. However, there is limited dispersal among geographical regions of the planet. A water-walker that originates in Europe can't easily get to Australia. Water walking will require a separate origination event in Australia, possibly from a very different creature. That is why evolutionists have a special fascination for remote islands, especially when they are created by volcanic activity. Suddenly, there is a landmass that didn't exist before, which is colonized by only a few creatures occupying a few niches. Vacant niches on remote islands must be filled by new origination events if they are to be filled at all. Of course, it is perfectly possible that new niches remain empty or become filled in a way that we didn't anticipate. We might well visit a remote island and discover no water walkers but flies that hover like helicopters and pluck those tasty dead and dying insects from the water surface. All of these predictions follow straightforwardly from viewing evolution as a natural process based on laws acting around us. None of them follows from creationism, except for a version of creationism in which God sets the laws in motion and doesn't intervene.

When we survey all of the water-walking creatures that exist on earth and in the fossil record, we find clear support for the haphazardness of evolution. How can I make this statement with such confidence when my personal experience is limited to sitting by the edge of my stream and gazing appreciatively at the living kinetic sculpture? Fortunately, other people have been studying water walkers, starting with the intrepid naturalists and collectors of Darwin's day and continuing with some of the smartest scientists, at the most elite universities, using the fanciest techniques available today. Even better, I can retrieve this knowledge at the touch of a button in today's electronic age. The infrastructure of modern science is awesome when you pause to think about it.

At home, in front of my laptop, connected by wireless and with a cup of coffee at my side, I start the timer on my digital watch and log onto an electronic database called Web of Science. I type "water strider" into the subject search field, and the database finds 272 references within seconds. I select a few options, click the print button, and abstracts for the references begin to emerge from my printer.

Each abstract is a concise summary of an article reporting scientific research that might have required years and many thousands of dollars to complete. Each article was critiqued by several experts before it could be published. The review process is not perfect, but it is one of the best systems on earth for holding people accountable for what they say. Like all scientists, I am routinely contacted by journal editors with requests to review manuscripts in my area of expertise. I am expected to do this without payment as part of my responsibility as a member of the scientific community. When I review a manuscript, I check it from stem to stern, from the basic justification for performing the research, to the way the study was designed, to the interpretation and statistical analysis of the results.

The abstracts I've printed are like the low power of a microscope. I can read each one in a few minutes and zoom in for a closer look, if I want, by reading the entire article. At the end of each article is a reference section that lists still more articles if I want an even closer look. Increasingly, entire articles are available electronically and can be downloaded as easily as the abstracts. Otherwise, I'll need to do it the old-fashioned way by visiting our science library. If our library doesn't have the journal or book, our ever-helpful librarians will get it from interlibrary loan within a few weeks. It's easy for scientists to take this system for granted, but it's awesome when you think about it.

As I peruse the abstracts, I learn that all sorts of creatures walk on water—mites, springtails (a tiny creature that usually lives in the soil), spiders, beetles, bugs, and ants. Some live on land but are adept at traversing small bodies of water to get to the other side or to skitter back to shore when they accidently fall in. Others have made the water surface their permanent home. Some made the transition so long ago that they have given rise to dynasties of many hundreds of species. Others are the first of their kind to make the transition. Every species has a geographical distribution that reflects where it originated and barriers to its dispersal. These patterns clearly reflect a haphazard evolutionary process. The creatures that walk on water are like the audience members of a television game show who are invited by happenstance to "Come on down!" and play the water-walking game.

FOR SPECIES THAT START PLAYING the game, chance plays a different role. Just as game-show contestants differ in their ability to provide the right answer, individuals playing the water-walking game differ in their abilities. Some have more hairs on their feet than others. Some have hairs that contact the water at a better angle. The best water walkers are more likely to survive and produce offspring that resemble themselves. Variation and selection. Variation and selection. Natural selection is like a blacksmith beating an implement into shape with repeated strikes of his hammer.

Once we reject the idea of special creation, it is remarkable to contemplate that something as perfect as the foot of a strider can be produced by the hammer blows of natural selection. Yet this is only one of many adaptations that striders must possess to survive and reproduce in their challenging environment. How do they avoid being eaten by fish that lurk underneath the surface of the water, like Bruce the shark in the movie *Jaws*? How do they avoid being swept downstream during floods? How do they survive the winter? How do females know the best place to lay eggs? It turns out that for some species, the best place to lay eggs is under the surface of the water. For this purpose only, females have a way of circumventing the adaptations that keep them *above* the surface during the rest of their lives.

One of the most amazing adaptations in striders is their ability to use ripples on the water surface as a new way of hearing and speaking. Our ability to hear is based on the existence of sound waves that travel through the air. The air is composed of gas molecules that are constantly colliding with one another. When a large object moves, it displaces the molecules in its vicinity, which displace other molecules in an expanding sphere. Eventually, the colliding molecules reach our ear, where they press gently against the membrane that we call our eardrum. That sets tiny bones in motion on the other side of our eardrum, which in turn press against another membrane in our inner ear. Ultimately, the whole Rube Goldberg device causes nerves to fire in the same pattern as the air molecules striking our eardrum. Then our brain goes to work, processing the information in ways that provide meaning. If it is night and the sound emanates from a snapping twig, we become fearful. If it is night and the sound emanates from our lover whispering in our ear, we become aroused. It is mind-boggling, when we think of it, that such complicated mechanical and neural machinery arose by the hammer blows of natural selection. Yet the evidence for the haphazard side of evolution is so compelling that there is simply no valid alternative

For water striders, ripples on the water surface provide a new medium of perception. Every movement results in ripples that radiate outward, packed with information about its source. A struggling insect, waiting to be eaten, will create one kind of ripple. A fish lurking below the surface will create another kind of ripple. A rival water strider or alluring member of the opposite sex will create still other kinds of ripples. The wind and swirling water will create still different kinds of ripples that need to be filtered out to detect the more meaningful ripples. All of those ripples emanate from their sources and cause the legs of the striders to move up and down, ever so gently, just as sound waves press against our eardrums. The undulating water causes the foot joints of the strider to flex, which cause nerves to fire in another mechanical Rube Goldberg device comparable to our ears. Then the nervous system of the strider goes to work, processing the information in a way that provides meaning.

If the ripples emanate from a prey organism, the strider skates directly toward it. It knows the direction because the expanding ripples do not strike the legs of the strider at exactly the same time, providing information about the direction of the source. The hammer blows of natural selection have caused the tiny differences in the up-and-down movement of the legs to become connected to the leg muscles in just the right way to cause the strider to head toward the struggling prey. A predatory fish rising up from underneath produces a bulge in the water that could never be produced by a struggling insect. The hammer blows of natural selection have caused this particular signal to result in the strider leaping out of the way like a trampoline artist.

FROM RECEIVING SIGNALS, it was only a small step toward *producing* them. Just as we create sounds in a meaningful pattern (speaking), striders create their own ripples by vibrating their legs to communicate with one another. The person who first discovered the speaking ability of striders is Stim Wilcox, who happens to work in my own department at Binghamton University. Stim's story shows how the pinball machine of life caused one man to become a scientist and how the scientific method enabled him to establish facts that are likely to last forever.

Stim's ancestors came to America from the British Isles and Europe several generations ago, and he had the kind of idyllic childhood spent outdoors that few children get to experience today. He loved to make things—especially weapons. In the days before video games, Stim made his own spears, slings, bows and arrows, and spear guns for catching fish while snorkeling off the coast of Massachusetts, where he spent his summers. Like me, Stim discovered in college that he could do what he already loved by becoming a field biologist. He ended up studying sound communication in insects that live under the water, which prepared him for his *aha!* experience with ripple communication. He was on a field expedition in Australia, sitting by the edge of a stream, when the ripples created by a surfacing beetle caught his eye. "Wouldn't it be a neat way to communicate, by ripples?" he commented idly to a colleague. Only a few days later, he observed a species of water strider vibrate its body and *knew* that it must be communicating by ripples. He then set about to prove it with a series of elegant experiments that became classics in the field of animal communication.

In one clever experiment, Stim painted the head of a dead male strider with latex paint. When the paint dried, he peeled it off, and it became a tiny mask that he could put on the head of a living male strider. Masked striders could move around and tell the difference between male and female striders, proving that they were not relying on vision. But were they relying on ripples? And how could Stim *prove* it?

Stim's lifelong love of making things came in handy. A galvanometer is an instrument for measuring an electric current. You might have one in your home for measuring the charge of your electric batteries. You touch the ends of a battery with wire probes, and the needle on the galvanometer rotates to indicate the charge. A galvanometer can be rigged to produce an electrical charge as the needle rotates, in addition to measuring a charge. Stim stuck a tiny Styrofoam ball obtained from inside a beanbag chair on the end of the needle of a galvanometer and floated the ball on the surface of the water. Ripples on the water made the Styrofoam ball rise and fall, which made the galvanometer produce a fluctuating electrical current that could be recorded on a computer. Stim had created a Rube Goldberg device for detecting ripples, much like the legs of a water strider. Even better, he could play the signal back through the galvanometer, causing the Styrofoam ball to create ripples by bouncing up and down in the precise pattern that had been recorded. Stim could actually talk to the striders like a modern-day Doctor Dolittle. Striders reacted to the bouncing ball, much as they reacted to one another.

Stim's next stroke of genius was to cause the striders themselves to vibrate at his command. He glued a tiny magnet onto the leg of a living strider and released it into an aquarium surrounded by a magnetic coil. Then he programmed a computer to generate a fluctuating magnetic field, which caused the magnet to vibrate up and down, causing the leg of the strider to beat out a ripple pattern on the water. Other striders in the vicinity responded in just the same way as they would to a normally signaling strider. It was the vibration, and not any other factor, that elicited the response.

I have recounted Stim's discovery in detail both to glorify and to demystify science. First, the demystification. Stim was just being a boy. His adventurous life as a field biologist was a continuation of his adventurous boyhood life outdoors. His scientific gadgetry was a continuation of his boyhood construction projects. Using a Styrofoam ball from a beanbag chair to make a ripple recorder was little different from using an old bicycle tire to make a spear gun. It was just plain fun.

Now for the glorification. Stim accomplished something magnificent and timeless with his experiments. With his miniature latex masks, he proved that striders do not require vision to move around and tell males from females. With his miniature magnets, he proved that they do rely on ripples. Anyone who disagrees is welcome to propose a different explanation, but it must be consistent with the observations. Like a chess master picking off his opponent's pieces, Stim's experiments excluded all plausible explanations but one. Perhaps another possibility will come along in the future. When it does, scientists will be happy to consider it. Until then, if anything deserves to be called a fact, it is that water striders use ripples as a medium of listening and talking, similar to the medium of sound waves.

SO MANY ADAPTATIONS PACKED INTO such a tiny creature, but now our story takes a dark turn. Even though natural selection has provided striders with tools for survival that seem heaven-sent, it has given them a social life from hell. They have the ability to talk, but all they have to say to one another is "Scram!" or "I'm going to mate with you—whether you like it or not!"

To learn more about the dark side of striders, I decided to zoom in on the work of one prolific scientist named Daphne Fairbairn. I didn't know Daphne personally, but evolutionists are a cooperative bunch, and she was happy to discuss her work with me, along with the story of how she became a scientist. Daphne's story shows how the pinball machine of life made it harder for a woman to become a scientist—although we can be thankful that this was more true in the past than in the present.

Daphne's ancestors were fishermen and boat builders on the

border between England and Scotland before emigrating to Canada in the early twentieth century. Her father grew up poor but had become a successful commercial artist by the time Daphne was born. Like Stim, Daphne became deeply attached to nature and adored visiting the family's cabin on the shore of a remote lake. She was her father's outdoor buddy and recalls becoming fascinated with examining the entrails of the fish they caught, demonstrating an early curiosity about how things work.

Even though Daphne's parents were loving, they were also deeply conservative. Girls should be beautiful and charming. They should plan for marriage and expect to be independent of their families by the age of eighteen. Higher education for women was a waste of the taxpayer's money. Daphne therefore received little encouragement from her family for academic pursuits. Fortunately, they lived close to her school, which could provide an outlet for her interests. She enrolled in every activity that she could fit in, not just science but also sports and theater, earning an award for outstanding contribution to school life and a scholarship to attend her local college (Carleton University in Ottawa). As strange as it might seem given our modern sensibilities, this was a source of embarrassment for her parents. They discouraged her from becoming a scientist, in part because they firmly believed that women were mentally inferior to men, which should be reflected in science more than any other occupation. Their single concession was to allow Daphne to remain at home while attending college — but only if she paid rent.

Daphne succeeded so well in college that she received a prestigious scholarship to attend graduate school at the University of British Columbia. Like Stim and me, she was now in the enviable position of doing as an adult what she had always loved as a child. She had also entered a social world in which her talents were admired rather than disparaged. She is now happily married to Derek Roff, another distinguished evolutionist, and it was second nature for them to encourage both their son and their daughter to pursue their aspirations.

Just as a microscope has high-power lenses for zooming in on detail, Daphne's research zooms in on the details of variation and selection. So far, I have said that the adaptations of water striders evolve by the hammer blows of natural selection. Daphne's research actually gives a blow-by-blow account.

In one heroic study conducted in the early 1990s, Daphne and her graduate student Richard Preziosi captured, marked, and photographed every adult strider along a 100-meter length of a stream near Montreal. Capturing them was the easy part, requiring only a quick flip of an insect net. Marking them involved applying little dots of enamel paint in various colors on various parts of the body, resulting in a unique code for each individual. To photograph them, they were placed in an apparatus that held them in a fixed position and at a fixed distance from the camera. In this fashion, more than 2000 striders were processed and released back into the stream.

That was just the beginning. During two years, at least once and sometimes twice a week, Daphne and Richard took a roll call of the striders by patrolling the 100-meter section of stream plus an additional 30 meters on each end. Striders don't move very far, so if a particular marked strider was alive, the chances were 76.0 percent that it would be spotted on a given day and 99.7 percent that it would be spotted over a two-week interval. Back in the lab, Daphne and Richard used the photographs to measure the dimensions of the striders in minute detail, including the length of the femur of every leg and the length of each of the three segments that make up total body length. They even measured penis length in males.

The purpose of all of this work was to observe the hammer blows of natural selection in action. If small striders are more likely to be eaten by predators, for example, then one by one, the smaller striders would fail to answer "Here!" on the weekly roll call, while the larger striders would remain.

From an evolutionary perspective, the only reason to survive is to reproduce. Daphne and Richard therefore needed to measure the reproductive success of both males and females. It's easy to record when a male is scoring, because he rides on the back of his conquest for a substantial period. In addition, Daphne and Richard periodically placed five males and five females in wading pools at the side of the stream to observe the mating success of each male in more detail.

The success of males in the pools corresponded closely to their success in the stream, enabling Daphne and Richard accurately to tell the studs from the duds.

To measure the reproductive success of females, Daphne and Richard periodically placed them in little cages within the stream and counted the eggs they laid over a three-day period. Additional experiments were conducted to make sure that caging females did not alter their egg-laying behavior. It's hard for me to imagine the amount of work that this must have required, several times for each of several hundred females throughout their egg-laying period.

God might or might not be recording our deeds in minute detail from on high, but a book of deeds exists for striders, thanks to the heroic efforts of Daphne and Richard. What are the results of their labors? In the Montreal area, striders become adults in summer and must bide their time until the following spring before they can mate and reproduce. Three-quarters of them perish during this long period, and it doesn't matter whether they are large or small, long-legged or short-legged. It's just a big lottery as far as the Grim Reaper is concerned. When the striders start reproducing in spring, however, the differences among individuals start to *make* a difference.

It turns out that smaller is better for survival during the reproductive season. One by one, the larger individuals don't show up for the weekly roll call. But larger is better as far as reproduction itself is concerned. Larger males are more likely to be riding on top of females, and larger females lay the most eggs. Male and female body size reflects hammer blows coming from different directions. Both would become smaller based only on survival, and both would become larger based only on reproduction. They end up somewhere in between, with the average female a bit larger than the average male.

Survival and reproduction do not depend on body size per se, however, but rather on the sizes of various *parts* of the body. Natural selection is so discerning that it is hammering each and every part of the body separately. It is not larger females that lay more eggs but females with larger abdomens. It is not larger males that are studs but small males with large penises and thick forearms, like a well-endowed Popeye. Total body size merely reflects the increases in these parts. When the parts are measured separately, as Daphne and Richard did with their painstaking measurement of the photographs, males and females are the same size in most respects except for those parts that are relevant to reproductive success. Males and females are also nearly identical in every respect before they become sexually mature. The hammer blows of natural selection beat them into the same shape, until they start playing the mating game.

The penis of a water strider is something to behold. It looks like a weapon, and it is. Its tip resembles a cross between a knife and a shoehorn, which is inserted into the female with the help of a thick leverlike midsection. Once inside, the male inflates a balloonlike structure covered with spines that can't feel good.

The mating behavior of the water strider consists of no courtship of any kind. A male simply leaps onto the back of a female and tries to insert his penis as fast as he can. If he succeeds, then he inflates his spiny balloon and can't be dislodged. Millions of generations of natural selection have perfected the penis of the male and the skill with which he performs his leap. In our phone conversation, Daphne described a particular leap that she recorded on high-speed videotape. In one frame, the male was a couple of centimeters away from the female and facing in the opposite direction. Two frames later, he was on top of her and properly positioned to insert his penis. It was that fast. Imagine the Rube Goldberg device that receives and processes information in just the right way to enable the male to perform such an acrobatic feat.

Female striders only need to mate now and then to have all the sperm they need. Beyond that, they want nothing to do with males, but the onslaught is endless. The female is larger and can shake them off some of the time, but they keep on coming. They are especially thick where the feeding is best. Her only choice is often to hide from them in marginal habitats or allow one to ride on her back so that the others will leave her in peace. The stream on my property might make me feel tranquil, but it's a treacherous world out there as far as the striders are concerned.

The animal world is full of tender love stories: ardent courtship, dazzling displays, nuptial gifts, devotion to offspring, even "till death do us part" monogamous pairings. The hammer blows of natural selection are capable of producing all of these forms, but in water striders, they led to a world where sexual harassment rules. I must confess that I am bothered by this conclusion, even though I am supposed to be an objective scientist. When I was playing the dating game, it took all my courage to approach someone and weeks to recover from even the most polite refusal. During my phone conversation with Daphne, I asked when it began to dawn on her that male water striders are such insensitive jerks. I can't be sure, but it seemed that she delivered her answer slowly and clearly, as if talking to someone a bit slow in the head.

"Well, David, when you're a reasonably attractive female, it begins to dawn on you around middle school."

THE PARABLE OF THE STRIDER comes with several messages. First, organisms are so well designed to survive and reproduce in their environments that their adaptations can take your breath away. For me, a scanning electron micrograph of a strider foot is as beautiful as the *Mona Lisa*. It's easy to understand why William Paley, writing before Darwin, would conclude that such exquisite design must signify a designer.

Second, as amazing as it might seem, these wonderful adaptations evolved by the hammer blows of natural selection. The evidence for this conclusion was compelling even in Darwin's day, based on evidence such as the haphazard side of evolution that caused such a motley crew of species to "Come on down!" to play the water-walking game. Today it is supported by the meticulous work of evolutionists such as Daphne Fairbairn, which provides a blow-by-blow account of natural selection in action.

Most important, natural selection can evolve adaptations that seem heaven-sent — and others that seem straight from hell. For millennia, theologians have pondered the problem of evil. How can the concept of a benign God be reconciled with the presence of so much evil in the world? This paradox does not exist for evolutionary theory.

The adaptations that seem heaven-sent and those that seem straight from hell evolve straightforwardly from the same process of natural selection. The parable of the strider tells us that evolution might, or might not, take us where we want to go — especially crucial to know as we contemplate how to make our own world a better place.

CHAPTER 4

The Parable of the Wasp

MY PROPERTY, LIKE A CITY, contains innumerable stories. Only a few steps from the stream where striders glide, I once encountered a completely different scene. Lost in thought, I stumbled into a cloud of insects hovering above the ground. Wasps! The eastern yellow jacket (*Vespula maculifrons*), to be precise. I retreated to a safe distance and watched their lazy arcs. Focusing on a single individual, I saw it descend slowly to the ground and disappear into the earth, like Alice down the rabbit hole. Focusing on the hole, I could see that all of the wasps in the cloud were in the process of entering in the same way. Other wasps were leaving the hole in a more determined fashion and flying out of sight. Still other wasps were presumably returning from a distance and joining the cloud, so that the cloud persisted even though the individuals composing the cloud were constantly changing. In the solitude of my property, I was observing something very similar to planes circling a major airport waiting to land.

Indeed, beneath my feet was a veritable city of wasps. Thousands of insects could be down there, constructing their homes, raising their young, and collecting food over an area of several square kilometers. This would be a *small* number as far as insect colonies go. Some termite and ant colonies have larger populations than any human city on earth. Even more amazing, if collective action is our gold standard, is that insect colonies work better than any human city on earth.

How is this possible, when an individual wasp has a brain the size of a mustard seed? In some of the largest and most successful ant colonies, the entire ant is only about the size of a mustard seed. Even more mysterious, how can the wasp colony beneath my feet be a miracle of collective action when the striders, only a few feet away, can't even say hello to one another in a civil fashion?

For millennia, people have used social insects as a metaphor to reflect on human society, although in very different ways. Plato thought that human society needs to be organized into classes, similar to social-insect castes. Christians and other religious believers have long compared their communities to beehives, not because their members divide labor but because they all work selflessly for the common good. That's why beehives are featured on the road signs in the Mormon-influenced state of Utah. The political theorist Bernard Mandeville (1670–1733) created a very different metaphor in his *Fable of the Bees,* which portrayed human society as a teeming beehive in which every individual is motivated by personal greed but the colony as a whole hums along harmoniously as a unit. This has become the guiding metaphor of our time, about which I will have much to say in future chapters.

These metaphors have had a powerful impact on human action throughout history, but none of them tells the true story of how harmonious social units, such as the wasps beneath my feet, can evolve by the same process that created the antisocial striders only a few feet away. The true story has emerged from painstaking research over the last century and is far more relevant for understanding the nature of human society than the previous metaphors we've relied on. Prepare to hear the real fable of the bees.

BEES, WASPS, AND ANTS belong to an enormous order of insects called the Hymenoptera, with more than 100,000 species described worldwide and many more unknown to science. The smallest species are smaller than the largest one-celled organism. The largest species are so large that they hunt tarantulas. Like striders, all of them are wonderfully adapted to survive and reproduce in their respective environments, thanks to the hammer blows of natural selection.

One of the most amazing adaptations of the Hymenoptera, shared with beetles (Coleoptera), butterflies and moths (Lepidoptera), and flies (Diptera), is the capacity for the young to be utterly different from the adults. A baby strider looks much like an adult and must subsist in the same way, but a baby wasp begins life as a soft-bodied larva and can live a completely different life from the adult.

The first wasps to appear in the fossil record, more than 200 million years ago, fed on plants as larvae, much like caterpillars, but some species made the transition to a predatory and parasitic lifestyle. This transition led to so many ways to survive and reproduce that the predatory and parasitic species now vastly outnumber the herbivorous species.

The main difference between a predator and a parasite is the size of its prey. A parasitic female wasp searches for victims much larger than herself, such as a caterpillar munching on a leaf, and inserts one or more eggs into its body with her needlelike ovipositor. Her young hatch into grubs that eat the victim from the inside out, saving the vital organs for last, before transforming into adults and flying away. This horrifying way of life (for the victim) was made vivid for a human audience by the movie *Alien*.

Some species of wasps allow their victims to remain active as they are being eaten, so that the victims protect themselves and eat food that ultimately will be used by the wasp grubs. The smallest wasp species live this way; smaller is better when it comes to approaching your victim and giving the fatal injection. The smallest known wasp species raises several offspring from a single *egg* of its insect host. That's why they're smaller than the largest single-celled organism.

Other wasp species paralyze their victims so that they become immobile, though still alive, as they are being devoured by the grubs. Immobilized prey cannot protect themselves, so the adult wasp must be large enough to drag its victim to a protected spot or to bury it before leaving in search of new victims. The giant wasp known as the tarantula hawk, which can reach two inches in length, lives this way. The adult females search for tarantulas by smell. Even tarantulas in their burrows are not safe, because they continuously shed molecules into the air that diffuse outward. Just as water striders' legs are exquisitely adapted to sense ripples, the antennae of female tarantula hawks are adapted to sense and follow the plume of molecules emanating from tarantulas.

Once a tarantula is found, the death match begins. The wasp circles the tarantula, which turns to face her with its fangs. Just as male striders are adapted to leap onto females in the blink of an eye, the female tarantula hawk senses the right moment to make her attack. They grapple, and she injects a neurotoxin into the tarantula that takes effect within seconds. Then she drags her immobilized victim into its own burrow or digs a burrow of her own. Depositing a single egg into the crypt with the victim, she fills in the burrow and leaves, never to return.

In virtually all wasp species, the males play no role in provisioning offspring and are adapted entirely to mate with females. We have already seen how the mating game causes male and females striders to become different from each other. In wasps, the differences can be even more extreme. When the hammer blows of natural selection beat males and females into different shapes, there is almost no limit to how different they can become. In many of the smallest wasp species, the males are blind and wingless. They mate with the females that emerge in their immediate vicinity and then die as the females fly off in search of victims for their young. In some species, local competition among males for females is so intense that the males have evolved weapons for lethal combat. A movie has not yet been produced that describes this horrifying scene, but imagine a world in which the only way for men to mate is to slaughter other men. Male tarantula hawks are physically not much different from females, but they are completely different behaviorally. Instead of searching for tarantulas, they defend territories against other males, perching on high spots and searching visually for females.

The ability to bury victims led to a key innovation. Instead of attacking victims large enough to raise an offspring, some wasp species evolved to kill smaller prey and bring them back to the same spot. Wasp species that live in this fashion have homing abilities that border on superhuman. Could you leave a spot in the woods, meander freely hither and yon, and find your way back to the exact same spot? Niko Tinbergen, who pioneered the study of animal behavior and shared the Nobel Prize with Konrad Lorenz and Karl von Frisch in 1973, showed how they do it. When they leave their burrow to search for prey, their tiny brains somehow remember the surrounding landmarks, which are used to relocate the burrow on their return. The way Tinbergen demonstrated this was simplicity itself. He surrounded a burrow with a ring of pinecones and then moved the ring a short distance while the female was away. Upon her return, she went to the center of the displaced ring. Moreover, the female was dumbfounded by the trick that had been played on her and was unable to locate her burrow only a few feet away. Natural selection had equipped her with an amazing ability to home on the basis of landmarks that don't move but not with the ability to solve a novel problem posed by moving landmarks.

Once some species of wasps could provision their young with small prey, other species evolved to exploit another nutritious food source: flower pollen and nectar. Pollen grains are tiny, but they can be collected with comblike appendages and aggregated into larger packages for the waiting grubs. In this fashion, wasps became bees. Pollen and nectar resources are so vast and diverse that thousands of bee species evolved from the first wasp species to cross this modest threshold.

Finally, from tunnels in the ground, it was another small step for wasps to become architects by creating protective structures of their own out of mud or plant fibers mixed with saliva to form something remarkably similar to paper. Some of the mud nests built by solitary wasps look like miniature Grecian urns. Most of us who have not become entirely citified have seen hornet nests dangling from tree branches or the much smaller open combs of the paper wasp, whose scientific name, *Polistes,* means "city founder," hanging by slender bases on the eaves of houses.

THESE DIVERSE LIFEWAYS REQUIRED tens of millions of years to evolve. As with the striders and other water-walking creatures, the vast hymenopteran order illustrates both the adaptive and the haphazard sides of evolution. If you are not already familiar with the ways of insects, then you might be amazed that so many adaptations can be bundled together into such tiny creatures and how physical and behavioral adaptations merge so seamlessly with one another. The ability of a female tarantula hawk antenna to sense a few molecules of tarantula odor, while ignoring the jillions of other molecules floating in the air, is a physical adaptation. Her abilities to follow the odor gradient to the tarantula and grapple with her formidable foe are behavioral. The neurotoxin that immobilizes the tarantula within seconds is physical. The behavioral becomes the physical when we study it closely enough. Every movement made by an organism is based on a physical environmental input, which initiates a physical chain of events inside the organism, which results in the physical movement of the organism — its behavioral output.

Even if you are familiar with the ways of insects, it is worth pausing to reflect with a new sense of wonder on the familiar. Take metamorphosis, for example. Everyone knows that caterpillars turn into butterflies, but what does this really mean? It means that some kinds of insects (but not others) have evolved the ability to be one kind of creature for part of their life and then to become a completely different creature, both physically and behaviorally, for the rest of their life. Talk about being born again.

These adaptations are so wondrous, when we take time to appreciate them, that we are tempted to call them miracles, just as parents often marvel at the miracle of their child's birth. The birth of your child is no less miraculous for being a product of evolution. In some ways, it is more miraculous to contemplate how something so wonderful could arise from laws acting around us, as Darwin put it, rather than directly by an intervening supernatural agent.

The haphazard side of evolution provides proof against creationism that was decisive even in Darwin's day, as we have already seen with the striders and other water-walking creatures. An omniscient God would have created the different kinds of hymenopterans in a single stroke, but evolution exhibits a property that in technical jargon is called *path dependence* and in more familiar terms could be called *you can't always get there from here*. Every adaptation must evolve from previous adaptations in a stepwise fashion. Wasps that provision their young with many small prey could evolve from wasps that provision their young with a single large prey, because that involved a single evolutionary step: the ability to return to the same burrow. Bees that provision their young with pollen and nectar could evolve from wasps that provision their young with many small prey, because that involved a single evolutionary step: the use of nectar and pollen rather than small prey. But wasps that provision their young with many small prey could not have evolved directly from parasitic wasps, and bees could not have evolved directly from wasps that provision their young with single large prey. That would involve too many evolutionary steps at the same time.

Path-dependence explains patterns of long-term evolution that could never be explained on the basis of an intervening God. The fossil record reveals vast stretches of time when there were no insects, only insects with direct development, only plant-feeding wasps, only wasps and no bees, and so on, as the step-by-step process of evolution unfolded. The taxonomic relationships among organisms provide a separate source of evidence supporting the fossil record. The first species to evolve a key adaptation, such as the ability to walk on water or to transform from a larval form to a completely different adult form, give rise to vast lineages of species, making it unlikely that the same key adaptation evolves in a separate lineage. That's why all species in some orders (such as Hymenoptera) have separate adult and larval forms, while none of the species in other orders (such as Hemiptera, which includes the striders) has the same capability.

Finally, the geographical distributions of species provide a third source of evidence supporting the other two sources. When a key adaptation evolves, the resulting lineage can spread only as far as barriers such as mountain ranges and oceans. Elsewhere, the same key adaptations must evolve separately if they are to evolve at all. It's little wonder that Darwin's ideas about identity by descent were accepted almost immediately, even as other aspects of his theory, such as the importance of natural selection, remained controversial. Never before had so many different aspects of the vast profusion of life on earth been explained on the basis of such a simple principle as the step-by-step process of evolution.

* * *

MOST INSECT SPECIES DO NOT form colonies, and for vast stretches of time, there were insects but no insect colonies on earth. Today insect colonies are so successful that they make up approximately half of the biomass of all insects. How could such a successful way of life be absent for so long? Because you can't always get there from here. The evolution of insect coloniality is heavily path-dependent and requires the right starting point.

The starting point for wasps, to the best of our knowledge, was a species of solitary wasp that provisioned its young with many small prey. I can find solitary wasp species on my property that still live like the ancestors of the social wasps. They are the ultimate in single moms and deadbeat dads. Every day and in every way, the males are designed exclusively to cruise for females. So many evolutionary steps would be required for them to share in parental care that it never happened. Wasp societies are sisterhoods, as we shall see, but in solitary wasps, all of the work falls to the single mom. She must build the home, find the food, and defend herself and her young from predators and marauders, including other females of her own species that try to invade her domicile, steal her provisions, and slaughter her young. These females are *tough*.

Given such a demanding life, the advantages of teaming up are tremendous. A group of females cooperating to raise their young could divide labor. Some could stay at home so that the young are tended and guarded at all times while others concentrate on foraging. The foragers could communicate to find food. If one female died, all would not be lost. Even large predators could be deterred by an angry swarm.

Given these advantages, the evolution of cooperative wasp colonies seems like a no-brainer, but two barriers must be overcome. First, there must be groups of females before they can evolve into cooperative groups. One way for this to happen is for unrelated females to begin sharing the same nest. Another way is for the offspring of a single female to remain in the same nest. The experts still argue over which is the most likely starting point, but we can regard this as a detail for our purposes.

Once we have groups, then we must weigh the costs and benefits of cooperation at two scales. A cooperative group might survive and reproduce better than an uncooperative group and solitary individuals. Within a group, however, cooperators might fare less well than individuals who behave in ways that we would call selfish in human terms.

These two comparisons underscore the fact that evolution is all about *differences* in survival and reproduction. Imagine a mutant plant that releases a toxin into the soil through its roots, to which it is somewhat immune. By poisoning its competitors more than itself, it survives and reproduces better than them and can be expected to evolve by natural selection. Poisoning counts as an adaptation in the evolutionary sense of the word. Never mind that the environment has been degraded and the poison-producing plant would itself be better off without the poison. Male water striders provide another example. Females and the population as a whole would fare much better with well-mannered males, but the jerks get the girls. In this fashion, traits that count as evil and self-destructive in human terms routinely evolve by natural selection. Evolution has much to say about the nature of human morality, but it is not as simple as equating "morality" with "adaptation."

The fact that evolution is all about differences reveals the main problem with the evolution of cooperation — not just in social insects but in all group-living species. Behaviors that are "for the good of the group" usually do not maximize one's relative advantage within the group. Imagine that you are an engineer assigned the task of designing two kitchen gadgets, a can opener and a knife sharpener. Now imagine that you must design a single gadget that performs both functions. You've probably seen all-in-one gadgets such as this, and they're often hilariously inefficient. Each task requires parts that interact in a certain way. Maximizing performance on one task reduces performance on other tasks as an inevitable by-product. Trade-offs are facts of life that engineers must cope with, whether they like it or not.

Trade-offs are a fact of life for evolution, no less than for engineers. Natural selection within groups will result in individuals that are streamlined to maximize their own relative advantage, no matter what the social cost or even their own individual cost over the long term, as we saw with the poison-producing plants. Call them knaves. Natural selection between groups will result in individuals that become model solid citizens, truly a part of something larger than themselves. When natural selection operates both within and between groups, the outcome can favor either knaves or solid citizens, depending on which level of selection is strongest. The eternal conflict between benefiting oneself and benefiting one's group, which suffuses religion and literature, also suffuses the biological world.

HIGHLY COOPERATIVE INSECT SOCIETIES EVOLVE when between-group selection trumps within-group selection. Numerous factors can stack the deck in favor of group selection. One is kinship. If group members are genetically related and if behaviors are based on genes, then cooperators will be clustered in some groups and noncooperators will be clustered in other groups. The segregation will not be complete unless group members are genetically identical, but the higher the degree of relatedness, the stronger between-group selection becomes and the weaker within-group selection becomes.

The environmental forces acting on whole groups as units provide another set of factors. In human life, collective disasters such as floods or attacks by enemies paradoxically bring out the best in people. They cooperate and derive pleasure from cooperating more than during peaceful times, when internal squabbles seem to take over. Recall the first pioneers to settle the Binghamton area, who spontaneously felt like family, only to lose the feeling when the town grew and people depended less on one another. Nonhuman species that live in groups experience the same differences. Some species are relatively secure from outside attacks and natural disasters that threaten the whole group. There is little to be gained from pulling together and much to be gained from besting your neighbor. In other species, the equivalent of September 11, 2001, or Hurricane Katrina can happen any day, making the profits of profiteering trivial compared with the benefits of collective action. In some species of ants, for example, hundreds of new colonies are established within an area that ultimately will be occupied by a single mature colony. As they grow, they fight among themselves until a single colony remains. Imagine a tournament among sports teams this intense. The teams plagued by internal conflict would drop out early. When collective action is this essential for survival, it's easy to appreciate how it can evolve despite the modest advantages of exploiting one's own group.

Systems of rewards and punishments that favor cooperation and inhibit cheating provide a third set of factors. Human solid citizens pay their taxes, for example, but it isn't voluntary. An elaborate machinery has been created to collect revenue from individuals to be used for the common good. The system isn't perfect, as we know only too well, but it works a lot better than if everyone gave according to the goodness of their hearts. Can you imagine the government collecting revenue in the same way that public radio stations conduct their pledge drives? It might seem amazing that social insects could evolve something comparable to a human tax code, but "policing" is a word that is often used by social-insect biologists to describe knaves that are punished for their misdeeds.

When these factors come together to make between-group selection a strong evolutionary force, something miraculous happens — in the sense of wondrous rather than supernaturally caused. Recall my description of our own hearing ability. An event at a distant location causes gas molecules to collide against one another. That chain reaction reaches our ear, which is structured to funnel the colliding molecules against our eardrum. The gentle vibration of the eardrum is transmitted by tiny bones to our inner ear, which triggers electrical impulses in our nervous system. Those impulses are constructed in just the right way to cause a response on our part that is appropriate to the distant event that started the whole thing off. This miracle of coordination is only one example of how single organisms are adapted to survive and reproduce in their environments, thanks to the hammer blows of natural selection.

When the group becomes the primary unit of selection, the interactions among group members become like the interactions among the parts of an individual organism. The social dance becomes as coordinated as the physiological dance. The members of the group become like organs, and the group becomes like an organism in its own right—a super-organism.

The concept of a group as a super-organism isn't new. It has a pedigree that stretches back to the ancient philosophers, before science existed as we know it now. Neither is it entirely benign. Most people want to live in harmonious groups, but they also want their own identity. The prospect of being a mere organ is fraught with ambivalence. We praise soldiers who sacrifice their lives in battle, but who wants to be like a skin cell that is routinely sloughed off? Then there is the terrifying prospect of super-organisms gobbling up smaller organisms and clashing with one another, like the brutal tournament among ant colonies. Nazi Germany regarded itself as a super-organism, and most of its atrocities were committed "for the good of the group." Isn't this a metaphor that we are better off doing without?

Evolution has much to say about the nature of human morality, but it is not as simple as equating "morality" with "group selection" any more than with "adaptation." For the moment, it is important to stress that super-organisms aren't a metaphor. They are a fact. They're really out there as part of the tangled bank, produced by the laws acting around us. They are also an essential part of the human evolutionary story. If we choose to deny their existence when we select metaphors to live by, we are denying reality. It might be part of human nature to deny reality when it isn't useful, but for the game of science, denying reality is against the rules.

THE CONCEPT OF SOCIAL-INSECT COLONIES as super-organisms also has a long pedigree and has been updated by my colleagues Bert Holldobler and Edward O. Wilson in their 2009 book *The Superorganism: The Beauty, Elegance, and Strangeness of Insect Societies.* Ed is world-famous and arguably the most influential evolutionist of our time. Bert is Ed's longtime collaborator on the study of social insects and shared the Pulitzer Prize with him for their previous book *The Ants.* Both books are remarkable for being literate and lavishly illustrated in addition to scientifically authoritative. The cover of *The*

Super-organism depicts the nest of the honey pot ant (*Myrmecocystus*), in which some members of the group become living food-storage vessels, hanging from the ceiling with hugely distended abdomens. The inside cover features a photograph of weaver ants (*Oecophylla*), which live arboreally in nests constructed from leaves. The adults form into a living net to draw the leaves together, which are then cemented by silk secreted from larvae held in the jaws of adults, like living glue guns. Even if we don't aspire to be storage vessels or glue guns in our own societies, we can appreciate that these insect societies are functioning as single coordinated units, similar to the coordination among the parts of individual organisms. A human city would do well to emulate the degree of coordination found in bodies and beehives, if not the specific mechanisms.

One reason that insect societies seem so strange to us involves path dependence, which dictates that all current adaptations must be derived from previous adaptations. That is why they often resemble Rube Goldberg devices, despite their impressive functionality, unlike what an engineer or a supernatural agent would design starting from scratch. Wasp societies are built from the parts made available by single-mom wasps and their kids, including one feature that seems surpassingly strange: the larvae feed the adults.

The scientist who worked out this strange story is James H. Hunt, and the story of how he became a scientist illustrates the importance of path dependence in our own affairs. Most of Jim's ancestors came from the hollows of Appalachia. Exactly when they arrived from Europe was not passed down through the generations. The rural poor are even less likely to become scientists than women are, but the pinball machine of life caused Jim's father to acquire an education, which made it easier for Jim to become a scientist.

Jim grew up in one of the first suburban developments that were springing up everywhere to accommodate returning GIs after World War II. Nature is scarcely to be found in modern suburban developments, but the first houses were located in old fields filled with wildflowers. Jim's earliest memories are of catching butterflies visiting the flowers by pinching their wings between his fingers, releasing them into the family's screen porch—and watching fascinated as they battered themselves to death against the screen. When he became a little older, he remembers catching honeybees in mayonnaise jars, filling his toy wagon with a layer of water, inverting the jars over the water—and watching fascinated as the bees drowned. As Jim humorously told me by phone, ask entomologists how they became interested in insects, and they typically began by torturing them as children!

Later, two of Jim's school friends happened to be children of wildlife-biology professors at North Carolina State College. One of the professors, Thomas Quay, nurtured Jim's interest in nature as a high school student. When Jim became a college student at North Carolina State, he became the assistant of the graduate students in Quay's lab and began to see that he could make a career out of what he already loved. He hung out with the grad students and assiduously copied their dress and mannerisms, just as an inner-city kid might try to enter a gang. His grades suffered—in fact, he didn't receive a single A—but his informal education was incomparably richer than his formal coursework.

When Jim was growing up, biology was still largely a matter of collecting, cataloging, and describing. Biologists specialized in a particular taxonomic group, such as birds or insects. By the time he became a graduate student, however, this information was increasingly being synthesized by evolutionary theory. The new fashion was to study all species as the manifestation of the laws acting around us, as I also discovered as a graduate student. The scientists who accomplished this transformation include a few who are known to the general public, such as Ed Wilson, and many others who are revered by those who understand their contributions, such as G. E. Hutchinson, Lawrence Slobodkin, Robert MacArthur, Daniel Janzen, David Lack, John Maynard Smith, John Vandermeer, Gordon Orians, Richard Root, and Robert Colwell.

Thanks to this generation of evolutionists and ecologists, Jim made the transition from studying birds to becoming interested in ecological communities of all sorts. Jim was always good at creating opportunities for himself, and before long he was at Harvard as a postdoctoral student, working on an ant-community project with one professor named Otto Solbrig but also rubbing elbows with Ed Wilson himself, who even forty years ago was preeminent in his field.

It was there that Jim had his epiphany about the importance of larvae feeding adults in the evolution of wasp societies, as he recounts in his book *The Evolution of Social Wasps*. The basic fact was known as a natural-history observation but had not been accorded any special significance. Jim now saw it as key in the transition from solitary to social life and therefore to the path-dependent organization of current wasp societies. Proving his conjecture became his goal for the rest of his career.

The larvae of solitary-wasp species have saliva for the same reason that we do: to lubricate their food and begin the digestion process. Everything they eat is brought to them by their mothers, but there is still a situation in which the mother takes back from her offspring. When food becomes scarce, she has two choices: either perish herself or eat her offspring's provisions and even her own offspring so that she can survive and reproduce another day. No matter how ghoulish it might seem to us, infanticide has evolved by natural selection in thousands of species of all kinds, including plants, insects, and primates, in yet another demonstration that the properties of all species can be understood by the same laws.

Given mothers who occasionally eat their offspring, how are offspring likely to evolve by the hammer blows of natural selection? To save their own skins by appeasing their hungry mothers with nutritious saliva! In this fashion, the paradoxical behavior of offspring feeding their mothers evolved in solitary-wasp species and became the starting condition for the evolution of sociality.

In social wasps such as *Polistes*, the city founder, one or a few females initiate a nest and begin to raise young. The first adults to emerge stay at the nest as workers rather than flying off to reproduce on their own. Getting them to stay was a key event in the evolution of wasp sociality. Jim realized that a solution to this problem might have been for the new adults to get a better meal at home than by flying away. Larval saliva became a *primary* resource for the workers, rather than an emergency meal during hard times.

In this path-dependent fashion, the larvae became a kind of

group stomach for the colony. It didn't necessarily have to happen that way, and it didn't happen that way in other lineages of social insects, where the larvae don't feed the adults, but it was the particular path taken in the social wasps.

Given that path, larvae feeding adults has become so foundational for the organization of wasp societies that preventing it is as fatal as separating your stomach from your esophagus. In one experiment, Jim diminished the larval saliva from colonies of the paper wasp *Polistes metricus* in the field by touching a pipette to their mouthparts, prompting them to regurgitate to him, much as they would to an adult wasp. As a control, he did an equivalent amount of poking and prodding in other colonies without removing the saliva. He did this only twice a week for three weeks and once a week for another three weeks, but the results were devastating. By the end of the season, the colonies with the saliva removed had produced no reproductive offspring at all. In experiments performed by other wasp researchers, when the older saliva-producing larvae are removed, the workers fly away, even though young larvae remain. When the saliva-producing larvae are allowed to remain but the workers are denied access, the workers fly away, and the queen dies. As with the solitary wasps that couldn't find their nests when Niko Tinbergen moved the ring of cones, social wasps are dumbfounded and unable to respond appropriately to novel alterations in their environment, despite being exquisitely responsive to other alterations more typical of their evolutionary past, such as fluctuations in the abundance of their prey.

This is only one example of a group-level adaptation in which the social dance among members of the group becomes comparable to the physiological dance among the parts of a single organism—or perhaps the dance that we perform in our own societies. The physical architecture of the nest is another miracle of coordination that takes place without any centralized planning. I wasn't so foolish as to dig up the yellow-jacket nest beneath my feet on my property, but I know from earlier intrepid naturalists that it is an elegant high-rise apartment building, consisting of layers of horizontal floors for raising the larvae enclosed within an outer dome, all made from paper manufactured by the workers and suspended like a piñata from the top of a cavity excavated by the workers so that there is a gap of air all around. A single entrance is located at the bottom of the nest, making it a fortress against intruders. It challenges the imagination how such an elaborate structure can be built by insects, until we realize that even more elaborate structures, such as our skeletons, are built by the cells of our bodies, also without any centralized planning.

Yellow-jacket nests are mere huts compared with the largest ant and termite colonies, which can rise taller than a person above the earth and below in vast cities of underground chambers. The famous leaf-cutting ants (*Atta*) of the tropics build some of the largest underground cities. These ants evolved in the New World tropics (Central and South America) and never dispersed to the Old World tropics (Africa and Asia), providing another demonstration of the haphazard side of evolution. They have abandoned the predatory ways of their ancestors and harvest vast quantities of leaves, which are taken underground, chewed, and used to cultivate a fungus that is eaten. In other words, these ants discovered agriculture long before our own species did. Their underground chambers and tunnels are so vast that the excavated soil piled aboveground can weigh 40 tons. In one study described in Bert and Ed's new book on super-organisms, a team of Brazilian scientists headed by Luiz Forti poured liquid cement into the entrance of a single colony, which flowed into the tunnels and chambers before hardening to make a solid mold; 6300 kilograms (6.3 tons) of cement and 8200 liters of water were required. When the earth was excavated around the mold, a vast network of tunnels, ducts, and chambers was revealed, extending as far as 8 meters underground.

Insect colonies have even evolved ways to evacuate their nests collectively. Fire ants, which have become major pests in the southern United States, often inhabit flood plains in their natural environment. When their nests become flooded, the colonies form themselves into living spheres that can be the size of a basketball, with the larvae and pupae at the center. The spheres float on the surface of the water and actually rotate so that no individual is submersed long enough to drown. I was so amazed by this description that I asked Ed Wilson to confirm it. He laughed and recalled seeing many spheres floating, like Noah's ark, on the flooded fields of Alabama where he grew up. Insect colonies have evolved the ability to walk on water collectively.

Beyond that, insect colonies have evolved to think as collective units. Earlier, I said that the behavioral becomes the physical when examined closely enough. Consider an individual organism deciding where to live. It could be you looking for an apartment or a fox looking for a den. An intelligent choice requires searching for options, gathering information about each option, comparing the information on the basis of certain criteria, and finally choosing the highest-ranked option. Research at the behavioral level has shown that animals such as foxes are pretty good at making the right decisions about where to live, what to eat, with whom to mate, and so on. Research at the neurobiological level shows how it happens in terms of physical interactions among neurons. The behavioral becomes the physical at the neurobiological level.

Research on honeybees and other social insects shows that colonies can also make smart decisions based on social interactions that are remarkably similar to the neuronal interactions within individuals. The concept of a group mind might sound like science fiction, but it has been demonstrated beyond a shadow of a doubt in social-insect colonies. Moreover, it is expected in terms of evolutionary theory at a fundamental level. After all, the reason a fox makes a smart decision is that a long period of natural selection favored those that receive and process information — listen and reflect — in just the right way. Those who made dumb choices were not among the ancestors of today's foxes. The same is true for social-insect colonies when natural selection operates at the group level.

HOW DOES THE TRUE FABLE of the bees compare with the metaphors about social insects that we have relied on in the past? Aristotle thought that human society needs to be organized into classes, similar to social-insect castes. Even though ancient Greece was the cradle of democracy, it was primarily a democracy of powerful men who kept slaves and deprived women and less powerful men of rights. It's true on the basis of design considerations that groups can often achieve collective goals best by dividing labor, but this does not necessarily require inequality or lack of equal opportunity. Indeed, the human path to sociality is based on egalitarianism, as we shall see.

How about the metaphor of Christians working selflessly for their communities, just as bees work with equal zeal to gather honey for their hives? Religious communities can indeed be considered the human equivalent of beehives, but the mechanisms that cause both to function so well as collective units are far more complex than zeal. An enduring religion bristles with control and coordination mechanisms to guide its members through their social dance. Moreover, when we appreciate the implications of path dependence, the specific mechanisms that cause beehives and religious groups to function well as collective units need not be the same. They aren't even the same for bees and wasps.

Finally, there is Bernard Mandeville's *Fable of the Bees*, which portrayed human society as a teeming beehive in which every individual is motivated by personal greed but the colony as a whole hums along harmoniously as a unit. Here is a passage from the fable, which was written in humorous verse.

> As Sharpers, Parasites, Pimps, Players,
> Pick-Pockets, Coiners, Quacks, Sooth-Sayers,
> And all those, that, in Enmity
> With down-right Working, cunningly
> Convert to their own Use the Labour
> Of their good-natur'd heedless Neighbour:
> These were called Knaves; but, bar the Name,
> The grave Industrious were the Same.
> All Trades and Places knew some Cheat,
> No Calling was without Deceit.

The sentiment of this passage has become the guiding metaphor of our time, but it is only half right when viewed in the light of the true fable of the bees. The right part concerns self-organization. It is true that individual bees and other social insects don't have the good of their group consciously in mind. They don't have *anything* consciously in mind in the human sense; they're only insects. Each member of a colony behaves according to a simple set of rules, such as a larval wasp that offers saliva when touched by an adult or a pipette held by Jim Hunt. The simple rules miraculously combine to produce colonies that function better than any human city without any centralized planning whatsoever.

The wrong part concerns individual greed. The true fable of the bees is about a conflict between levels of selection. Within-group selection produces individuals that are designed to benefit themselves at the expense of their neighbors, the essence of knavery. Between-group selection produces individuals that behave for the good of their groups, the essence of solid-citizenry. It isn't just wrong to say that knaves are no different from solid citizens and that a well-functioning society can emerge spontaneously from interactions among knaves. It's as wrong as it can possibly be. The kind of self-organization that evolves by group selection should never be confused with individual greed, for bee colonies or for human societies.

THE TRUE FABLE OF THE BEES has a prequel. For hundreds of millions of years during the early history of the earth, life consisted only of bacterial cells. Then a new kind of cell arose with a nucleus and other structures such as mitochondria and chloroplasts that are called organelles because they work together to make the cell function, just as our organs work together to make our bodies function at a larger scale.

Nucleated cells were thought to evolve by small mutational steps from bacterial cells until a biologist named Lynn Margulis proposed a radical alternative in the 1960s. To be precise, she revived a theory that had been proposed in the early 1900s and dismissed as crazy. The theory was that nucleated cells are groups of bacterial cells that evolved to be so well integrated that they became super-organisms in their own right. In other words, according to Lynn, everything that I have recounted for the evolution of social-insect colonies from solitary insects took place during an earlier epoch of time for the evolution of nucleated cells from bacterial cells.

Lynn's theory of the symbiotic cell is now widely accepted, and she is justly honored as one of the great biologists of our time. Once nucleated cells evolved, the process repeated itself for the evolution of multicellular organisms, as the evolutionists John Maynard Smith and Eörs Szathmáry recount in their books *The Major Transitions in Evolution* and *The Origins of Life: From the Birth of Life to the Origin of Language.* Amazingly, everything that we recognize as an individual organism today is a highly integrated society of lower-level organisms. The concepts of "organism" and "society" have become indistinguishable from each other.

Whenever I visit my tree house and property, I return refreshed. When we consider what the creatures on my property have to teach us about ourselves, we become intellectually refreshed. The themes that we have touched on in the last two chapters have been pondered by philosophers, religious sages, and storytellers throughout the ages. More recently, they have been pondered by theorists of all stripes, with and without the trappings of science. Never, until now, have they been pondered in a way that makes the human condition so continuous with the rest of life, based on the same laws acting around us. We are lucky to be alive at this moment of intellectual history. Can we use our new scientific understanding to improve the quality of human life in a practical sense? To answer that question, I must return to my city of Binghamton, New York.

CHAPTER 5

The Maps

WHAT IS A HUMAN CITY? Is it like a pool of water striders, colliding with one another in pursuit of their own goals? Or is it like a wasp colony, working together, without necessarily knowing it, for the common good? If a city is not functioning well as a collective unit, can it be made to function better with enough scientific and evolutionary know-how?

One thing is for sure: I can't answer these questions for my city of Binghamton unless I listen and reflect on it in the right way. Recall how narrowly the striders perceive their world. Wasps are the same, even though they process information for the good of their group rather than themselves. My sensory and processing abilities are the same ones that evolved to help us survive and reproduce in small groups on the African savannah. I simply don't have the equipment to perceive a whole modern city and to understand how it works, let alone how to improve it. As the Brooklynites like to say, *fuhgeddaboudit*.

Science and technology have become absolutely essential to listen and reflect—acquire and process information—in the modern world. We need them to peer into outer space, to see something as small as the super-hydrophobic hairs of a strider leg, or to behold what happens inside our brains when we think and feel. Now I need science and technology to listen and reflect on my city of Binghamton. I also need evolutionary theory to ask the right questions. I need to study my city in the way that Daphne Fairbairn studies water striders, Jim Hunt studies wasps, Tim Clutton-Brock studies meerkats, Peter and Rosemary Grant study the Galapagos finches, and Jane Goodall studies chimps.

Throughout my career, I have danced from subject to subject, organism to organism, as if disciplinary boundaries didn't exist. That's what so special about evolutionary theory. When it comes to the nuts and bolts of studying any particular subject and organism, however, there is always a lot to learn. When I studied mites that ride on the backs of beetles, I had to learn about mite and beetle taxonomy. When I studied how sunfish become adapted to different environments within the same lake, I had to learn the newest morphological measurement techniques. When I studied religion, I had to immerse myself in the world of religious scholarship. I always started as a novice, with much to learn from the experts. Why should the experts bother with a novice who waltzes in out of nowhere asking to learn from them? Because I was posing a new set of questions informed by evolutionary theory. That was my contribution to the relationship.

Now I was leaving my comfort zone one more time to become a novice on the subject of community-based research. I would need to pick up new techniques and work with people who didn't know me from Adam. Most of them wouldn't be scientists, and some wouldn't care a rat's ass that I'm a professor. Some would claim to believe evolution and others to disbelieve it. Almost everyone would associate evolution with dinosaurs, fossils, and human origins. They would start out mystified by how evolution might relate to the city of Binghamton and why I was asking for their attention. There was as much humor potential as scientific potential in my fool's errand.

My first goal was to measure the inclination of people to act as knaves or solid citizens at the scale of the whole city. Were they more like water striders or more like wasps? One convenient aspect of studying people, as opposed to other creatures, is that you can directly ask them questions. How would you answer the following questions about yourself on a scale of 1 (this doesn't describe me at all) to 5 (this describes me exactly):

- I think it is important to help other people.
- I resolve conflicts without anyone getting hurt.
- I tell the truth even when it is not easy.
- I am helping to make my community a better place.
- I am trying to help solve social problems.
- I am developing respect for other people.
- I am sensitive to the needs and feelings of others.
- I am serving others in the community.

These questions are part of a survey called the Developmental Assets Profile (DAP), created by a nonprofit organization called Search Institute (www.search-institute.org) that has been promoting healthy kids and communities for more than fifty years. The DAP is designed to measure a number of internal assets (the way individuals are) and external assets (the social environments inhabited by individuals) that contribute to healthy development. Across the country, schools and other organizations have used the DAP to measure and improve the assets of their youth and communities. I didn't need to start from scratch in my own effort. As a newbie in the field of community-based research, I could rely on the expertise of veteran organizations such as Search Institute.

If you are the skeptical sort, your mind might be starting to flood with questions about now. Can you really measure how nice and nasty people are just by asking them? Especially schoolkids, who are probably bored silly and couldn't care less about telling you, even if they knew? Wouldn't some of them lie about how nice they are because it is the socially acceptable answer? And who's nice or nasty all the time, anyway? Aren't most of us nice under some conditions and nasty under others? Surveys like this are as common as air pollution and road noise. Haven't they already delivered their dubious insights by now? What's new about giving someone else's survey to another group of kids?

These are all fine questions that require answers. Skepticism is part of the scientific method. Self-report surveys have their problems, but they are a useful way to begin. I intended to validate the results with other methods. As for using someone else's survey on another group of kids, I had bigger plans.

Questions about nastiness and niceness, knaves and solid citizens, needn't be confined to kids. I wanted to know about the whole life cycle, but public-school students were the best place to begin for three reasons. First, there is an outpouring of interest in helping our children develop into healthy and productive adults, which is expressed at the level of scientific research in addition to parents, schools, and communities. Second, insofar as adults are influenced by their upbringing, the best way to improve their civic virtues is to begin when they are children. Third, if you want to administer a survey to thousands of people, there is no better person to know than the superintendent of the public school system.

That is why I found myself at the front entrance of the old Christopher Columbus School on a gray February day. Originally one of the city's schools, it now houses the school administration and classes for students who need special services, including those who made so much trouble in the city's current high school that they had to be removed. It is located in a section of the city that has seen better days. Downtrodden neighborhoods such as this one were invisible to me in my previous life because I had no reason to visit them and good reasons to avoid them. Now I briefly worried about the welfare of my car as I signed in at the front desk attended by a security guard. I am confident and outgoing when I'm in my element but can become shy and withdrawn otherwise. I might want to make a difference in my city in some abstract and intellectual sense, but I had to admit that some growth on my part would be required before I could feel comfortable mingling with the very people I was trying to understand with my survey.

Peggy Wozniak, Binghamton's school superintendent, is a wonderful person. So wonderful that whenever I mention her name around town, the most likely response is "Peggy! She's a wonderful person." Tall and fit, with silver hair—she has run marathons and still runs—the most noticeable thing about her is her open smile. Peggy didn't become a scientist, but here's how the pinball machine of life caused her to become a school superintendent.

Peggy is mostly of Polish descent and was born in the house built by her grandfather, a tailor from Warsaw who emigrated to Buffalo,

New York, in the 1880s. Her neighborhood was so ethnic when she grew up that Polish was still the primary language spoken. Peggy didn't realize until later how poor her family and her neighborhood were. It didn't seem to matter, because everyone was roughly equal. The Wozniaks, however, were known for their planning and upward mobility. Peggy's mother became a teacher, and her father took business courses and became an accountant. Their marital relationship was unusually egalitarian for its time. Peggy's father was the chief cook and even became the main domestic when he retired early because of health problems.

Peggy thrived in school and college and decided to make her career in elementary education. She also made the most of her young adulthood in the early 1970s, traveling throughout the United States and Canada with friends by hitchhiking and by train. She was tempted to settle in Southern California but returned home to tend her ailing mother and became a sixth-grade teacher at a middle school on the outskirts of Buffalo. Others might regard this as a disappointment, but she described it as a great experience, thanks in part to an enlightened principal who gave Peggy and her fellow teachers enough freedom to pursue their own ideas and develop into a strong team.

Peggy realized her dream of moving to Southern California after her mother passed away. She lived a carefree life with three other women from Buffalo, picking up work where they could find it. A job as a substitute teacher turned into full-time employment at an "alternative" high school, which was really a repository for the kids everyone else had given up on. Ever ebullient, Peggy regarded this as her most rewarding teaching experience. She was encouraged to go into administration, took the necessary training, and gradually worked her way up the ladder at a variety of school districts, some that were predominantly Latino and Asian. Along the way, she married a husband who, like her parents, was fully comfortable with an egalitarian relationship. Finally, the pull of family drew her back east, and she chose the Binghamton superintendent position among several options, as much for its challenges as for its opportunities.

As soon as Peggy arrived in Binghamton, she started to plug into the many civic organizations that I had only heard by name: the Binghamton Rotary Club, the Broome County Urban League, the Regional Advisory Board for Lifetime Healthcare Companies. I know how Peggy would answer those questions on my survey listed above, and she wouldn't be lying.

As I listened to Peggy's story, I was struck by how her energy, friendliness, and openness to experience turned everything she encountered into an opportunity. I didn't realize what it meant to be a school superintendent or how much she is honored within her profession until I did a little background checking. Peggy supervises 1100 employees, manages a budget of $91 million, and was just entered into the Commission on Independent Colleges and University's Hall of Fame for K-12 educators who have made a difference in New York. This was the silver-haired person who greeted me with a firm handshake and a broad smile in her office at Christopher Columbus School on Binghamton's Hawley Street.

PEGGY WAS THRILLED THAT AN egghead professor who appeared out of nowhere wanted to make her schools and the city of Binghamton a better place. She already knew about Search Institute and its Developmental Assets Profile. She had even worked with them at previous times during her career. *Of course* she would help me administer the DAP to the Binghamton students, especially since I was offering to do most of the work without imposing an additional strain on her budget. She also agreed to help me combine the information from the DAP with other information about the students in the school records, with due respect to their privacy, of course. She assigned me to someone named Doug Stento to help me get the job done.

Doug was the first native of Binghamton I had met since starting the Binghamton Neighborhood Project, as I was beginning to call it. His grandfather emigrated directly to Binghamton from Italy during the boom times of the Endicott-Johnson shoe company, which was such a draw for European immigrants that some of them stepped off the boat onto Ellis Island knowing only the English phrase "Which way EJ?" Whole communities emigrated, not just individuals. When

Doug's grandfather decided to get married, he returned to Italy and was paired with a girl so young that they had to travel to Brazil to get married before returning to Binghamton.

Instead of working directly for EJ, Doug's grandfather started a construction company, which Doug's father built into a successful business. Doug was the first member of the family to attend college, but only thanks to a clever manipulation by his father. At the age of fifteen, Doug announced that he had no intention of going to college but instead planned on taking over the family business. "In that case," said his father, "you will begin as I did." Doug was assigned to a construction crew that summer, and the foreman was instructed to treat him like anyone else, even though he was the boss's son. On the first day, Doug was assigned to a crew of older men digging a ditch with pickaxes. It seemed to Doug that they were working too slowly, and he was eager to show what he was made of. Thirty minutes later, Doug was exhausted, and one of the older men said, "We have seven and a half hours left to go, son." Doug learned two things from his summer's experience: first, a respect for the working man; second, a strong desire to go to college.

When Doug started to think about a career in college, he enjoyed teaching but also wanted a higher standard of living. Once again, his father came through with some good advice, comparing a teaching job to the fabled tortoise racing the hare. Another career might offer more money and glamour over the short term, but a teaching job offered security and a good pension. Doug took his father's advice and never regretted it. He became one of the first teachers trained in health and education. This was during the early 1970s, and states were only beginning to require subjects such as sexuality, mental health, substance abuse, and nutrition to be taught in public school. Jobs were plentiful, and Doug intended to leave Binghamton, but just as he was making his decision, his father had a heart attack, and Doug took a job in his hometown so he could help out with the family.

Teaching subjects such as sexuality in school was controversial, and Doug started a revolt among the conservative members of the community when he chose to use a rigorous freshman college textbook in his high school classes, rather than the previous antiquated textbook written at a fifth-grade level. Aided by an organization outside the community, local conservatives demanded that the book be banned and claimed that Doug was encouraging his students to take drugs, have sex, and masturbate. Doug was required to defend himself at a raucous school-board meeting with an angry and well-organized audience. He patiently explained that the previous textbook was an insult to the students' intelligence and that the new textbook was a resource for class discussions that were highly relevant to the lives of the students. Then, to Doug's surprise, one of his students rose from the audience, praised his class, and concluded by saying, "As a healthy young man of seventeen, I learned how to masturbate long before entering Mr. Stento's class." The room burst into laughter, and Doug carried the day.

I didn't realize until I started talking with public-school teachers how many hats they are required to wear. Not only did Doug teach biology and earth science in addition to health, but he coached sports teams, first wrestling and then track and football. He loved being a coach because it allowed him to interact with his students on a different plane from being a mere teacher. He became a surrogate father to some of the most underprivileged youth and helped them attend college on sports scholarships, their only conceivable chance for a higher education. Doug had dozens of stories to tell about his students, some triumphant and some tragic, recalling names and dates as if they were yesterday.

After meeting Doug and Peggy, I was starting to feel inadequate. I had lived in Binghamton for more than twenty years without lifting a finger to help anyone. Compared with these two titans of civic virtue, what did I have to offer with my theories and my survey, fretting about my car on my first visit to the Columbus school?

AS I BRAVELY STARTED my new line of research, my novice status weighed heavily on me. Everything was new; nothing was routine. I wanted to add questions to Search Institute's DAP, but how many, and which should I add? Some items, such as religious affiliation, had more than five possible answers. How could I fit these into the format of an answer sheet designed to be scanned by a computer that provides only five possible answers for each question? I still remember sitting at the printing office, giving the survey a final scan, and noticing only then that two questions were numbered 52, which would have caused all subsequent questions to be misnumbered on the answer sheets. How many times had I scrutinized the survey without noticing that mistake? When at last the 3000 copies of the survey were printed, I rushed over to the printing office and carried the heavy box back to my office on a hot spring day with an absurd feeling of accomplishment.

Before I printed the survey, I had to get it approved by my university's human-subject review board. All human research, no matter how innocent, is rigorously policed in this manner. A training course that I am required to take at three-year intervals describes some disasters of unregulated research that make policing necessary. One professor in New York City wrote fake letters to dozens of restaurants stating that he had experienced food poisoning at them, setting off a panic of throwing out food and contacting lawyers. What was he thinking? A sociologist studying homosexuality made dates with men in public restrooms, used the information to discover where they lived, and then visited their homes disguised as a repairman. By this means, he discovered that homosexuals are not an isolated group but come from all walks of life and are frequently family men. Perhaps there were even a few pastors and senators among them. I find this information fascinating, but the way it was gathered certainly can't be condoned.

My survey was tame compared with these whoppers. Information from school records that was added to the survey included such things as age, grade level, eligibility for free lunch (an indicator of poverty), and residential location. Student identity was protected by replacing the name of each student with an arbitrary ID number. No individual scores or residential locations were to be reported, only statistical averages. With these precautions, we received the blessing of our human-subject review board.

Doug also had work to do on his end. A school system is not like an army, where everyone is expected to follow the orders of superiors. The teachers are already overworked with their normal duties, and class periods are already filled with other activities. When someone in Doug's position asks a teacher to administer a survey that will require the bulk of a time period, it is more like a request than a demand and must be carefully negotiated, even when he is acting on behalf of the school superintendent. Besides, this was not just a survey but *yet another* survey. Teachers are asked to poke and prod their students with surveys all the time, usually without any discernible benefits. Doug therefore had to exercise all his social networking skills to cajole his colleagues into taking part. If it weren't for his sterling reputation, the Binghamton Neighborhood Project would never have become airborne.

AT LAST, THE DAY ARRIVED when the survey was to be given to the students. I packaged them into folders for each class, dropped them up off at Doug's office in the morning, and retrieved them at the end of the day. By now, I had become so bent on avoiding mistakes that I had to do everything myself, even the mundane tasks that could have been assigned to a student or a secretary. The stack of answer sheets was more than a foot tall, but I examined each and every one of them before taking them to the computer center to be scanned. This proved to be necessary because numerous students failed to follow the instructions. Some failed to write their last names first. Others wrote their names in the top row but didn't fill in the corresponding circles required for their names to be read by the computer. Others wrote in ink despite being clearly instructed to write in pencil. Some forgot to indicate their gender, and it wasn't always obvious to guess from their names. Every time I caught a mistake, I laboriously erased it and filled in the correct information, sometimes recopying the entire sheet. Then there were a few jokers who entered the names of celebrities, such as A-Rod, and had to be discarded entirely. Just as a telescope lens requires countless hours of polishing, I polished my data until every last blemish was removed.

Once the answer sheets were cleaned to the best of my ability, they were ready to be scanned at our computer center, in the same way that multiple-choice exams are processed. Within minutes, my foot-high stack was converted into an electronic file and a printout on oversized computer paper. With mounting excitement, I e-mailed the electronic file to our Geographical Information Systems (GIS) center. GIS is a sophisticated technology that has developed over the past few decades to visualize and analyze spatial information. MapQuest, Google Earth, and the navigators in your car are based on GIS technology, which can create maps and link them with other information for any location in the world. Thanks to GIS technology, I can create maps of Binghamton that are linked to my survey data. A map of the residential locations of the students who took my survey, for example, would show a standard map of the city and a scattering of the thousands of points obtained from my electronic file. These points can be color-coded to indicate how each student responded to the various survey items, such as those listed at the beginning of this chapter. Insofar as the DAP gathers trustworthy information, I was on the verge of seeing, for the first time, how kids who report being knaves and solid citizens are geographically distributed among the neighborhoods of Binghamton, New York.

An especially striking visualization of GIS data is made possible by a technique called kriging. Imagine picking a random location in the city of Binghamton. Chances are that a student does not reside at that exact location, but so many took my survey that numerous students are sure to live close by. With kriging, the computer estimates a value of an imaginary student at that location based on the values of real students at nearby locations. When this is repeated for all locations, the scattering of color-coded points is converted into a continuous landscape, with hills and valleys, looking exactly like a topographic map.

I'll always remember the moment I first saw the krig map for the questions listed at the beginning of this chapter, which strive to measure nastiness and niceness, knaves and solid citizens. If every neighborhood had exactly the same mix, then the krig map of the city would look like a flat plain. Instead, it looked like the Himalayas, with tall peaks representing neighborhoods in which most of the students reported being solid citizens and deep valleys in which most of the students reported being knaves, or at least lacking a sense of civic virtue. Some of the hills and valleys of civic virtue corresponded to real hills and valleys. My own house is on a physical hill, for example, and I was pleased to see that it also sits on a civic hill that slopes down to a valley by the river. Other neighborhoods by the river were civic hills, despite being situated in a physical valley. The downtown area of Binghamton looked like Death Valley, but volcanoes of civic virtue rose here and there. It was intoxicating to roam the map with my eyes, trying to relate the hills and valleys to what little I knew about the various neighborhoods. What were the social counterparts of geological forces that brought them into existence?

This was only one of many maps that I could generate from my survey data and link to other information about my city of Binghamton. I couldn't wait to show them to Peggy and Doug. I had the GIS center print a poster-size version of the civic-virtue map, laminated it in plastic, and before long was rolling it out on the table in Peggy's office.

Peggy and Doug immediately began roaming the map with their own eyes, commenting excitedly about what the hills and valleys might mean on the basis of their own vast experience. Doug had roamed the neighborhoods for his entire life, and both of them interacted with students and their parents from the neighborhoods on a daily basis. Both had devoted their hearts and minds to the welfare of Binghamton rather than to their own narrow concerns. Even for them, the egghead professor and local deadbeat had provided a new way of seeing.

CHAPTER 6

Quantifying Halloween

EVERYONE FAMILIAR WITH Binghamton loved to examine my maps, but what did they really mean? Do kids really know their degree of civic virtue? Even if they do, will they accurately report it when asked? If the survey was total garbage—if all of the students were giving the socially acceptable answers, for example—then everyone would receive a high score, and there would be no variation. As it turned out, not only did the students vary in how they answered the survey, but high- and low-scoring individuals were clustered into different neighborhoods. The rugged topography of my maps demanded an explanation.

As for people knowing their degree of civic virtue, let's pause to reflect on how different people are from one another. Terms such as *nice* and *nasty* only scratch the surface of our differences. A more complex taxonomy is needed, but at the end of the day, people differ enormously in how self- or other-oriented they are. Another survey that I have used in past research is based on the writings of Niccolò Machiavelli, whose name has come to represent a kind of person who manipulates others for his or her own personal gain. Here are some items from the Machiavelli survey:

- Never tell anyone the real reason you did something unless it's useful to do so.
- The best way to handle people is to tell them what they want to hear.

- The biggest differences between criminals and other people is that criminals are stupid enough to get caught.
- It's safest to assume that all people have a vicious streak and it will come out when they are given a chance.

If these statements strike you as a bit shocking, then you're not a high-Mach, as they are affectionately called in the literature. High-Machs are *proud* of these values. My point is that people are so different from one another that they're like different species in an ecosystem. It's easy to tell the difference between a cougar and a moose, and it's easy to tell the difference between a narcissist and a Good Samaritan. To a crude approximation, just asking is good enough.

Instead of the metaphor of listening and seeing, a better metaphor for my first maps would be a blind man's cane. Something was going on, but more tapping would be required to know what it was. Could I confirm and extend the results of the survey by gathering more data from completely different sources? Fall was approaching, and perhaps the crisp air prompted my epiphany. Halloween! Everyone remembers as a kid that some neighborhoods are better than others for trick-or-treating. From an adult perspective, decorating one's house, buying treats, and staying at home are an expression of interest in one's neighborhood. Might variation among neighborhoods on Halloween compare with the hills and valleys of my GIS maps based on the DAP? That would confirm and extend the results of my survey from a completely different source. Best of all, the information could be collected on a single night. We could count the kids, weigh the candy, score the decorations, and count the smashed pumpkins and toilet-papered trees the day after.

Call me a nerd, but I get especially excited by the prospect of quantification. After all, in order to create a GIS map for Halloween, I must convert it into a table of numbers to be fed into the maw of a computer. Some people find quantification distasteful, as if the true essence of something like Halloween can't be turned into numbers. On the contrary, in my quest to understand the nature of civic virtue, I was studying the wellspring from which Halloween, and much else, emerges.

I would need help to gather my numbers. Fortunately, just as Sherlock Holmes had his Baker Street Irregulars, I had a willing cadre of EvoS students. They would not be surprised by the idea of quantifying Halloween, and at least some of them would be charmed by the prospect of standing on a street corner, clipboard in hand, counting the little ghosts and witches, princesses and pirates, as they went by.

I WAS ALSO FORTUNATE to have a new graduate student to help me, Dan O'Brien, the one who'd insisted that I attend the Saint Patrick's Day parade. His Irish ancestors on his father's side and Italian ancestors on his mother's side were working-class folk who settled into ethnic enclaves in the New York City area. His parents met as math teachers at a Catholic school and later switched to computer-oriented careers. When Dan told me about his upbringing, I was struck by the degree to which he was nurtured and coached by his parents, in stark contrast to the boys Doug Stento worked so hard to nurture. Dan's mother didn't tolerate lazing around; when Dan had nothing to do, he was given educational games. He read voraciously without needing to be encouraged. His math skills were so heavily nurtured that he was doing long division and multiplication by the first grade. As he progressed through school, he memorized the countries and capitals of the world. He was also an avid sports fan, and his uncle used to show him off as a sports-trivia prodigy at the age of six.

Perhaps the most telling story that Dan told me about his parents was about when they decided to have a third child. Dan was eleven, and his sister Tania was eight. Their dad gently broached the subject during an outing, and they expressed enthusiasm for a new sibling. Dan is convinced that they were being meaningfully consulted and in some sense had been allowed to make such an important decision. With the arrival of Liam, they were now a family of five.

After Dan graduated from high school, he raced through Oberlin College in three years, double-majoring in math and biology. For all his brilliance, he wasn't interested in achieving in a conventional sense and had a fatalistic streak, assuming that opportunity would find him. He spent two years teaching middle and high school in Mexico and paradoxically became more involved in the Catholic religion than he had been inspired to be by his Irish-Italian parents. Eventually, he decided that he should be using his intellectual talents and applied to some programs on the basis of what had interested him in college. His path to Binghamton was pure serendipity, another example of the pinball machine of life. It was just my good luck that this consummate good kid, with an engine like a Maserati's purring under his hood, was now joining me in my quest to quantify Halloween.

Working with our GIS map, Dan and I identified neighborhoods that appeared to be high, medium, and low for neighborhood quality according to the DAP. Then we developed a system for scoring Halloween decorations, including whether the house, the lawn, or both were decorated, how many pumpkins, whether they were carved, and whether they were real or fakes brought down from the attic every year. A few days before Halloween, we started cruising the streets of our designated neighborhoods and scoring the houses. Some had no decorations at all, and others were so elaborate that I wondered what had possessed the owners — tombstones, spider webs, life-size zombies, even a ladder leaning against one house with a pair of pants and shoes sticking out of the ground, as if a workman had fallen off and embedded himself in the soil. For Halloween night, we organized crews of EvoS students to stand on designated street corners and count trick-or-treaters as they passed by, along with estimates of their age and other information, such as whether they were accompanied by an adult. It was such a whimsical project that everyone was in a festive mood when the night of the 31st finally arrived.

Who is not charmed by the sight of children dressed in their costumes, going from door to door as their parents hover protectively nearby? Counting them as they went by on my assigned street corner brought back memories of when I accompanied my own kids, Katie and Tamar, years ago. First we visited Mrs. Benyi, our neighbor on the right, who made them special treats, then the houses in our immediate neighborhood, and finally, by car, other neighborhoods that were known to have the best loot. Some of their Halloween costumes are still in our attic. One year, Tamar had the idea of going as a bathroom sink. We made a realistic-looking sink out of cardboard and a wooden frame representing the mirror. It hung on Tamar's shoulders so that she was looking at you through the frame when you opened the door. You in the bathroom mirror—what could be scarier than that? The sink costume was such a hit that the next year she decided to go as a toilet. We made a realistic-looking toilet with a flush mechanism. You put candy into the toilet, pressed the flusher, and the candy disappeared through a trap door into a hidden compartment. People loved flushing their candy down the toilet so much that Tamar got twice as much as the other kids.

After Binghamton's kids returned home to spread out and organize their loot on their living-room floors, Dan and I began toting up our numbers to see how the celebration of Halloween corresponded with our GIS maps based on the DAP. The lowest-quality neighborhoods identified by the DAP also scored lower for Halloween decorations. That tasty result gave us faith that the hills and valleys of our maps based on the DAP were real. The kids weren't just making it up.

Quantifying Halloween was fun but had an obvious flaw. Some people object to Halloween on religious grounds. They and their neighborhoods might be virtuous in their own way, just not expressed in the form of Halloween. Luckily, this flaw had a corrective: Christmas. The bias for Christmas should be the opposite of that for Halloween, with Christians showing up in force and pagans, Jews, and Muslims shuttering their windows. On the other hand, if holiday decorations are primarily an expression of neighborliness, then the same neighborhoods that light up for Halloween should also light up for Christmas. Certain households might avoid certain holidays for various reasons, or even avoid all holidays, but they would not obscure the general trends for neighborliness. Scientific progress seldom consists of a single definitive experiment; typically, the evidence accumulates in a brick-by-brick fashion. Christmas decorations would provide a third brick in the foundation that Dan and I were laying for the Binghamton Neighborhood Project.

We decided to mount an even more ambitious effort for Christmas than for Halloween. Instead of measuring selected neighborhoods, we would measure the entire city. First, we obtained an electronic file of all street names in Binghamton from the GIS center. Then we put out a call for volunteers to help us take a census of houses over a four-day period from December 16 to December 20. The EvoS Irregulars rose to the occasion, even though they were busy with their own holiday preparations. Dan and I developed a simple system for scoring a house in less than a minute, which ranged from 0 for no decorations whatsoever to 4 for both lighted and nonlighted displays on both house and lawn. We were careful to make sure that a high score need not depend on wealth. Even humble homeowners can put a string of lights on their shrubbery or a wreath on their door. Finally, we made a special category 5 for "exceptional displays" that seemed to be vying for a Guinness record. The volunteers were formed into pairs who drove along streets assigned to them, recording each house address and assigning a decoration score. We couldn't record every street, but the streets that we did record were randomly selected so they lay like pickup sticks across the entire city.

Most people enjoy driving or walking around at night during the Christmas season, admiring the decorations. Some are tasteful, and others are silly, but if all of the outside decorations were suddenly to disappear — all of the wreaths, lights, candy canes, Santas, crèches, snowmen, and sleighs — and if the only decorations were *inside* the houses, it wouldn't be Christmas anymore. For me, driving down randomly assigned streets and scoring each house only heightened the pleasure. My partner was Charles Sontag, one of my graduate students, who was studying toad tadpoles and showing that they cooperate to find food, much like the social insects. That's right — toad tadpoles have civic virtue. Our random assignments took us down streets that I would never have visited otherwise. I especially appreciated one home in a humble neighborhood that had a ring of lights, an illuminated carousel, an illuminated star, and a wreath on the porch, along with the following lawn decorations: an inflatable Santa, snowman, and giant snow globe, one illuminated tree, one free-standing illuminated reindeer, and one illuminated sleigh with reindeers. We didn't count the old washing machine at the curb waiting to be carted away. That house received a 5. I also enjoyed a house on the very fringe of the city, on a hill with a fine view of downtown, with an illuminated Santa's sleigh and a string of lights hung on the clothesline, as if he was taking off and leaving a trail of stardust.

As soon as the census was completed, Dan and I worked feverishly with Holly Kelleher, an exceptionally dedicated EvoS student, to key the data into the computer. Kevin Heard, who managed the GIS center, generated within hours a krig map of holiday-decoration scores for the city of Binghamton. I was so proud that I used it as my electronic holiday card for the year. The hills and valleys representing neighborhoods with high and low decoration scores were displayed in shades of red and the locations of houses with exceptional displays were shown in green. It might have been the only holiday card in the world that included a "methods" section in addition to the traditional wish for a happy and prosperous new year.

OK, I know what you're thinking. I'm a freak, or at the very least an incurable nerd, but I couldn't help boasting about what we had accomplished. *In six days, we had measured an entire city without spending a single penny other than for gasoline.* You can't understand something if you can't see it. I was developing a capacity for seeing my city of Binghamton, New York. It was as if the scales were falling from my eyes. A formal statistical analysis would be necessary, but to my eyes, the hills and valleys of holiday decorations bore a strong resemblance to the hills and valleys of neighborhood quality reported by the public-school students on the DAP. On a clear night, I could probably measure it from an aerial photograph: the more nurturing neighborhoods actually glowed more brightly during the holiday season.

THE MORE ADEPT DAN AND I became at working with our EvoS Irregulars and mastering the mechanics of this kind of research, the more we were able to see, or at least tap with our cane. Each study reinforced and extended the results of the previous studies, like a solid foundation of bricks.

Using a technique called the lost-letter method invented by the great social psychologist Stanley Milgram in the 1960s, we had the Irregulars drop stamped, addressed envelopes at 200 randomly determined locations around the city. Each envelope was addressed to a fictitious "Job Search Committee" at BU's biology department to make it seem as if the letter was a job application. The return address listed a name that was gender-neutral and a fake street address that was coded to indicate the drop location. What was the likelihood that someone passing the envelope on the sidewalk would commit a small act of kindness by picking it up and mailing it? Would neighborhoods differ in their return rates, and would those differences correspond to the DAP, holiday decorations, and other differences that we were accumulating in our database?

Details such as "Job Search Committee" and a gender-neutral name in the return address reflected careful thought about the design of this study to avoid potential biases. It never occurred to us, however, that times had changed since Milgram invented the method in the 1960s and that our experiment would trigger the U.S. homeland security system. Evidently, postal workers noticed the unusual number of letters going to the same address and became suspicious when they noticed that the return addresses were fake. They *really* began to worry when they opened one and discovered an ominous blank page. I had told the secretary of my department to expect an onslaught of letters but hadn't thought of telling my department chair and good friend, John Titus, who was therefore surprised to receive a call from the regional postmaster inquiring about the matter. John's reply was " I don't know, but I know whom to ask!"

Holly Kelleher and Monica Lee, two irrepressible EvoS undergraduate students, began an ambitious study of garage sales. Shouldn't *they* reflect the same social forces that account for the hills and valleys of our GIS maps? Working for independent-study credits in addition to their own interest, they set to work recording the addresses of past garage sales from the archives of our local newspaper and visiting current garage sales with a survey that they developed. Holly complained that her apartment was filling up with bargains that she couldn't resist but had no use for after she brought them home. Even I came back from one expedition with an ermine stole for only five dollars, which ended up draped uselessly over one of the computers in my lab.

Dan got his sister, Tania, then an art major at Cornell University, into the act by having her take photographs of the neighborhoods. What would people who knew nothing about Binghamton think about the neighborhoods based only on the photographs, and how would this compare with what the actual residents think about their neighborhoods?

Just as social psychologists such as Stanley Milgram invented clever techniques such as the lost-letter method, economists have invented a set of techniques called experimental games that have become all the rage in the scientific study of human social behavior. In the Dictator game, for example, one person is given an amount of money and allowed to share it with another person, no strings attached. In the Ultimatum game, the second person can accept or refuse an offer and by refusing causes the first person to get nothing also. In the Prisoner's Dilemma game, two people decide whether to cooperate when there is a temptation to cheat. Each game is a microcosm of human social interactions that can be played in a carefully controlled fashion, with factors manipulated in almost infinite variety. For example, the Prisoner's Dilemma game can be played once or repeatedly, the decision to cooperate or cheat can be simultaneous or sequential, and the capacity for punishing cheaters can be present or absent. An entire field called experimental and behavioral economics has developed around these games, about which I will have much to say in future chapters. As an egghead professor, I was familiar with this arsenal of newly developed techniques. As a reborn community planner, I was in a position to apply them to my city of Binghamton, New York. How would the forces that created the hills and valleys of my GIS maps, the social equivalent of the geological forces that create real hills and valleys, look when measured by the sensitive barometers of experimental economics games? Dan and I prepared to find out by asking Peggy's permission to play experimental economics games with the same students who had filled out the DAP. Brick by brick, the Binghamton Neighborhood Project was taking shape nicely.

Sherlock Holmes amazed Watson with his deductions. When he explained his reasoning, however, Watson regarded the same deductions as obvious, much to Holmes's irritation. What Holmes knew, and Watson had to be told, was the web of causality that connected an observation, such as the scratches on the back of a pocket watch, to the causes, such as a drunken man having difficulty inserting the wind-up key. Like almost everyone else in my city, I had spent my first twenty years bombarded by observations without knowledge of causes, like Watson. Now I was beginning to feel a bit more like Holmes. The web of causation connecting surveys, holiday decorations, lost letters, garage sales, photographs of neighborhoods, and experimental games was even beginning to seem elementary.

CHAPTER 7

We Are Now Entering the Noosphere

IT WAS 2009, the year of Darwin, the 200th anniversary of his birth and the 150th anniversary of the publication of *Origin of Species*. The whole world celebrated, and not just because of the round numbers. Abraham Lincoln was born on the exact same day and year as Darwin, but his round number was not celebrated nearly as much, even in America. Instead, the Darwin celebrations reflected a dawning awareness that evolution matters more than most people think. We need evolutionary theory to understand the world around us. Not just nature but the nature of humanity. Not just academic understanding but practical understanding that can help to solve some of the most pressing problems of our age. Call it a great awakening.

As someone who was already trying to change the way the planet thinks about evolution, I was becoming more popular than an Irish band on Saint Patrick's Day. And I found the offers hard to resist. As Alfred P. Doolittle told Henry Higgins in *My Fair Lady*, "I'm willing to tell you! I'm wanting to tell you! I'm waiting to tell you!"

Before I knew it, I had accepted thirty invitations, a veritable world tour, including five different trips overseas in addition to those around the United States. One workshop titled "Do Institutions Evolve?" would be held at a villa in the hills of Tuscany just outside Florence, Italy. What person in their right mind would refuse an invitation like that? Another workshop titled "Why Aren't the Social Sciences Darwinian?" was scheduled for May in Cambridge, England, where Darwin went to college. In July, the entire city of Cambridge was celebrating Darwin with a five-day festival, including two forums on religion that I couldn't resist. I would even be able to visit Darwin's dorm room, which had been re-created as if he might return at any minute.

Anne, my evolutionist wife who works as hard as I do, was not amused. How exactly was I going to fit thirty trips into my already busy schedule, not to speak of the turbulence it would create in our shared life? How would I teach my courses and carry out my other faculty duties? How would I manage hosting the ten speakers per semester who come through Binghamton in our EvoS seminar series? How would I pay attention to my graduate students? One of them was already fond of saying that I had a supply of Batman smoke bombs that I threw down to make myself disappear. How about our semblance of a home life, including our big vegetable garden? Would I be around to help start the seedlings in April and transplant them in May? Even I ached at the thought that my visits to our tree house and property would become less frequent and that the Binghamton Neighborhood Project might suffer. Anne started to rib me by calling me "Your Darwinness." I protested that she should be glad that I was trying to change the way the planet thinks about evolution. I was flattered that she wanted me around, but wasn't life easier in some respects when I was away? No, she replied, it was hard to adjust to the constant coming and going. As the saloon girl in the movie *Blazing Saddles* sang about men, "They're always coming and going and going and coming and always too soon."

Of all the trips that I couldn't refuse, two were especially enticing. The first was a small workshop organized by the John Templeton Foundation, which has almost single-handedly funded research on science in relation to religion and other "big questions," as it puts it on its Web site. The workshop was titled "*Homo symbolicus:* The Dawn of Language, Imagination, and Spirituality." It would be held in late January at Cape Town, South Africa, and would include a field trip to the Blombos Cave on the coast of the Indian Ocean, where some of the earliest evidence for the dawn of culture in our species—from approximately 100,000 years ago—had been excavated.

The second event was a five-day conference titled "Biological Evolution: Facts and Theories," organized by the Vatican, to be held at the Pontifical Gregorian University in Rome in early March. The Vatican was using the year of Darwin to conduct a thorough review of evolutionary theory in relation to Catholic theology and had invited an A-list of evolutionists to help them out. I was privileged to be among them.

These two events were irresistible by themselves but doubly so when considered together. In the space of three months, I would trace the 100,000-year human journey from the first shell beads and lines etched in stone on the southern tip of Africa to one of the greatest inflorescences of modern human culture, the Catholic church in Rome. Even though I would be leaving my city of Binghamton, the journey I would be tracing was highly relevant to the Binghamton Neighborhood Project. Cities exist — as does our hope for improving them — thanks only to our capacity for the rapid behavioral change that we call culture.

SOMEONE BEFORE ME HAD TRACED the same route in the opposite direction. Pierre Teilhard de Chardin (1881–1955) was a Jesuit priest and paleontologist during a time when science was regarded as a suitable path to God. He was part of the team that discovered Peking Man, one of the first missing links connecting *Homo sapiens* to the apes, which was an international sensation in 1929 and today is classified as the extinct hominid species *Homo erectus.* He did most of his work in China but also visited South Africa to consult with colleagues such as Raymond Dart, who discovered an earlier missing link that became known as *Australopithecus africanus.* Teilhard's best-known work is *The Phenomenon of Man,* written in 1940 and published shortly after his death in 1959. The introduction by Sir Julian Huxley, one of the greatest evolutionists of his day and grandson of Thomas "Darwin's bulldog" Huxley, indicates the respect that Teilhard commanded among scientists. The most remarkable thing about *The Phenomenon of Man,* however, is its spiritual quality. Teilhard claimed to provide a 100-percent scientific account of humanity in a way that affirmed and strengthened, rather than threatening, his religious faith. He liberally used words such as "spirit" and "soul" and ended his book with these words: "even in the view of a mere biologist, the human epic resembles nothing so much as the way of the Cross."

Today, Teilhard is read almost exclusively for his spiritual quality. He has been forgotten by scientists and is virtually never mentioned in current discussions of evolution in relation to human affairs. Most of my colleagues would be embarrassed to discuss the spiritual side of Teilhard, as if there is something wrong with mixing spirituality and science. In part, this is because almost everyone today has become cynical about the prospects for improving the human condition and especially the role of science in the enterprise.

Sir Julian Huxley, in contrast, was a passionate humanist who felt that mankind must take charge of its own destiny. His many books include *Religion without Revelation* (1927, 1979), *Evolutionary Ethics* (1943), and *Essays of a Humanist* (1964), in addition to *Evolution: The Modern Synthesis* (1942, 2010), which literally defined the field of evolutionary biology for the ensuing decades. Here are two passages from Huxley's humanistic work:

> There is no separate supernatural realm: all phenomena are part of one natural process of evolution. There is no basic cleavage between science and religion....I believe that [a] drastic reorganization of our pattern of religious thought is now becoming necessary, from a god-centered to an evolutionary-centered pattern.
>
> Many people assert that this abandonment of the god hypothesis means the abandonment of all religion and all moral sanctions. This is simply not true. But it does mean, once our relief at jettisoning an outdated piece of ideological furniture is over, that we must construct something to take its place.

It's no wonder that Huxley praised Teilhard's own grand vision in *The Phenomenon of Man*, even though Teilhard would certainly disagree with the notion of his Catholic faith as an outdated piece of ideological furniture.

How on earth did this kind of expansive optimism lead to the cynicism and limited expectations of today? Why have Teilhard and the humanistic side of Huxley been forgotten by professional evolutionists, who continue to celebrate Huxley as one of the fathers of the modern synthesis? A large part of the answer is contained in this third quote from Huxley, written in 1941:

> The lowest strata are reproducing too fast. Therefore... they must not have too easy access to relief or hospital treatment lest the removal of the last check on natural selection should make it too easy for children to be produced or to survive; long unemployment should be a ground for sterilization.

This passage sounds horrifying to most of us today, certainly to myself. Even more horrifying is the fact that Huxley had lots of company. It was acceptable at that time for passionate humanists such as Huxley to argue that mankind should take charge of its destiny in this particular way. More horrifying still, their talk was not idle and led to social policies on both sides of the Atlantic that can only be looked back upon with shame.

Thus, Huxley and his compatriots were largely responsible for their own demise. Their vision became a pariah concept known as social Darwinism. Against this background, we can begin to appreciate why Darwin's theory applied to humanity was not a matter of smooth, continuous progress; why it became restricted to the biological sciences; why most people interested in human-related subjects wanted nothing to do with it; why most evolutionists were willing to respect the boundary by concentrating on the birds and bees, leaving human improvement to others; why a fresh look at evolution in relation to human affairs didn't gather steam until late in the twentieth century—and why only now is the rest of the world prepared to take notice during the year of Darwin.

When I decided actually to read *The Phenomenon of Man* in preparation for the Vatican conference, I thought I was paying my dues as a good scholar but that I was unlikely to find much of contemporary interest. I was wrong. Once I grew used to Teilhard's vocabulary (example: "noosphere") and compressed sentences (example: "The consciousness of each of us is evolution looking at itself and reflecting upon itself"), I read the entire section of the book on human evolution in a single sitting. I was able to do this for only one reason: Just as a chord played on one instrument causes the corresponding strings of a nearby instrument to vibrate, what Teilhard was saying resonated with my own understanding of the human evolutionary story with little need for alteration. In some respects, he seemed to be still ahead of his time, for his science in addition to his spirituality, and I am happy to convey his message. Fasten your seatbelt—we are about to enter the Noosphere.

ONE OF THE BEST WAYS of conveying the true nature of science is by telling the stories of scientists, and Teilhard's story rivals Darwin's in its panoramic scope. Like Darwin, Teilhard was born into an affluent family and reveled in natural history as a boy. The family estate was located in the French province of Auvergne, which offered abundant wildlife, and Teilhard's father was an avid natural historian. Teilhard also exhibited a spiritual and mystical streak from an early age that he attributed to his mother. At the same time that Teilhard was collecting pebbles and rocks as a little boy, he was also pondering the frailty of life. Life was indeed frail in those days, even for the affluent. Political instabilities threatened old social orders, and diseases could strike anyone down, including several of Teilhard's own siblings during his lifetime.

It was natural for such a studious and pious lad to become a Jesuit priest, and it was natural at the time for the Jesuits to regard science as a legitimate pathway to God. Teilhard's Jesuit training took him to first to England and then to Egypt, where fossil shark teeth were intermixed with the artifacts of an ancient civilization. From Egypt, he returned to England, where he began to ponder evolution not only through his naturalistic pursuits but also through philosophical tracts such as Henri Bergson's *Creative Evolution*. From the beginning, Teilhard was reaching for a concept of evolution that would explain the totality of experience, the natural *and* the spiritual. He was following in the footsteps of philosophers such as Bergson, but his path was leading him toward conflict with his own church. When the liberal Pope Leo XIII was succeeded by the more conservative Pope Pius X, Bergson's *Creative Evolution* was placed on the Vatican's *Index of Forbidden Works*.

After his ordination in 1911, Teilhard trained to become a fully professional geologist and paleontologist, earning his doctorate from the Sorbonne in 1922, but first he had to endure the horror of World War I as a stretcher bearer. Once again, he was voyaging to far-off locations such as North Africa, only this time as a helpless observer of a clash of modern civilizations. Throughout the war, he risked his life to carry the wounded and the dead off the battlefields, for which he was awarded the medal of the Legion of Honor in 1921.

Here is an excerpt from a letter written by Teilhard during the war, explaining the paradoxical fact that wounded soldiers often want nothing more than to return to the front:

> The front cannot but attract us because it is, in one way, the extreme boundary between what one is already aware of, and what is still in the process of formation. Not only does one see there things that you experience nowhere else, but one also sees emerge from within one an underlying stream of clarity, energy, and freedom that is to be found hardly anywhere else in ordinary life and the new form that the soul then takes on is that of the individual living the quasi-collective life of all men, fulfilling a function far higher than that of the individual, and becoming fully conscious of this new state. It goes without saying that at the front you no longer look on things in the same way as you do in the rear; if you did, the sights you see and the life you lead would be more than you could bear. This exaltation is accompanied by a certain pain. Nevertheless, it is indeed an exaltation. And that's why one likes the front in spite of everything, and misses it.

Teilhard was writing neither as a scientist nor as a priest in this passage but merely as an acute observer. His ability to see beyond the individual and to use a word such as "soul" in a way that has nothing to do with supernatural agency would become a hallmark of his scientific worldview.

Teilhard took his final vows in 1918, but his writing from the battlefield was already beginning to trouble his Jesuit superiors. Teilhard was convinced that he saw a deeper truth behind current Catholic doctrine. Remaining true to his religion required becoming heterodox. Moreover, his growing reputation in the world of science and his eloquence as a writer were causing him to be heard. Even young Jesuits were eager to learn his strange new creed. Teilhard had a quality that pulled people toward him, like iron filings to a magnet.

But Teilhard was supposed to be obedient to his Jesuit order, which in turn was supposed to be obedient to the Vatican. Science was viewed as a legitimate path to God, but some paths had locked gates, especially when they seemed to challenge sacred doctrines such as original sin. To make matters more difficult for Teilhard, what counted as acceptable varied with the papal administration. Just as a piece of legislation might sail through a liberal political administration but become doomed when conservatives come into power, Teilhard might be allowed to work and write as he wished under one pope but become ominously censured under another. The Vatican required permission for every major decision, such as publishing a paper, attending a conference, or joining an expedition. Previously published work might come to the attention of a zealous conservative faction, eager to make an example of him, at any time. Throughout his life, Teilhard struggled to remain an obedient servant of his church without sacrificing his scientific integrity or spiritual vision, which he was convinced was truer to the church than the dogma forced upon him.

To limit his influence, the church sent him to China as a kind of intellectual Siberia, as far from the centers of Western science as possible. But by banishing him from the centers of existing science, they unwittingly placed him in the middle of the action with respect to new science. As Teilhard traversed the vast Asian continent on his collecting expeditions, he observed geology, nature, and culture at a scale rivaling Darwin's voyage around the world on the *Beagle*. Even better, Teilhard combined all of this with his vast knowledge of Catholic theology and a personal spiritual quality so strong that it was hardened, rather than broken, by the cataclysm of World War I, personal tragedies such as the death of his beloved siblings, and his status as marginal within his own church.

Much as Darwin became famous among scientists on the basis of what he accomplished on the *Beagle*, Teilhard's international reputation among scientists grew from his exile in China, the exact opposite of what the church intended. The discovery of Peking Man created a sensation and allowed Teilhard to ponder the mystery of human origins more deeply than ever before. Although only one skull was unearthed, the cultural remains of Peking Man were more plentiful, mingling with the fossils of extinct animals among the still more ancient rocks.

Teilhard resisted what must have been an extreme temptation to leave the church, which forced him to sign a statement repudiating his ideas on original sin (ironically, during the same week that the *Scopes* Monkey Trial began in America), refused to let him accept a professorship at the Sorbonne that would have been the zenith of his academic career, refused to let him publish his spiritual work for his entire life, and even refused to let him spend his final years in France. Teilhard died in America and is buried in a quiet spot in the Hudson River Valley, only a few hours from Binghamton. His final quiet act of rebellion was to place his unpublished manuscripts in the hands of friends and beyond the reach of the mighty hand of the church.

When *The Phenomenon of Man* was published in 1959, Teilhard's spiritual and scientific flames could at last burn brightly together. Then his scientific flame began to wane and ultimately flickered out. Science has no central hierarchy, no capacity to exile its members, to force them to sign confessions, or to censure publications that challenge dogma—except, perhaps, when the dogma is so widespread that a centralized authority isn't needed. Ironically, whatever happened in a decentralized fashion to silence Teilhard among scientists was more effective than the centralized efforts to silence Teilhard within the church.

* * *

LIKE DARWIN'S, TEILHARD'S WORLDVIEW was suffused with the immensity of time and space. A photograph from one of Teilhard's expeditions shows an arid landscape in northwestern China that seems almost without life, with barren mountains thrusting out of the earth's crust, decomposing into sand filling the valley below. Since Teilhard, like Darwin, was trained in geology in addition to zoology, he had a deep appreciation of the purely physical forces that shaped the surface of the earth for billions of years before the first spark of life.

Teilhard was not the slightest bit tempted to attribute supernatural agency to the origin of life, despite his spiritual nature and Jesuit training. He assumed that life was a purely physical process but one that was qualitatively different from what came before. He used the metaphor of water completely changing its properties when it is brought to a boil. With life came a new kind of diversity in the thousands of species adapting to their environments. Seen in the immensity of time and space, living creatures spread over the surface of the planet and formed a kind of a skin, as Teilhard put it. Most places on earth are not like the barren landscape in the photograph; they are cloaked in vegetation rooted in soil that includes the remains of past life mixed with the physical earth. Current life breathes from an atmosphere conditioned by past life. The word *biosphere* had already been coined to describe the influence of life on earth, and Teilhard adopted the term with pleasure.

Now for Teilhard's own contribution. He asks the reader to imagine excavating layers of soil. Deep down, there is only the physical earth. Closer to the surface, organic materials begin to appear. Then, still closer to the surface, human artifacts start to appear. At first, they are barely present, such as flakes of stones chipped from rocks to make tools. Then they become more abundant. In the immensity of space and time, the artifacts of human activity spread over the surface of the planet and form a kind of a skin, like the skin of life that preceded it. A word is needed for this human skin. The *noosphere*.

Now for an exceptional act of brilliance. Teilhard did not view humans as merely a highly successful species. He imagined humanity as a new evolutionary process, capable of generating a diversity of cultural forms, just as life is capable of generating a diversity of organic forms. That makes the origin of our species as momentous, in its own way, as the origin of life.

As with the origin of life, Teilhard was not the slightest bit tempted to attribute human origins to a divine spark. "Man came silently into the world," as he put it, a species like any other. A chance combination of biological adaptations led to the metamorphosis. The convergence might have been serendipitous, even highly improbable, but once accomplished, it literally took on a life of its own. The term *noosphere* therefore has two meanings: the physical skin of the human presence on earth and the new process of evolution that Teilhard loosely referred to as "thought."

The new process necessarily relied on different mechanisms from biological evolution, but its outcome was essentially the same. Teilhard was adamant that *human cultural diversity is like biological diversity.* He asked the reader to imagine the biological tree of life branching over a period of hundreds of millions of years. Then one of its tips becomes a new evolutionary process that starts branching at an incomparably faster rate, overtopping many of the previous branches as human cultures spread over the earth and displace other species. Teilhard did not pass moral judgment on the replacement of biological diversity with human cultural diversity. His main point was to stress that both biological and cultural diversity obey the same laws of natural history. Culture did not free humanity from evolution. Culture was evolution at warp speed.

Next, Teilhard described the long-term arc of cultural evolution as resulting in coalescence in addition to a diversity of forms. The earliest hunter-gatherer cultures were "grains of thought" that merged with other grains to form ever larger societies. Looking into the future, he envisioned a single global society, which he called the Omega Point, which would also be a form of supreme consciousness — the process of evolution reflecting fully on itself.

THE TEMPLETON FOUNDATION WORKSHOP PROVIDED a perfect opportunity for me to compare Teilhard's vision with current scientific knowledge about the dawn of culture in our species. In less than a day, I traveled from Binghamton to Cape Town, South Africa, a trip that would have required weeks on shipboard for Teilhard. Templeton Foundation workshops are an intoxicating blend of intellectual and social interactions among people from diverse disciplines, at a pleasant location and punctuated with good food and drink. This one included archeologists, anthropologists, primatologists, linguists, philosophers, and evolutionists such as myself. Two days of conversation were followed by the visit to Blombos Cave, several hundred kilometers from Cape Town on a high bluff overlooking the Indian Ocean. As the cave was being prepared for visitors by Christopher Henshilwood, who heads the archeological team excavating the site, I marveled at the fact that I was sitting on the exact same spot as my distant ancestors, anatomically the same species as myself but just beginning to wear ornaments such as shell beads and using pigments such as ochre. This was the very moment that mankind was transforming from a mere species to a new process of evolution, as Teilhard would have put it.

It was easy to see why the mere species would choose this location, now a nature preserve and almost as wild today as it was back then. The coast was uninhabited as far as I could see in both directions. The sky and the sea were a kaleidoscope of blue and white—blue sky, white clouds, blue sea in a dazzling variety of hues, and white froth as the surf crashed against the rocky shore. The cave would have afforded protection against the big predators that were a constant threat to our ancestors but now were removed from the landscape. Braving the surf would be treacherous, for my ancestors no less than for myself, but numerous quiet tide pools were carved into the rocks, which teemed with shellfish and beckoned my companions and me to use them as natural Jacuzzis. Furry mammals about the size of woodchucks scurried among the rocks. They are called rock hyraxes and, amazingly, are more closely related to elephants than to woodchucks. Life is so malleable that the same ancestral species can be molded into forms as different as an elephant and a rock hyrax, depending on the hammer blows of natural selection. My ancestors certainly enjoyed feasting on rock hyraxes, whose bones are mingled with the fish bones, mollusk shells, and ashes from the campfires inside the Blombos Cave.

Actually, this was not necessarily the scene that my ancestors gazed upon from this spot. During the last 100,000 years, climate change caused the sea level to drop and the shore to recede so far into the distance that some of my ancestors living during this period would have gazed upon an African savannah rather than the sea. For part of this time, the cave was completely swallowed by a sand dune, only to be exposed again when the changing climate once again brought the sea to the doorstep of the cave. One of my companions directed my attention to the two giant rock formations on the shore that I had been gazing at all along. They were clearly different from each other. One was ancient rock formed during the early history of the earth. The other was sandstone, made from the compressed sand of the dunes that engulfed the cave only 70,000 years ago. The difference was obvious after it was pointed out to me, but unlike Darwin and Teilhard, my worldview is not suffused with the vastness of space and time. For me, my physical surroundings seem so solid that they must be eternal. I must stretch my imagination to see them as a point on an arc of continuous change, stretching into the past and the future as far as the eye can see, like the uninhabited coastline in front of me.

AT LAST, THE EXCAVATION SITE within the Blombos Cave is ready to receive visitors. All archeological sites, including the excavation that preceded the construction of the university's downtown building in Binghamton, are handled with extreme care. This one is handled with even more care than usual, as one of a handful of sites that can shed light on the dawn of culture in our species. We are instructed to walk carefully along some planks and down a ladder into a pit with a vertical wall of sand in front of us. The wall reflects the chronology of time. Barring physical disturbance, the lower you go, the earlier the material was deposited inside the cave.

Just as a microtome shaves ultrathin slices of tissue for inspection, the excavation involves shaving slices of sand from the vertical wall. It is an unbelievably meticulous process, involving cataloging every fragment that might be of possible interest. The current face of the wall is studded with objects that I yearn to identify. This was exactly how Teilhard asked his readers to imagine the physical noosphere: human artifacts beginning to mingle with biological material and the physical matrix of the earth.

Forget about the movie scenes of Indiana Jones entering an ancient temple and yanking the precious object from the grasp of a rotting mummy. Precious objects are jutting from the face of the sand wall, but they are waiting for the *next* slice, which might require years. Christopher points out features of interest, including the bones of their food and dark layers of ash from their fires. The two most spectacular finds are shell beads and a block of ochre, a mineral used by indigenous people around the world to make red pigment, with lines etched in a design. Even the most humble objects contain clues to the lives of our ancestors, however, and it is surprising how the clues can add up to a convincing story based on clever detective work. Christopher points out a red horizontal line in the wall that indicates the presence of ochre. He speculates that this might have been a spot where the stone was ground to make the pigment, with some of the powder falling onto the cave floor. One member of our group speculates that it might be a leather garment dyed with ochre that rotted to become the thin red line in the sand. Christopher is intrigued with this idea, which hadn't occurred to him. It's easy enough to test the hypothesis by analyzing the soil for organic compounds indicative of animal skin. In this fashion, the story of our ancestors at the dawn of culture can be pieced together with more certainty than you might think.

Our current knowledge of the dawn of culture in our species is based not only on clues from sites such as the Blombos Cave but also from all of the scientific disciplines that were represented at the Templeton Foundation workshop. Teilhard would be pleased. Especially pleasing would be the union of two themes that he stressed separately but did not put together: reflection and cooperation.

For Teilhard, the vital spark that transformed us from a mere species to a new evolutionary process is the capacity for *reflection,* which he described this way:

> From our experimental point of view, reflection is, as the word indicates, the power acquired by a consciousness to turn in upon itself, to take possession of itself as of an object endowed with its own particular consistence and value: no longer merely to know, but to know oneself; no longer merely to know, but to know that one knows. By this individualization of himself in the depths of himself, the living element, which heretofore had been spread out and divided over a diffuse circle of perceptions and activities, was constituted for the first time as a centre in the form of a point at which all the impressions and experiences knit themselves together and fuse into a unity that is conscious of its own organization.

Teilhard's writing is dense but hypnotic in its precision, poetic while remaining descriptive in a literal sense. We might want to decompress what he is saying, but there is nothing that is obviously wrong. In fact, contemporary books such as Terrence Deacon's *The Symbolic Species* confirm and elaborate upon Teilhard's densely stated thesis.

The capacity for reflection allows us to imagine new worlds and then step into them, as Teilhard describes in a continuation of the same passage:

> Now the consequences of such a transformation are immense, visible as clearly in nature as any of the facts recorded by physics or astronomy. The being who is the object of his own reflection, in consequence of that very doubling back upon himself, becomes in a flash able to raise himself into a new sphere. In reality, another world is born. Abstraction, logic, reasoned choice and inventions, mathematics, art, calculation of space and time, anxieties and dreams of love—all these activities of inner life are nothing else than the effervescence of the newly-formed centre as it explodes upon itself.

Once again, this is an accurate summary of Terry Deacon's thesis that we are unique in our capacity for symbolic thought, which allows us to create imaginary worlds and then step into them.

It is clear from these passages that Teilhard regarded reflection as an *individual* capacity, similar to his own prodigious capacity for reflection. He also appreciated the importance of groups in passages such as this one:

> Throughout living phyla, at all events among the higher animals where we can follow the process more easily, social development is a process that comes relatively late. It is an achievement of maturity. In man, for reasons closely connected with his power of reflection, this transformation is accelerated. As far back as we can meet them, our great-great-ancestors are to be found *in groups* and gathered around the fire.

Even in this passage, Teilhard implies that the individual capacity for reflection came first and accelerated social development. The contemporary evolutionist Michael Tomasello developed the same thesis in his book *The Cultural Origins of Human Cognition*. Mike has changed his mind in his more recent books *Why We Cooperate* and *Origins of Human Communication*, however, as part of a rapidly growing consensus that *cooperative groups came first*. The human capacity to reflect could not have evolved, and cannot exist in its current form, without trustworthy social partners.

Teilhard himself was a titan of reflection as an individual, but he was building on the ideas of Darwin and hundreds of other thinkers who preceded him. His development was nurtured by both individuals and social institutions. Every expedition that he took required extensive cooperation. His reflections would be useless if he couldn't transmit them, requiring more cooperation. The failure of his church to cooperate almost put an end to his reflection. Teilhard was not an individual. He was a node in a vast system of cooperation, including his immediate social partners and his culture, which had become structured over a period of centuries to facilitate the cultural practices that we call science. As a thought experiment, imagine that Teilhard had been given plenty of food, water, and affection as a boy but no access to any intellectual resources whatsoever. If he could write

The Phenomenon of Man under those conditions, that would be individual reflection.

As for modern reflection, so also for its rudiments in our ancestors. Our closest living relatives — chimps, bonobos, gorillas, orangutans, and gibbons — are extremely smart, but their particular form of intelligence is predicated on the fact that they can't necessarily trust their neighbors. Male chimps cooperate to hunt for food or patrol their territory, but they also are obsessed with achieving social dominance within their group. Female chimps also compete with one another to monopolize the best resources for themselves and their kin within the group. A baby chimp can't leave its mother to play with the other chimps; it might get beaten up or killed.

Modern human social life can get this dysfunctional. Think of arms races among superpowers, blood feuds in tribal societies, and bitter political disputes in which the only thing that matters is to beat one's opponent. The kind of reflection that Teilhard had in mind comes to a screeching halt under these conditions, no matter how smart people remain in other respects. For reflection to get started in the first place, there had to be an atmosphere of trust.

That atmosphere was not created by everyone suddenly becoming nice but by the ability to thwart the ambitions of others easily. My evolutionist colleague Christopher Boehm calls this "reverse dominance" in his book *Hierarchy in the Forest*. In a typical primate dominance hierarchy, the meanest individual or coalition manages to intimidate the others and monopolize the resources. In a typical small-scale human group, including hunter-gatherer societies around the world, the meanest individuals and coalitions are ridiculed, punished, expelled, or even executed unless they change their ways and fit in with the rest of the group. We are an aggressively egalitarian species, and our passion for equality is manifested whenever we exist in small groups with a relatively even balance of power among the members.

Equality is the requirement for a major transition, as we learned from the parable of the wasp. As soon as individuals could no longer succeed at the expense of their neighbors, collective survival and reproduction became the primary means of natural selection. Only then could our ancestors begin to share freely what they learned, develop an inventory of symbols with shared meaning, and otherwise make the transition from just another species with a fixed repertoire of behaviors to an open-ended evolutionary process.

Seeing cooperation as a precondition for reflection and reflection as a form of cooperation fits Teilhard's broad vision even better than reflection as an individual capacity that came first. After all, Teilhard thought that reflection *becomes* a kind of distributed consciousness at the Omega Point. Now we can say that it *begins* as a form of distributed consciousness at the scale of very small groups.

AS I COMPARE *The Phenomenon of Man* with current scientific knowledge, I am struck by how often Teilhard gets it right and in some respects is still ahead of his time. Inside and outside the ivory tower, there are two major conceptions of human nature that are both wrong in their own ways. The first is to regard ourselves as exclusively the product of genetic evolution, giving us a fixed nature that cannot change except by future genetic evolution. This is the specter of genetic determinism that so many people find threatening because it implies an incapacity for change. It is profoundly wrong to the extent that human mentality and culture also count as evolutionary processes, capable of producing new forms that never existed in the past. Many of my evolutionist colleagues still think primarily in terms of genetic evolution. Those who fully appreciate the import of the statement that "there is more to evolution than genetic evolution" are still in the minority. Teilhard was way ahead of his time when he described humanity as a phylum of diverse and rapidly evolving psychological and cultural forms, while remaining a single biological species.

The other major conception of human nature that is wrong in its own way is to suppose that evolution explains our physical bodies and a few basic instincts but has nothing to say about our rich behavioral and cultural diversity. In contrast, Teilhard insisted again and again that humanity was still bound by the rules of evolution, as in this passage:

> There is no need for me to emphasize the reality, diversity and continual germination of human collective unities, at any rate potentially divergent; such as the birth, multiplication and evolution of nations, states and civilizations. We see the spectacle on every hand, its vicissitudes fill the annals of the peoples. But there is one thing that must not be forgotten if we want to enter into and appreciate the drama. However hominised the events, the history of mankind in this rationalized form really does prolong—though in its own way and degree—the organic movements of life. It is *still* natural history through the phenomena of social ramification that it relates.

In other words, *human cultural diversity is fundamentally like biological diversity,* a statement that I made at the beginning of this book as the essence of the evolutionary paradigm. It will be a great day in the future when everyone follows Teilhard's lead by avoiding both the error of genetic determinism and the error of regarding culture as a liberation from evolution.

Teilhard was also right about the speed of cultural evolution and the coalescence of cultures at ever-larger scales. He appreciated the importance of path dependence in cultural evolution—the fact that you can't always get there from here—which I emphasized for genetic evolution in my parable of the wasp. He even anticipated Jared Diamond's magnificent book *Guns, Germs, and Steel* by observing that European culture was capable of more rapid evolution than traditional Chinese culture. It is wrong for Teilhard to be read only for his spiritual message. Let his scientific flame burn brightly again!

AND WHAT ABOUT HIS SPIRITUAL message? How could Teilhard remain so strong in his Christian faith without requiring a God interested in the affairs of people or even a single divine spark anywhere in his entire story? Because he regarded Christianity as the leading edge of cultural coalescence, not in the sense of conquest but in the sense of expanding the human capacity for love. The following passage toward the end of *The Phenomenon of Man* helped me to prepare for my own journey from the Blombos Cave in January to the Vatican in March 2009:

> It is relatively easy to build up a theory of the world. But it is beyond the powers of an individual to provoke artificially the birth of a religion. Plato, Spinoza and Hegel were able to elaborate views which compete in amplitude with the perspectives of the Incarnation. Yet none of these metaphysical systems advanced beyond the limits of an ideology. Each in turn has perhaps brought light to men's minds, but without ever succeeding in begetting life. What to the eyes of a "naturalist" comprises the importance and the enigma of the Christian phenomenon is its existence-value and reality-value.

In other words, religions *live* in a way that philosophies don't. They are cultural life forms that grow, replicate, and adapt on their own, which makes them *more real*. Teilhard continues:

> Christianity is in the first place real by virtue of the spontaneous amplitude of the movement it has managed to create in mankind. It addresses itself to every man and to every class of man, and from the start it took its place as one of the most vigorous and fruitful currents the noosphere has ever known. Whether we adhere to it or break off from it, we are surely obliged to admit that its stamp and its enduring influence are apparent in every corner of the earth today. It is doubtless a quantitative value of life if measured by its radius of action; but it is still more a qualitative value which expresses itself—like all biological progress—by the appearance of a specifically new state of consciousness. I am thinking here of Christian love.

By the time the Vatican conference rolled around, the year of Darwin was killing me. I returned from each trip to catch up frantically on my work, only to leave for the next trip further behind. Maintaining a semblance of order at home requires both Anne and me at the best of times. Now the accumulating mail began to resemble snowdrifts, and the dust bunnies under the bed grew so large that Anne started to call them wooly mammoths. She was tolerant but didn't spare me her "I told you so" look at choice moments. Plane flights became precious opportunities to work. I wouldn't try to sleep on a transatlantic flight, just doze long enough to continue work.

I arrived at the Rome airport punchy with fatigue and didn't appreciate where I was until the taxi passed the Colosseum on the way to my hotel. My jaw dropped open, as if I were a country bumpkin. Everyone has heard of the Colosseum, but I was overwhelmed by its size. This was the place that held 50,000 spectators, that was flooded to hold mock naval battles, where people and animals were killed by the thousands as a form of entertainment, including the early Christians, who really were fed to the lions. Right next to the Colosseum was the Arch of Constantine, the Roman emperor who converted to Christianity and attributed his military successes to the protection of the Christian God. Never mind that Jesus preached to turn the other cheek.

The conference was held at the Pontifical Gregorian University in downtown Rome, founded more than 450 years ago by the Jesuits, Teilhard's order. It was organized at the highest level, and the speakers were actually scheduled to meet the pope, although that event had to be canceled at the last minute. Many of the people attending the conference wore religious vestments signifying their order and their rank within it. To me, they looked as exotic as Obi-Wan Kenobi from *Star Wars*. Sprinkled among them were my own evolutionist colleagues—lots of them, since this conference was going to include five whole days' worth of talks. When the Vatican decides to evaluate evolutionary theory, it goes all the way.

During one of the intermissions in the cavernous hall outside the auditorium, I was surprised to be approached by a young man in the ankle-length robe of a student priest who spoke with an American accent and was from Binghamton, New York. Then a German priest named Emerich Sumser introduced himself to say that he had read and enjoyed my book *Darwin's Cathedral* as part of his doctoral dissertation. In neither case should I have been surprised. Of course, student priests come to Rome from around the world for training, including my city of Binghamton, and Catholic priests are among the best scholars and scientists in the world, as we saw in the case of Teilhard.

Yet I was also surprised when one of my evolutionist colleagues who is also a knowledgeable Catholic pointed out a cardinal, or a "red hat," as he called him, in the audience.

"Do you know who that is?"

No, I replied dumbly.

"That's the head of the Congregation for the Doctrine of the Faith. Do you know what that used to be called?"

No, I replied dumbly.

"The Holy Office. Do you know what that used to be called?"

No, I replied dumbly.

"The Inquisition."

Whoa! I was impressed. The church was still keeping a watchful eye on scientific inquiry, blocking some paths and allowing access to others, just as in the days of Teilhard. The amazing thing about Teilhard was that he could retain his idealism about Christian love, which he would describe as the soul of his church, despite a lifetime battling its bureaucracy. He referred to his final futile trip to Rome to seek permission to publish his work as "stroking the whiskers of the tiger."

THE CONFERENCE BEGAN with the nitty-gritty facts of biological evolution—paleontological evidence, molecular evidence, speciation, development, complexity. Lynn Margulis was there to talk about her symbiotic-cell theory, the prequel to the parable of the wasp. Then attention was focused on the origin and evolution of our species, including cultural evolution. The last two days were devoted to philosophical and theological implications of evolution, with talks by scholars and theologians from the Vatican and the Gregorian in addition to my evolutionist brethren. Translators were present to translate Italian into English and English into Italian through earphones that could be obtained in the lobby.

In my own talk, I gave a whirlwind tour of major transitions—when evolution goes the way of the wasp—and its implications for human biological and cultural evolution. I ended with a discussion of religion as a product of evolution and the current relevance of Teilhard's ideas. I said that Teilhard got a lot of things right, as I have related in this chapter, but that he got one thing wrong, which I have saved for the end of this chapter. Teilhard portrayed the Omega Point as the inevitable outcome of cultural evolution. Our current knowledge does not allow us to be so sanguine. Regardless of whether evolution is biological or cultural, small-scale or large-scale, it can always go the way of the strider or the way of the wasp. The only way to reach the Omega Point is by becoming wise managers of evolutionary processes. Left unattended, cultural evolution will take us where we don't want to go.

I was not the only person to discuss Teilhard at the Vatican conference. His name surfaced repeatedly, even with pride that a Jesuit priest had contributed so substantially to the history of evolutionary thought. Yet, judging from the conference, the Vatican seemed no closer to accepting the full implications of Teilhard's ideas than when he was alive. The Catholic scientists, scholars, and theologians were far too sophisticated to accept American-style creationism and intelligent design, which they would regard as the *Beverly Hillbillies* version of theology. They seemed willing to accept what the evolutionists had to say about biological evolution. When it came to humans, however, they were still committed to divine intervention and were wrestling with the age-old question of how so much evil can exist in a world created by a benign and all-powerful God. As an individual, Teilhard had managed to transcend this formulation to achieve a deeper understanding of his religion, like a caterpillar metamorphosing into a butterfly. As an organization, the Catholic church might never be able to accomplish the same transformation. Cultural evolution, like biological evolution, is path-dependent. You can't always get there from here.

CHAPTER 8

The Parable of the Immune System

PIERRE TEILHARD DE CHARDIN discovered a path between two constellations of beliefs, each of which is wrong in its own way. The first is centered on the fact of genetic innateness. In some respects, we are a biological species like any other. Anatomically, we are so similar to the great apes that even Linnaeus, who invented our current system of biological classification in the 1700s, placed us in the same family with them. Today, molecular techniques allow us to specify our ancestry with even more precision. Astonishingly, chimps and bonobos are genetically more closely related to us than they are to gorillas. To be precise, the ancestral species that branched to give rise to modern chimps, bonobos, and humans lived approximately 6 million years ago. The ancestral species that branched to give rise to modern gorillas, chimps, bonobos, and humans lived approximately 8 million years ago. Orangutans and gibbons, the two other great-ape species, branched off even earlier.

All of this is correct, but it becomes wrong when it leads to the conclusion that we are inflexible. This is the specter of genetic determinism that so many people find threatening about evolution. If we are products of genetic evolution, and if genetic evolution operates very slowly, doesn't that mean that we are stuck with our current "natures," that we have an incapacity for change over the time scales that matter for human affairs? Religious believers and nonbelievers alike cherish the thought of human potential—that the future need not be like the past, that we can become much better as both individuals and societies, that there can be such a thing as a path to enlightenment. Accepting the theory of evolution becomes a horrifying prospect when it seems to deny human potential—unless you are already a cynic, in which case, you merely smile and say, "I told you so."

The second constellation of beliefs is centered on the fact of human flexibility. Who cares if chimps and bonobos are genetically more closely related to us than they are to gorillas? No other great apes are even remotely like us in our behavioral and cultural diversity. That's why they're still living in the last remnants of tropical forests and we're the ones cutting down the forests with chain saws and machines the size of dinosaurs. That's why we speak hundreds of different languages and otherwise become so culturally different that we must struggle to comprehend one another. That's why we are overwhelmed by art and sit quietly in lecture halls to receive the information that is contained in our cultures, not in our genes.

All true, but it becomes wrong when it leads to the conclusion that we have mysteriously become liberated from evolution, that we can understand and improve the human condition without any knowledge of evolution. This is the ideal of social constructivism that so many people rush to embrace because it affirms the value of human potential. Alas, it is one thing to affirm the value of something and another to achieve it. Social constructivists often have their hearts in the right place, but they arrive at the construction site without a toolbox.

Like the prophets of old, Teilhard foresaw a great truth when he described human flexibility as a rapid process of evolution, following its own rules in some respects but otherwise obeying the rules that govern all evolutionary processes. Also like the prophets of old, Teilhard pointed the way without providing the details. It is remarkable that he could even point the way, given the rudimentary state of knowledge about evolution during his lifetime.

Today, we are overwhelmed with details. Not only has evolutionary science become sophisticated, but all branches of science and scholarship trained on humanity have become sophisticated, regardless of whether the E-word is used. This tremendous accumulation of knowledge has done little to alter the two constellations of beliefs outlined above, which still clash like the irresistible force striking the immovable object. Very few people have discovered Teilhard's middle way, even among my own evolutionist colleagues. Some scientific social constructivism is required. Perhaps we can turn Teilhard's footpath into a durable highway that anyone can follow, using the details that were unavailable to him.

IN MY PARABLES of the strider and the wasp, I drew on the biological tangled bank to reflect on themes that apply equally to the human tangled bank. It is time once again to draw from the well of nature. This time, it is unnecessary to visit my property, because what I am about to describe exists in each and every one of us. It is the vertebrate immune system.

The immune system evolved by genetic evolution to protect us from parasites and diseases. It does its job so well that the magnitude of its task can be easily overlooked. The big predators that threatened our ancestors and that today are restricted to game parks, zoos, and our imagination are nothing compared with the little predators that remain all around us. Every breath we take, every morsel of food we eat, every time we physically touch something brings us into contact with millions of microbes that can potentially use us as a meal. Don't even think about getting rid of them with disinfectants and antibiotics. With generation times as short as twenty minutes, microbes evolve so fast that killing 99 percent of them, as boasted on the bottles of mouthwashes and disinfectants, merely removes the competition for the 1 percent of survivors that can really do us damage.

Precisely because diseases and parasites can evolve so fast, the vertebrate immune system has evolved a remarkable capacity to keep pace with them by forming antibodies at random and selecting for those that successfully fight the disease of the day inside our bodies. In other words, the vertebrate immune system possesses exactly the same combination of genetic innateness and flexibility that Teilhard imagined for human behavior and culture. If we understand how the immune system works, we can begin to understand how our behavioral and cultural flexibility works — a precondition for improving the quality of life at any scale, from individuals, to neighborhoods, to nations.

The immune system is so complex that an entire branch of science is devoted to understanding it, requiring years of specialized training to become a professional. Current scientific research is still unlocking its secrets. Here is how my colleague Steve Frank begins the first chapter of his book *Immunology and Evolution of Infectious Disease:*

> "The CLTs destroy host cells when their TRCs bind matching MHC-peptide complexes." This sort of jargon-filled sentence dominates discussions of the immune response to parasites. I had initially intended this book to avoid such jargon, so that any reasonably trained biologist could read any chapter without getting caught up in the technical terms. I failed — the quoted sentence comes from a later section in this chapter.

Fortunately, a book with the modest title *How the Immune System Works* does present the big picture of the immune system in a way that anyone can understand. The author, Lauren Sompayrac, received his PhD in particle physics before switching to medical science, specializing in tumor viruses. After retiring from the University of Colorado's department of molecular, cellular, and developmental biology, he began to write modestly titled books such as *How the Immune System Works, How Pathogenic Viruses Work,* and *How Cancer Works.* If there was a Nobel Prize for writing clearly and elegantly about complex topics, I would nominate Lauren Sompayrac.

One reason the immune system seems hard to understand is that it is such a team effort. Lauren asks us to imagine focusing on a single player in a football game, such as a tight end, who runs down the field and then stops. His behavior doesn't make any sense until you see it in the context of the whole game: he took two defenders with him, leaving the running back uncovered to catch the pass and run for a touchdown. The immune system includes dozens of players — specialized cell types — that communicate with one another in their elaborate defense of the body. In fact, the more one learns about the immune system, the more it begins to resemble a social-insect colony inhabiting our body as its environment. Just like the specialized castes of an ant colony, the specialized cells of the immune system are mobile but remain connected through a chemical signaling system that operates completely beneath our conscious awareness.

Let's focus on one of the star players of the immune-system team, the macrophage. It is a giant among cells, like a 300-pound linebacker. Like an amoeba, it is capable of crawling around and engulfing foreign objects with blobby extensions of itself called pseudopods (from the Latin for "false feet"). There is nothing random about its search. It has an elaborate sensory system for hunting its prey. The cover of *How the Immune System Works* shows a photograph of a macrophage about to engulf two bacteria. Once again, the tools of science are required to make something so tiny visible to the naked eye. The single macrophage cell looks like a flat sheet many hundreds of times larger than the two bacterial cells. The portion closest to the bacteria is elevated above the surface by little extensions that look like legs, and the pseudopod is extending toward the bacteria like a long snout. With its legs and snout, it appears as if the macrophage has a front end, and indeed it does, but only for the moment. Any other edge of the sheet might also transform into legs and a snout as soon as it approaches a victim. Producers of horror films, take note: this is one monster that has not yet stalked the silver screen.

If the photo were a video, the pseudopod would flow around the two bacteria, encasing them in a pouch called a vesicle. Then another vesicle containing powerful chemicals would approach and merge with the first vesicle, digesting the bacteria. The chemicals are so powerful that they would also digest the macrophage if they weren't isolated within the vesicles, like acid in a glass bottle.

Billions of macrophages and other members of the immune system team are roaming our bodies at any particular time. Our bodies seem solid to us, but they are like a porous sponge to the immune-system cells. For example, the cells that make up our capillaries, our tiniest blood vessels, are arranged like shingles that an immune system cell can easily squeeze through to enter the body.

Because members of the immune-system team communicate with

one another, an injury as simple as a splinter initiates a symphony of signals that cause members of the team to rush to the site. The immune-system cells already present at the site release chemicals that increase blood flow to the area, make the vessels more porous so that fluid from the capillaries can leak out into the tissues, and stimulates the nerves, which we subjectively feel as pain. Other chemicals, called cytokines, diffuse outward and recruit additional immune-system cells to the site—just like the alarm pheromones emitted by an angry wasp colony.

The more one learns about the mechanistic details of the immune system, the more mind-boggling it becomes. Another member of the team is a killer cell called a neutrophil. About 20 billion neutrophil cells are circulating in our blood at any time. Each one lives only a few days, so millions are also being created at any moment. How neutrophils get from the blood to a wound site makes a fascinating story. Normally, they are swept along at a high speed of about 1000 microns per second. When macrophages at a wound site release cytokines, the internal surface of the surrounding blood vessels become stickier, like Velcro, to the circulating neutrophils. As they roll along more slowly, receptors on the surface of the neutrophils are able to detect the signs of a battle in the vicinity, and they wedge their way through the capillaries to the site.

It gets even more complex. Different signaling molecules make the internal surface of the blood vessels sticky for different specialized members of the immune-system team. To use a human analogy, police, firefighters, and ambulance teams are specialized to cope with different emergencies. If you have a heart attack, you don't want all three to show up at your door. The specialized signaling system of the immune system causes the right members of the team to slow down inside the blood vessels while everyone else sweeps by. As Lauren describes it, "this whole business is like a mail system in which there are trillions of packages (immune system cells) that must be delivered to their correct destinations."

EVERYTHING THAT I HAVE DESCRIBED so far, and much more, is part of what's called the innate immune system, which evolved over hundreds of millions of years. Even sea urchins have elements of our innate immune system. Despite its fabulous complexity and sophistication, the innate immune system can also be baffled when presented with a genuinely new problem for which it has not been prepared by genetic evolution. Recall the elegant experiment by Niko Tinbergen, in which he moved the ring of pinecones surrounding the burrow of the solitary wasp while she was away foraging. Even though she was exquisitely adapted to survive and reproduce in her environment, she was dumbfounded because genetic evolution did not prepare her to solve this particular problem. Or recall Jim Hunt's simple experiment in which he prevented adult social wasps from being fed by the larvae. The entire colony went limp, like separating your esophagus from your stomach. Both of these examples illustrate the paradoxical concept of *rigid flexibility.* Genetically evolved adaptations are often marvelously flexible but only at solving the problems that resulted in their evolution in the first place. The same adaptations prove to be rigid when confronted with a new problem that was not part of the "environment of evolutionary adaptedness (EEA)," to use a bit of technical jargon.

The immune system faces new problems all the time in the form of disease organisms that have evolved new methods of attack. For example, macrophages are genetically adapted to sniff out bacteria on the basis of components of bacteria cell walls that differ from the host cell walls. A new disease strain that manages to disguise itself can bypass this defense. It wins the evolutionary arms race for a while, until genetic evolution produces a new strain of host organism that can detect the new disease strain. This coevolutionary arms race between hosts and their diseases has been going on since time immemorial.

Antibodies provide a powerful new weapon against diseases that goes beyond the innate immune system. By producing antibodies at random and selecting those that successfully bind to disease organisms, a single host organism can protect itself from a new disease organism that has bypassed the defenses of its innate immune system. The rapid evolution of antibodies inside each organism partially replaces the more cumbersome process of genetic evolution eliminating susceptible hosts and favoring resistant hosts. For this reason, the formation and selection of antibodies is called the adaptive component of the immune system.

Using a variation-and-selection process to fight diseases is brilliant, but how is it actually accomplished mechanistically? An antibody is a molecule that functions like a hand with an arm. The hand part grabs onto the surface of a disease organism or any other surface recognized as foreign to the host organism. The arm part connects to the surface of immune-system cells. The ability of the hand to grab onto a foreign surface depends on its shape. Because there are so many foreign surfaces — and new ones are evolving all the time — the challenge for the immune system is to produce a huge variety of hand shapes to find the ones that work. To be precise, about 100 million hand shapes are required to grab onto any conceivable surface. How does the immune system produce 100 million different hand shapes?

Imagine that you have five large bowls in front of you, each filled with numbered balls. The balls in the first bowl are numbered 1 through 40. The balls in the second bowl are numbered 41 through 80, and so on, up to 161 to 200 for the fifth bowl. Now imagine picking a ball at random from each bowl. Their numbers might be 8, 42, 106, 141, 198. Now pick again. The numbers might be 16, 55, 83, 124, and 171. How many possible combinations of numbers are there? 102,400,000! Welcome to the world of combinatorial explosions.

That's how the immune system generates 100 million different hand shapes. The antibodies are produced by a specialized member of the immune system team called B-cells. Each immature B-cell contains multiple versions of certain genes, like the different-numbered balls in each bowl. As a given B-cell matures, it randomly selects one version from each gene, like selecting a ball from each bowl. Thereafter, it faithfully replicates its nearly unique combination. In this fashion, immature B-cells that are genetically identical are transformed by combinatorial magic into mature cells so diverse that their antibodies can latch onto virtually any conceivable organic molecule on foreign surfaces. Brilliant!

So much for the variation part of the variation-and-selection process. How about the selection part? Each B-cell displays its antibodies on its own surface, with the hand part facing outward. When the hands latch onto a foreign surface, it acts as a signal for the B-cell to grow and reproduce. Each division requires about 12 hours, but the magic of exponential growth ensures that a single cell can produce 20,000 cells in a week. Now this army of B-cells starts to produce its particular antibody and pump them out into the body unattached. When you study the immune system, you need to get used to large numbers. One hundred million different antibodies. Three billion B-cells circulating in our body at any one time, each "fishing" for a foreign substance to latch onto. Two hundred thousand B-cells from a single B-cell that makes a "catch." And now, each B-cell pumping out 2000 unattached antibody molecules *per second!*

When an unattached antibody molecule latches onto the surface of a disease organism with its hand, it makes its arm available to connect to the surface of a macrophage or another member of the immune-system team, which does the dirty work of killing the disease organism. This coordinated attack is usually successful at removing the disease organism from our body. The B-cells pumping out the antibody soon die, and the immune system returns to its resting state, with 3 billion B-cells bearing 100 million different antibodies patiently "fishing" for new disease organisms.

Once we develop antibodies to a disease organism, we remain protected for the rest of our lives. Our immune system is able to mount a much faster response to a given disease when it subsequently enters our body, compared with the first time. This does not happen by chance but requires a special memory built into the immune system by genetic evolution.

There are two additional types of B-cells, besides the short-lived type that I have already described. The second type lives much longer and takes up residence in the bone marrow, pumping out moderate amounts of the antibody even after the disease organism is no longer present. The third type does not produce antibodies but stands ready to proliferate and begin antibody production as soon as the disease reenters the body. Together, they provide long-term protection against diseases that struck once and are likely to strike again.

My account of the immune system is a simplified version of Lauren's simplified version. I haven't even mentioned the subsystems that are specialized to fight different categories of disease organisms, such as bacteria, viruses, and large-bodied parasitic worms. I have only described the tip of the iceberg. One implication of all this complexity is that the adaptive immune system doesn't replace the innate immune system. It is an add-on that utterly depends on the innate immune system to function. Lauren makes this point by returning to a football analogy. The adaptive immune system is like a quarterback who cannot possibly function without the coach and the rest of the team.

Because the adaptive immune system is so dependent on the innate immune system, it has not entirely escaped the problem of rigid flexibility. Despite its amazing variation-and-selection ability, it can still be dumbfounded in environments that depart from the EEA ("environment of evolutionary adaptedness"). If you suffer from asthma, hay fever, irritable bowel syndrome, certain kinds of diabetes, and many other modern ailments, then your problem is the result not of a disease organism but of a dumbfounded immune system.

Throughout its long history, the vertebrate immune system has been confronted with an onslaught of disease organisms. They weren't welcome, but they were always there. The immune system could therefore rely on their presence to develop appropriate antibodies, just as solitary wasps can rely on the presence of landmarks to find their way home. Whenever something is a constant part of the environment, adaptations can rely on their presence to function properly.

For the first time in the 200-million-year history of the vertebrate immune system, modern medicine and hygiene have created an environment in which these ancient companions are largely absent. At first, this seems like an unambiguous blessing, until we reckon with the arsenal of the immune system, poised to unleash its firepower. If the usual signals that it evolved to detect are absent, there is no telling what it will interpret as a signal and what kind of friendly fire it will unleash on the body that it is supposed to protect. The most exquisite adaptations to one environment can go spectacularly wrong in a different environment, even when they employ their own variation-and-selection processes. Modern medical science is only beginning to absorb this sobering message, which accounts for a large fraction of what ails us in modern life.

WHAT CAN THE PARABLE OF the immune system teach us about our capacity for behavioral and cultural change? How can it help us follow Teilhard's prophetic lead by steering a middle course between genetic determinism and social constructivism? How can we use the knowledge to change our behavioral and cultural practices for the better?

First, even though parasites and diseases are only part of the environment inhabited by a large organism such as a vertebrate, the problems that they pose for survival and reproduction are mind-bogglingly complex.

Second, the ways in which the immune system protects the body from parasites and diseases are profoundly behavioral and social. Earlier, I said that the behavioral becomes the physical when examined closely enough. That's true with a vengeance for the immune system, which fundamentally involves specialized cells moving and communicating with one another, just like the specialized castes of a social-insect colony. If that's not behavioral and social, what would be? Some aspects of the immune system even qualify as cultural, in the sense of the transmission of information across cell generations.

Third, the adaptive repertoire of all organisms has the paradoxical property of rigid flexibility that I have already emphasized for solitary wasps, social wasps, and now the immune system. The concept of rigid flexibility seems like a fusion of opposites, like a Zen koan, but it is one of the most important tools in the evolutionary tool kit and becomes fully intuitive with practice. Organisms can be amazingly flexible and intelligent in response to their accustomed environment and yet dumbfounded by the slightest change in their accustomed environment. That's because their flexibility is a product of evolution, which is only capable of adapting organisms to their accustomed environment—the EEA.

Fourth, the adaptive component of the immune system represents a special kind of adaptation that evolved by genetic evolution but is also itself an evolutionary process. This kind of adaptation makes flexibility less rigid. If a new disease organism arrived from Mars and invaded your body, your immune system could probably handle it, even though it has never been encountered during the entire history of the immune system. That's because the variation-and-selection process employed by the immune system enables it to adapt rapidly to its current environment, just like the variation-and-selection process of genetic evolution.

Fifth, in another fusion of opposites, the variation-and-selection process of the adaptive immune system is, and must be, richly innate. Antibody variation doesn't happen by itself. It requires an elaborate mechanism that creates 100 million combinations out of a much smaller number of genes. Selection doesn't happen by itself, either. It requires elaborate mechanisms to recognize the most successful antibodies and link their success to the reproduction of the B-cells that produce them.

Sixth, the adaptive component of the immune system supplements but does not replace the innate component. Most diseases and parasites are dispatched by the innate component. Even when the adaptive component is required to recognize a disease that has managed to fly beneath the radar of the innate component, it merely tags the disease for destruction by the innate component.

These points prepare us to think about other variation-and-selection processes that have evolved by genetic evolution. In particular, the tradition of behaviorism in psychology focuses on the open-ended behavioral flexibility of animals — not just humans but the rats and pigeons that B. F. Skinner made famous for their ability to learn anything that he wanted them to do in his Skinner boxes. He even trained pigeons to play Ping-Pong, a game that assuredly was never part of their EEA. In a famous paper, "Selection by Consequences," Skinner explicitly described the learning ability of animals as a product of genetic evolution and an evolutionary process in its own right. However, he also thought that "learning" had largely replaced "instinct" in our species, and he scorned the idea of studying how learning actually takes place inside the head. By analogy, this would be like saying that the adaptive component of the immune system has largely replaced the innate component and that we can understand what we need to know about the immune system without bothering with the mechanistic details.

Seventh, the immune system is profoundly cooperative. It is not even remotely like a single agent that makes all of the decisions and does all of the work by itself. It requires a team of agents that perform different functions and are in constant communication with one another. This fact goes a long way toward explaining why our species took variation-and-selection processes to new heights. Other species can learn as individuals, but a variation-and-selection process at the level of groups requires a degree of cooperation that most species lack. In most group-living species, members of a group cooperate to a degree but are also their chief rivals. These species also have culture to a degree. In fact, we are only beginning to discover the extent to which creatures such as chimps, lions, and crows have cultural traditions that adapt them to their environments. You might be surprised to see crows on this list, but remember that we don't need to travel to Africa or study our closest primate relatives to learn the lessons of the tangled bank. We don't even need to travel to my property to study crows, because they have moved into our cities. I am hearing their raucous calls from my Binghamton home as I write these very words.

In the last chapter, I made the foundational point that cooperation came first in human evolution. Even though we seem most distinct in our mental abilities, making it tempting to speculate that an individual ability to reflect came first (as Teilhard put it), there is nothing individual about the ability to reflect. A major evolutionary transition was required for our ancestors to trust members of their group sufficiently to share the information required for symbolic thought to evolve as a mental capacity. The parable of the immune system allows me to make the same point now with even more force. The human major transition created something that was new on the face of the earth: an intelligent primate that was also sufficiently cooperative to evolve a variation-and-selection process at the group level. *Human groups are like the immune system*. We are only beginning to appreciate the implications of this statement.

Eighth, even the adaptive component of the immune system does not escape the problem of rigid flexibility. A single change in the environment, such as removing parasites and diseases that have always been present, can cause the immune system to fire on the very organism that it is designed to protect. It is beyond the capacity of the immune system to solve this problem. The only solutions are to alter the environment perceived by the immune system to avoid friendly fire or wait for genetic evolution to adapt the immune system to its new environment.

This final lesson is sobering enough for the actual immune system but even more sobering for the other variation-and-selection processes that are like the immune system. Earlier, I said that religious believers and nonbelievers alike cherish the thought of human potential, that the future need not be like the past, that we can become much better as both individuals and societies, that there can be such a thing as a path to enlightenment. I'm here to testify that these dreams can come true, thanks to the fact that human groups are like the immune system. Alone among all species, we have the capacity to imagine new worlds and make them a reality. But beware. Our marvelous capacity for change is merely rigidly flexible. A single change in the environment can cause us to become as dumbfounded as a solitary wasp whose ring of pinecones has been displaced. As limp as a wasp colony when the larvae can't feed the adults. As tormented as a body being destroyed by its own immune system. We wouldn't even know what was happening, for lack of a theory that makes sense of what is in front of our faces.

CHAPTER 9

The Reflection

EVEN THOUGH the Binghamton Neighborhood Project is one of my newest endeavors, I am beginning to regard it as my anchor to reality. Pierre Teilhard de Chardin can have his Omega Point. You are welcome to believe that Jesus could raise the dead. I'll settle for raising the valleys of my GIS maps into hills. When I scan my eyes over the hills and valleys of my GIS maps, I'm not just trying to improve my city of Binghamton, although that would be good enough. I'm pondering the deepest and most philosophical questions that can be asked about the human condition. If I succeed at raising the valleys into hills, it will be because I have learned something general that can be applied to the valleys of any human population. When I leave Binghamton to rub shoulders with the philosophers, intellectuals, and scientists of the world, it's the Binghamton Neighborhood Project that I want to talk about—and they want to listen. The clarity with which I can examine the human condition in the real world is like exchanging Leeuwenhoek's microscope for the most recent SEM model.

Now that we are nearing the midpoint of this book, it is useful to take stock and reflect on its full meaning. I began with the ancient tension between rationality and faith, including the fight between the rabbi and his wife about why the geese shrieked. Then I said that science, the flowering of rationality, can become a way of listening and reflecting on the human condition, like religion and literature.

Listening, seeing, and all other forms of perception are ways of gathering information about the world. Reflecting is the processing of the information to give it meaning. Meaning results in sustaining action. All creatures have evolved to listen and reflect in ways that enable them to survive and reproduce in their environments. The cycle of listening and reflecting leading to sustaining action has been perfected by countless generations of natural selection.

The adaptations that evolve by genetic evolution are breathtaking in their sophistication, but they are only good at what they are designed to do. In another Zen-like fusion of opposites, one must be blind in order to see. In scientific plain talk, there is so much information to gather and so many ways to process it that both perception and reflection must be highly selective. Our genetically evolved perception and reflection abilities equip us to survive in small groups living off the land. They don't even remotely equip us to solve the problems of modern human existence. Unless we use the tools of science to listen and reflect on the human condition, *fuhgeddaboudit*, as the Brooklynites like to say.

This point is easiest to make for perception. It's a no-brainer that we need the tools of science to see tiny objects, distant objects, big diffuse objects such as a city, and entire sensory modalities such as electrical fields that other creatures perceive as plainly as we see and hear. We are so utterly dependent on these extensions of our perceptual abilities that we need to remind ourselves that they are not a gift of God or a gift of genetic evolution. They are a gift of cultural evolution operating over a process of many thousands of years.

The same point is more novel and therefore more potent when made for reflection. What do we *do* with all of this information? How do we give it meaning that leads to sustaining action? Sustaining for whom? Me? Us? Everyone? Religions, stories, philosophies, ideologies, and theories of all sorts are meaning systems that begin by interpreting the world and end by motivating action. They are not a gift of God or a gift of genetic evolution. They are a gift of cultural evolution operating over a process of many thousands of years, just like the modern technologies that extend our perceptual abilities. Meaning systems emerge, mutate, recombine, and jostle with one another for prominence. For every survivor, there are hundreds of failures known only to historians or permanently lost in the mist of time. The process is still taking place all around us, as we can see as soon as we know what to look for.

Among this crowded field of meaning systems, science and evolutionary theory are relative newcomers. All meaning systems, including the most fundamentalist religions, make statements about the world that are regarded and defended as facts. These "facts" frequently have no evidential basis and even flaunt what is known to be true—not only for religions but also for nonreligious meaning systems such as political ideologies. Modern life is awash with falsehoods ardently defended as facts, such as the claims that surround America's debate over health care. At first, it seems that we have all gone crazy, until we realize that the primary function of a meaning system is to motivate sustaining action. There is nothing like a putative fact for motivating action. If I dislike what you are doing, I can call you sick or immature. If I think that a woman's place is in the home, I can claim that she is mentally inferior and that it is abnormal for her to become sexually aroused. It's a Darwinian world, for meaning systems no less than for biological organisms, as Teilhard was among the first to appreciate. Putative facts are so good at motivating action that their actual veracity seldom stands in the way in the push-and-shove among meaning systems. When it comes to truth and consequences, consequences dictate truth for most surviving meaning systems. If a putative fact is actually true, so much the better. If not, whatever, as long as it causes us to do the right thing.

Science is a cultural system, only a few centuries old, that uniquely values the truth for its own sake. I do not say this to glorify science. In fact, science's obsession with the truth makes it incomplete as a meaning system, because science by itself does not tell us what to do with the facts that have been established. Moreover, scientists are not immune to letting consequences dictate truth in their own minds. Those claims about women that I described in the last paragraph were received scientific wisdom during the nineteenth century. What scientists say should never be accepted just because they call themselves scientists or even if it seems to have been affirmed by scientific research. It is always necessary to jump up and down on the scaffold of knowledge to make sure that it is solid. If you are skeptical about a scientific claim, then jump up and down on it as hard as you can until you expose a weakness or convince yourself that it is solid.

Science works, to the degree that it does, because scientists are held accountable for what they say more than almost any other cultural system. If I want to say that somebody's health-care plan is going to kill your granny, I'm free to do so on the public stage, despite screams of protests that it's a bald-faced lie. If I try to do something comparable as a scientist or a scholar, I'm outta there. I'll be accused of incompetence, my papers won't be published, and I won't get a job. If I willfully falsify information, I'll be excluded in the same way that the most grievous sinners are excluded from their churches.

Like a strong religion, science and scholarship offer plenty of structure to implement their ideals. Religious believers aren't just exhorted to obey the dictates of their faith; they are locked into a system that makes it difficult to do otherwise. In just the same way, scientists and scholars are locked into a system that makes it difficult not to seek the truth. Something as simple as publishing a paper in a peer-reviewed journal is a cut gem of accountability, as I described in the parable of the strider. I am happy to call science a religion that worships truth as its god.

THE CONFLICT BETWEEN RATIONALITY AND faith exists whenever a meaning system departs from factual reality on its way toward motivating sustaining action. That's when a rationalist, such as Isaac Bashevis Singer's mother, can scornfully insist that dead geese don't shriek and decisively settle the issue with the evidence of windpipes. That's when her rabbi husband can legitimately worry that cold logic is tearing down faith. Sustaining action is a good thing, at least when it doesn't interfere with the sustaining action of others. A meaning system that leads to sustaining action is an object of great value, even if it departs from factual reality along the way. If facts are used to tear down faith in an enduring meaning system, a murder of sorts is being committed. That is the essence of the ancient tension between rationality and faith.

Can the conflict be resolved? Yes, although it would be for the first time in history. What's needed is a meaning system that respects factual knowledge as never before, in the same way that scientists and scholars do, and then uses factual knowledge to implement its values. This is different from science as usually practiced, which is silent about values. The new meaning system would require a more self-conscious and explicit consideration of values than ever before, in addition to a greater respect for factual knowledge. Most meaning systems are not fully self-conscious about their values, but the new meaning system would need to be.

Stories about animals, human origins, and human nature that have been told throughout the ages are good examples of useful falsehoods. I refer not only to children's stories such as Aesop's fables and religious stories such as the garden of Eden and original sin but also secular stories for adults, such as Mandeville's fable of the bees, Hobbes's war of all against all, Rousseau's noble savage, Freud's primal horde, and the portrayal of human nature sometimes called *Homo economicus*. All of these made enough sense to pass as fact when they were coined but in retrospect can be seen as putative truths dictated by consequences. Authentic scientific knowledge about the biological world, human origins, and human nature has incomparably more to tell us about ourselves than these fictions.

Thinking of ourselves as part of the tangled bank is so new that most of what we have learned was discovered during the last twenty years and is still unknown to the vast majority of scientists and scholars studying humanity, not to speak of the general public and experts who run our governments and economies. Nevertheless, a theory this useful for understanding the human condition must also be useful for improving it. The reason everyone believes in physics and chemistry is not that they are better supported by evidence than evolution but that they are more useful on a daily basis. They are a form of authentic knowledge that contributes to durable meaning systems, so we embrace them as warmly as we embrace useful fictions. Evolutionary science will eventually prove so useful on a daily basis that we will wonder how we survived without it. I'm here to make that day come sooner rather than later, starting with my own city of Binghamton.

The first phase of the Binghamton Neighborhood Project was long on perception and short on reflection. It was necessary to gather information and visualize it in the form of GIS maps before we could process it. Scientific information processing requires statistical analysis, which is seldom regarded as comparable to philosophical reflection. The more I analyzed my data with the help of modern statistics, however, the more I appreciated Plato's parable of the cave wall. Plato imagined a group of people chained so that they were facing a cave wall with a fire behind them. They can see the shadows of events taking place inside the cave projected onto the wall, but they cannot directly see the events. The shadows are as close as they can get to reality. For Plato, a philosopher is someone who becomes freed from the chains and can directly perceive the world as it is, rather than its distorted projections.

In my case, reality was a complicated web of causal relationships that resulted in the hills and valleys of my GIS maps. The shadows were the correlations among the variables in my database. The purpose of scientific analysis, with the help of statistics, is to infer the causal relationships from the correlations, as the philosopher does. A correlation does not automatically signify causation. If A goes up when B goes up, that doesn't necessarily mean that A caused B to go up. It could be the reverse, a third factor that caused both A and B to go up, or an even more complicated web of causal interactions. On the other hand, something happened to make A and B rise together. With enough information and ingenuity, the causal web can be revealed. That is the goal of statistics, no matter how technical it gets.

LAPTOP COMPUTERS MAKE IT POSSIBLE to analyze data anywhere, but my preferred location is the university's new downtown center, by the Chenango River where Native American long houses used to be, a stone's throw from Confluence Park. The downtown center was constructed to give the university a greater presence in the community and to house the departments that are most dedicated to helping people in a pragmatic sense, such as the department of human development, the department of public administration, and the department of social work. These are regarded as blue-collar departments at most universities, compared with white-collar departments such as physics and economics. There is no Nobel Prize for social work, and the very name has a proletariat ring. When I decided that the trenches are the best places to dig for gold, I developed a newfound respect for these disciplines and the faculty in these departments. My goal was now the same as theirs, and it wasn't obvious what my white-collar academic credentials had to contribute. Just as Doug Stento learned some lessons when he was assigned to the ditch-digging crew of his father's construction company, I had some lessons to learn from my new colleagues in these so-called blue-collar departments. I was therefore honored when they offered to provide me with a second office in the new building where they would be housed.

The building itself is a sight to behold. Earlier, I said that Binghamton's most durable and beautiful buildings were built during the nineteenth century, making it seem that we are squatting on the ruins of an earlier civilization. The downtown center rivals the older buildings in both respects. It can even be said to have a body and a soul.

The body of the building uses the most recent technology to survive the elements and efficiently regulate its internal environment, like a single organism or a wasp colony. It recognizes the magnetized signature of my university ID card from a distance, automatically raising the gate of the parking lot and unlocking the entrance doors, like an immune system recognizing a cell as one of its own. Motion detectors conserve energy by turning out the lights when moving objects are absent, sometimes plunging me into darkness in my office when I am lost in thought. Temperature is conserved and regulated with the newest technology and ancient techniques such as the strategic placement of eaves to provide shade from the sun when it is high in the sky during the summer and to admit its warming rays when low in the sky in winter. It even has an impermeable skin that extends underground to protect it from periodic flooding.

The building is aesthetically pleasing in addition to being functional. It has an uplifting quality similar to a church, which means that it has a soul as Teilhard used the word. Entering the building takes you into a central atrium 54 feet tall with skylights and the entire facing wall of glass, making you feel half indoors and half outdoors. A kiosk to your left sells food and coffee, with tables and chairs arranged as at an outdoor café. An entrance to your right welcomes you to the "Information Commons," which is what libraries have become. Books and journals printed on paper have become so insignificant that they are located in one corner of the Information Commons, on shelves ingeniously mounted on tracks so that they can be pressed together to save space and rolled apart to create an aisle only when a particular book or magazine is needed. I have never seen it used even once. The rest of the space is devoted to computers, illustrating the degree to which they have replaced words printed on paper, for better or for worse. Most of the computers are arranged in circular clusters with partitions between the computers and a petal-like awning above each chair, making it seem that you are sitting underneath a flower. The arrangement creates a feeling of privacy for each person using a computer, even though the room is crowded and the wall separating the Information Commons from the atrium is glass, allowing everyone to see in and out.

A room for holding meetings with members of the community includes computers connected to the Internet, a large conference table to sit around, attractive upholstered chairs for more casual conversations, and even a corner with toys and books for young children. Potted plants line the big picture windows facing the street, affording the pleasant distraction of watching people pass by. I feel good when I am in this room, and so does everyone else. Nobody has yet said, "Can't we move into a stuffy windowless room with lower ceilings and fluorescent lighting that makes us look as if we have jaundice?"

The building even has an iconography, similar to the symbols of a church or a Masonic temple. The seal of the university is inlaid in marble on the floor as you enter the atrium. Curves of colored marble radiate outward from the university seal, like ripples from a stone thrown into a pond, which continue to radiate outside the building in the form of textured concrete. The curves symbolize the optimistic hope that the downtown building will have benefits that radiate outward into the community. The outlines of the nineteenth-century buildings that used to stand on the spot are also inlaid in colored marble to give a sense of history.

A lot of money, planning, and love went into the construction of this building. If you think it is a waste of the taxpayer's money, I beg to differ. At the conference in Tuscany that I attended during the year of Darwin, I had dinner in a house that was built a century before Columbus discovered America. One reason parts of Europe are so beautiful is that the human constructions are so durable and blend so harmoniously with the natural landscape. Contrast these with the buildings built by communist regimes after World War II, which were so narrowly utilitarian and penny pinching that they sapped the human soul and started falling apart almost immediately. I would like to think that the richest and most powerful nation on earth can construct buildings that lift the spirit and will look as good during the next century as Binghamton's churches and nineteenth-century buildings look today. I'm especially gratified by the fact that this uplifting building was not created for the rich and powerful but for anyone in the city of Binghamton who wishes to pass through its doors. Many of the people who use it are the first in their families to attend college.

My own office in the downtown building is not prestigious. In fact, it is windowless and located next to the restrooms, from which I periodically hear the whir of the electric hand dryers. Nevertheless, I love it and regard it as my lean, mean analysis machine. My laptop plugs into a 21-inch monitor with a keyboard adjusted to just the right height for typing. My office chair is perfectly adjusted. Stapler, tape dispenser, scissors, paper clips, and stationery are neatly arranged within arm's reach. The four windowless walls enable me to create my own visual environment with art. One wall displays an old map of Binghamton created in 1840, as if from a hot-air balloon, with each building minutely drawn with the precision of a draftsman's hand. Below it hang two of my GIS maps, mounted as if they are artwork. On another wall is a photograph taken during the early twentieth century of the buildings lining the Chenango River, with signs advertising stores and products that no longer exist. The wall that I face while working at the computer has a poster-size photograph of the Andaman Islands, a tropical paradise. A brightly painted rowboat in the foreground floats on water so clear that it seems to be suspended in the air. The two oars stick out like the wings of an ungainly insect. In the background are some rocky islands and sailing ships against the ocean horizon. This photograph reminds me that I am on a voyage of discovery as exciting, in its own way, as the age of exploration.

AS I BEGAN TO ANALYZE my data, I was guided by the prediction that Binghamton is a Darwinian world at two levels. First, there is a rapid variation-and-selection process similar to antibody evolution in the immune system. To a large extent, people are good at figuring out what works in their particular situation and doing it. Second, there is something comparable to the innate immune system that evolved by genetic evolution and makes the rapid variation-and-selection process possible. That "something" is likely to be complex. It goes way beyond the conscious weighing of alternatives to include subconscious processes that must be discovered scientifically to be understood.

That "something" might go beyond the individual. As we have seen, the immune system is profoundly a matter of teamwork. The behavioral variation-and-selection process might be better understood as a coordinated group than as a single reasoning agent. The teamwork might also take place largely beneath conscious awareness. We could be playing our respective roles every waking moment without the slightest idea of what we are doing.

In the past, the statement "It's a Darwinian world" evoked images of nature red in tooth and claw, as if life in a Darwinian world must be nasty, brutish, and short. I hope that my parables of the strider and the wasp have permanently put these associations to rest. It's not as if evolution makes everything nice, as the parable of the strider attests. Instead, evolution can potentially explain the full spectrum of behaviors, from the nastiest to the nicest, as the two parables taken together attest.

The behaviors that we associate with nastiness and niceness coexist in the human tangled bank for the same reason that species coexist in the biological tangled bank: because each is adapted to different environmental circumstances. If we can understand the environmental circumstances favoring niceness, or its "niche," to use an ecological term, we can attempt to expand the niche for niceness and shrink the niche for nastiness, raising the valleys of my GIS maps into hills.

What is the niche for niceness? Part of the answer depends on the difficulty of the task. If I want to move my coffee mug, I can do it myself. You would be foolish to offer your help, because you would only get in the way. Many tasks don't require cooperation, and niceness is not such a virtue that we should make tasks difficult just so that we can cooperate. Whenever we can make life as simple as moving a coffee mug, let's do so!

Moving a piano is a different matter, not to speak of a really difficult task such as building the downtown building. *That* required cooperation. Whenever cooperation is required to get something done, there is a single overriding criterion for it to evolve. The cooperators must confine their interactions to one another and avoid interacting with the noncooperators. Cooperation, niceness, and solid citizenry in all their forms tend to be vulnerable to noncooperation, nastiness, and knavery in all their forms. Those who give must get to be sustained over the long term. When this happens, evolution goes the way of the wasp. When those who don't give get, then evolution goes the way of the strider.

That was the first prediction that I attempted to test with my data. The Developmental Assets Profile (DAP) that we administered to the public-school students included items that asked about one's own niceness, such as "I am serving others in my community." Also included were items that asked about social support from various sources such as family ("I have parents/guardians who help me succeed"), neighborhood ("I have neighbors who help watch out for me"), school ("I have a school that cares about kids and encourages them"), religion ("I am involved in a religious group or activity"), and extracurricular activities ("I am involved in a sport or creative things such as music"). The statistical method of multiple regression enabled me to show that the most prosocial kids in Binghamton also received the most social support. Those who gave were indeed getting. Moreover, the most prosocial kids were bathed in social support from multiple sources. Family, neighborhood, school, religion, and extracurricular activities all made separate contributions that added up. People are like buildings. What they become depends on how much is wisely invested.

The results didn't have to turn out that way. It is perfectly possible to imagine a kid who reports getting a lot but has no tendency to give. It's also easy to imagine a kid who heroically gives without getting anything. All combinations of individual prosociality and social support are possible in principle. If they were completely uncorrelated, then my prediction based on what to expect in a Darwinian world would have failed, and I would need to go back to the drawing board. Instead, my prediction was confirmed, and I had taken one solid step forward in my analysis.

The degree to which highly prosocial kids in Binghamton were matched to their social environment was impressive, as we can see by comparing it with the matching that occurs among kin in simple genetic models. These models are based on the assumption that behaviors are coded directly by genes. If I behave nicely, it is because I have a gene for niceness. I tend to share the same genes with my relatives because we both got them from the same common ancestor. That automatically creates a correlation between giving and getting that increases with the degree of relatedness. If we are identical twins and I am nice, you are certain to be nice. If we are siblings with the same mother and father, the probability is 0.5 that you have my nice gene which is identical by descent. If we are completely unrelated, then the correlation between giving and getting declines to zero.

Human niceness includes nepotism, just as with any other species, but also is extended to nonrelatives much more than it is in other species, so something else must create a correlation between giving and getting when we interact with nonrelatives. There has been plenty of theorizing, but before now, the correlation has never actually been measured in any human population, much less a population size of a whole city. The strength of the correlation that we measured was about 0.7. Astonishingly, the correlation between giving and getting for school-age kids in the city of Binghamton was greater than the nepotistic niceness expected among full siblings in a strictly genetic model.

THE KIDS WHO SCORED LOW on prosociality should not necessarily be regarded as antisocial. Imagine that you are a kid who is unlucky enough to inhabit a harsh social environment. Nobody is giving to you, so what should you do? You only have four choices: (1) leave your social environment if you can, (2) attempt to improve your social environment, (3) turn off your own niceness to protect yourself, or (4) remain nice and suffer the consequences. Might you feel tempted to choose option 3? In this fashion, many of the kids who contribute to the strong correlation by not giving and not getting might be perfectly capable of giving, if only they can find their way to a more nurturing social environment.

On the other hand, suppose that your environment was so harsh, from such an early age, that you have become like a turtle afraid to poke your head out of your shell. Then you are lucky enough to find a more nurturing social environment. Can you immediately cast off your shell? Possibly not. Some survival skills acquired early in life might not be that easy to change later in life, especially if they were orchestrated by complex mechanisms operating beneath conscious awareness, similar to the innate component of the immune system.

Finally, even though some kids who score low on prosociality might be like turtles in their shells, others might be like sharks and parasites after all. Many species inhabit the biological tangled bank, and we should expect the human tangled bank to be similarly diverse. You needn't be a scientist to appreciate that it's potentially a dangerous world out there. Some people are prepared to take advantage of you, no matter how nice you are. Predatory and parasitic strategies are especially effective when they are rare. That's when the cooperators drop their guard. Again, any given social strategy need not be deliberate and might be caused by complex psychological mechanisms that operate completely beneath conscious awareness. Social predators and parasites might feel perfectly virtuous, for example, even though their behavior tells a different story.

Identifying these various "species" would require lots of data and sophisticated statistical techniques. One technique, called hierarchical linear modeling (HLM), didn't even exist when I got my degree but has become essential for analyzing units that are nested within one another, such as individuals who live in the neighborhoods that in turn make up a city. What can be attributed to variation among neighborhoods, as opposed to variation among individuals?

I'd like to think that I'm an old dog who can learn new tricks. On the other hand, there was Dan O'Brien, my ace graduate student, with his mathematical aptitude purring like the engine of a Maserati under his hood. In what seemed to me like no time, Dan took the most advanced training offered by the psychology department and became the statistical specialist for the Binghamton Neighborhood Project. I could keep up with and contribute to the analysis at an intuitive level, but he took care of the nuts and bolts. Hooray for division of labor!

I had already shown that kids who report being highly prosocial also report that their neighborhoods are supportive. That is an individual-level correlation. Dan was able to demonstrate an additional neighborhood-level correlation. In other words, if you are a highly prosocial kid in the city of Binghamton, not only do *you* report that your neighborhood is supportive, but the *other kids* in your neighborhood also report that it is supportive, and *their* estimation statistically contributes to *your* prosociality over and above *your own* estimation. This result helped affirm our faith that the survey was measuring something real. If the kids were just making stuff up, then what other kids in your neighborhood say should not be predictive of your own response.

When people see the hills and valleys of prosociality on my GIS maps, they often ask if it's just a matter of being rich or poor. My data included a few items indicative of wealth, such as parents' education and whether a kid qualified for a free school-lunch program, but much more information was available electronically from other sources. Thanks to GIS technology, I can add any spatially based information to my database. The U.S. Census Bureau, for example, has a wealth of information about all neighborhoods in America, including median income. With a few expert mouse clicks, this information could be added to our database, and we could compare the hills and valleys of median income with the hills and valleys of prosociality.

A superficial comparison revealed a positive correlation. On a graph with median income on the x axis and prosociality on the y axis, the cluster of points had a positive slope, although there was also a lot of scatter. When we included median income and social support in the same analysis, however, a different result emerged. Median income correlated with individual prosociality only insofar as it contributed to social support. The web of causal relationships that we were beginning to discern through the correlations looked something like this: The prosociality of a kid depends on social support. Money can be used for lots of things. When money is used for social support, kids become more prosocial. When it is used for other things, it has no effect on prosociality. Money can even have a *corrosive* effect on prosociality, as we shall see.

IN CHAPTER 6, I DESCRIBED our whimsical studies involving holiday decorations, lost letters, and garage sales, which we performed to validate the survey. The numbers from these studies were lined up in neat columns alongside the rest of our data, like new battalions joining a vast military parade. They broadly confirmed our survey results, giving us a high degree of confidence that the hills and valleys of our GIS maps were real. In the very same neighborhoods that the kids rated as unsupportive, people stepped over the stamped envelopes on the sidewalk and didn't bother to put up decorations during the holidays.

The experimental economics games that we played with the students brought the hills and valleys into sharper focus. Imagine that you are a Binghamton High School student walking into your health class. You are told that today we have two visitors from Binghamton University who want to play a game with us. In walk Dan O'Brien and Omar Eldakar, two cool-looking young adults, who explain the rules of the game. You will be paired with another member of your class and can choose whether or not to cooperate. If both cooperate, each of you gets $30. If neither cooperates, each of you gets only $15. If one cooperates and the other doesn't, the noncooperator gets $45 and the cooperator gets $10. Finally, one person must announce his or her decision first. Playing the game therefore requires three decisions. What will you do if you go first? What will you do if you go second and your partner has decided to cooperate? What will you do if you go second and your partner has decided not to cooperate? You are asked to write your decisions and a brief description of your rationale on a piece of paper.

Depending on what type of student you are, you might be intrigued, or you might be looking at the clock, wondering when this is going to end. Then you're told that the game will be played for real money. After everyone indicates his or her decision, the papers will be collected, two will be chosen at random, one will be chosen at random to go first, and the two students will be paid real money on the basis of their decisions. Now you're interested, even if you were looking at the clock before.

A year earlier, you took a survey in another class. The survey wasn't mentioned today, and you have no reason to remember it, but what you indicated on the survey is combined with how you played the game, the longitude and latitude of your residential location, and many other columns of numbers inside my laptop.

This experiment enabled Dan and me to dissect cooperation into different components that could be examined separately. Deciding to cooperate as the first mover requires an element of *trust*, because you are making yourself vulnerable to defection. If you think that your partner will take advantage of you, then you will elect not to cooperate, earning $15 rather than $10. If you trust your partner, then you will elect to cooperate to get $30.

Deciding to cooperate as a second mover if your partner cooperates requires an element of *trustworthiness*. There is no uncertainty about what your partner will do. You know that you can get $45 rather than $30 if you want, but should you do the right thing by reciprocating? Your decision is likely to depend heavily on details such as whether your decision will be made public, whether you will have an ongoing relationship with your partner, and so on. The beauty of the game is that it can be played under many different conditions in which these details are systematically altered. In our case, Dan and Omar picked the two papers at the end of the class period, so all of the students knew that their choices would be revealed if they were chosen to receive actual money.

Finally, deciding to cooperate as a second mover even when the first mover has defected requires an element of *unconditional generosity*. People are frequently exhorted to turn the other cheek, and perhaps some really do. Dan and I didn't need to speculate, because our experiment would give us hard numbers, at least within the microcosm of the game.

As with our original survey, Dan and I subjected the data to a thorough inspection before adding them to our database. If the choices made no sense (such as defecting as a first mover and cooperating in response to defection as a second mover), or if the verbal comments suggested that the student didn't understand the game, the data were removed. Our accumulating database already included math scores from a state-mandated test. A comparison of math scores for the students who played the game revealed that those who were dropped from the analysis also had low math scores, providing independent evidence that they probably just didn't get the game.

Once the data had been inspected, the analysis could begin. Patterns emerged among how the students played the game, how they responded to our survey a year earlier, and information about their neighborhoods from the U.S. Census. A web of relationships was causing these variables, gathered at completely different times and in different ways, to vary in concert with one another. Our job was to infer the causal web from the correlations, like the unchained prisoners in Plato's cave.

Of the 130 students who were retained in the analysis, 63 percent were trusting as first movers, 66 percent were trustworthy in response to a cooperative first mover, and 15 percent were unconditionally generous in response to a defecting first mover. Dan and I argue over whether the cheek turners were really motivated by generosity, but in any case, their numbers were too small for statistical analysis, so we focused on trust and trustworthiness.

With respect to trust, we discovered that kids who elect to cooperate as first movers tend to come from neighborhoods that are *high* in quality and *low* in median income. Wealth actually had a toxic effect on this component of prosociality, instead of the merely neutral effect that we measured in our survey. Here's how we interpreted this result from our evolutionary perspective. Recall the two requirements for cooperation to evolve. First, people must be engaged in some activity that requires cooperation in the first place. It can't be like lifting a coffee mug; it must be like moving a piano or constructing the downtown building. Second, given the need to cooperate, there must be enough social structure so that the solid citizens can profit from one another and avoid being exploited by knaves. Rich kids can get what they want with a credit card, which is as easy for them as lifting a coffee mug. Poor kids have to cooperate to get what they need; for them, life is like moving a piano. However, a nice neighborhood is required to provide the social structure, even when cooperation is needed. If that was the actual web of causal relationships, it would explain the correlations that we observed in our analysis. Of course, other causal webs might also explain the same correlations. Inferring reality from the shadows on the cave wall is not an easy matter.

Another result was that kids on the free-lunch program tended to be untrusting, even though the most trusting kids came from low-income neighborhoods. We regarded this as broadly consistent with our interpretation. Too much poverty can be toxic for trust, in addition to too much wealth.

The results for trust did not apply to trustworthiness, which at first didn't seem to correlate with anything in our database. Dan and I didn't know what to make of this. When you're a scientist, you need a theory to point you in the right direction, but your search always becomes a fishing expedition at some level. As Einstein famously said, if we knew what we were doing, it wouldn't be called research. Dan started to sort through the many variables from the U.S. Census and the city that had become a part of our ever-accumulating database. Two variables correlated with trustworthiness at a level that was statistically highly significant: the kids who cooperated as second movers in response to a cooperative first mover tended to come from neighborhoods with low population density and high mixed land use, such as residential housing located close to businesses. We also received our first indication of a cultural difference: the ethnic heterogeneity of a neighborhood had a strong negative effect on trust for the white kids but not for any of the ethnic minorities.

Each time we discovered a new correlation, we needed to incorporate it into our causal story, changing the story as needed to remain consistent with the correlations. We have names for this kind of reasoning in everyday language, such as having 20/20 hindsight and being a Monday-morning quarterback. These disparaging names reflect the fact that it is always easy to explain something in hindsight and that such explanations are by no means necessarily the right ones. Is science merely a case of 20/20 hindsight? No, because it is done again and again. Each round of hindsight leads to a new set of expectations that is tested by a search for new correlations. With every cycle of hypothesis formation and testing, erroneous explanations are weeded out, and the explanations that most successfully approximate the actual web of causal relationships survive. Science is a self-conscious variation-and-selection process designed to discover the truth of the matter. As with any evolutionary process, many cycles of variation and selection, or "generations," are required for adaptation to occur.

Dan and I were only a few generations into our scientific inquiry, and here was our best guess about the causal web on the basis of our existing correlations: as before, there must be a reason to cooperate and a social structure that protects the solid citizens from the knaves. The new correlations that we discovered involving population density, mixed land use, and ethnic heterogeneity make sense in terms of these two basic ingredients. Crowded and ethnically heterogeneous neighborhoods might well erode social structure, as other studies in the scientific literature suggest. As for the cultural difference, Binghamton is still a largely white city, and the most heterogeneous neighborhood is still 50 percent white. The other ethnic groups are not in a position to have their majority status threatened. One problem with purely residential neighborhoods is that there isn't much to do. A mixed-use neighborhood might provide greater opportunities for cooperative social interactions. Future research would be required to confirm this interpretation or to discard it in favor of a better one.

HOWEVER PROVISIONAL OUR UNDERSTANDING OF the causal web, one thing was for sure. When the students were in their classrooms playing the game, their behavior was influenced by events that took place in their neighborhoods previously in time. Otherwise, there would be no correlation between the neighborhood variables and their behavior in the classroom. The study that Dan initiated using photographs taken by his sister, Tania, introduced a new set of possibilities. Imagine that you are a college student at Binghamton University who is told by your instructor that someone will be visiting class today to play a game with you. In walks Dan O'Brien, who distributes a short survey asking where you are from and how familiar you are with the city of Binghamton. Let's say that you are from Long Island and know nothing about Binghamton, other than having visited the bars that line State Street. After waiting a few minutes for everyone to answer these questions, Dan shows a PowerPoint presentation that begins with a montage of photographs of Binghamton neighborhoods that were chosen to reflect the full range of neighborhood quality, as reported on our survey by the public-school students who actually live in the neighborhoods. After showing the montage, Dan displays the photographs for each neighborhood in randomized order for a period of thirty seconds. During this brief period, you are asked to study the photographs and answer the same questions about neighborhood quality that the public-school students answered on the survey. How will your assessment, based on a mere glance at each neighborhood, correspond to the assessment of the kids who actually live there?

When Dan conducted this study, the correlation that emerged was amazingly strong. Whatever it is about a neighborhood that accounts for its quality, it can be accurately perceived by viewing the photograph of the neighborhood for only a few seconds. Then Dan performed a second version of the experiment that produced an even more striking result. Instead of asking the college students to assess the neighborhoods from the photographs, he said that they were going to play a game with a high school student from each neighborhood that involved deciding whether or not to cooperate. Dan then outlined the same game that the high school students had previously played in their classrooms and had the college students make the three decisions about whether to cooperate as first mover, as second mover given that the first mover cooperated, and as second mover given that the first mover had failed to cooperate. Best of all, the college students were informed that the game was not hypothetical. After all of the photographs had been displayed, the responses would be collected. The responses for one college student would be chosen at random, one neighborhood would be chosen at random, and the game would go forward between that college student and a high school student from that neighborhood, based on our previous study in the high school. The college student would then be paid real money on the basis of his or her decision.

When you're a scientist, a cloud of data points on a graph surrounding a regression line that rises or declines steeply is an object of great beauty, because it means that you have discovered a strong correlation that is statistically highly significant. The more tightly the data points cluster around the regression line, the less likely it is that the association could have happened by chance. When Dan created a graph with neighborhood quality as assessed by the high school students on the *x* axis and the likelihood of a college student offering to cooperate as first mover on the *y* axis, I wanted to frame it and mount it on my wall like a trophy. The regression line rose steeply, and the data points were tightly clustered around it. The graph showed that the perception of a neighborhood based on a photograph has an immediate effect on the tendency to cooperate. This immediate effect is in addition to whatever long-term effects a neighborhood might have on those who actually live in it.

Like Watson before and after being educated by Holmes, the more

I pondered this astounding result, the more reasonable and even obvious in retrospect it became. Every species in the biological tangled bank is adapted to assess the safety of its environment and proceed cautiously or with confidence accordingly. A turtle withdraws into its shell at the sight of a predator. Eons of natural selection have caused the assessment of danger to be fast and automatic, operating largely beneath conscious awareness. The turtle that thought long and hard before withdrawing into its shell was not among the ancestors of current turtles. People are just like other animals in their automatic and largely subliminal assessment of threat. They might also think long and hard, but that would be an add-on to their older assessment abilities. Why shouldn't a college student become wary at the sight of a blighted neighborhood, even in a photograph, and withdraw from cooperating in the same way that a turtle withdraws into its shell?

Astounding or obvious, our new result was full of implications for understanding the psychological and social forces that create the hills and valleys of our GIS maps. I was especially struck by the thought that kids living in high- and low-quality neighborhoods might experience completely different faces of human nature, a smile or a frown, even in people they were meeting for the first time.

THE DOWNTOWN BUILDING IS A sizable investment by my university and the State of New York to understand and improve the quality of life in Binghamton. It is a purely secular building but has a church-like quality that should be taken seriously, and it is occupied by dozens of faculty representing entire disciplines dedicated to understanding and improving the human condition in a practical sense. Given such an outpouring of effort, goodwill, and expertise, what do I have to offer in my windowless office?

One thing that I have to offer is the ever-expanding database that I am assembling with the help of Dan and others. Information is everywhere, of course, but it is a unique vision of the Binghamton Neighborhood Project to combine it into a single integrated database. What would it be like for every sector of the city to merge its information in a way that could concentrate its firepower on any particular subject, as Dan and I were beginning to do? What would happen if this information were made available to every academic discipline inside the ivory tower so that they could contribute to every practical problem outside the ivory tower? That was my vision for the Binghamton Neighborhood Project, which I was beginning to describe as a "whole-university/whole-city" approach to community-based research.

To most of my colleagues, the scope of my vision is like the vision of world peace. Of course, but let's be practical. They have their hands full with their own columns of numbers and don't see the need for a thousand times more. Yet Dan and I were showing that our approach was eminently practical. It's not as if the whole thing must be built before it can be used, like the downtown building. Instead, it could be built grain by grain, like a termite colony. EvoS had already laid the groundwork for the "whole university" part from modest beginnings. Dan and I were making progress on the "whole city" part and profiting from the very start. It was merely a matter of adding to our database as need and opportunity arose and encouraging others to do the same. In this organic fashion, we were on our way toward creating an edifice of information as impressive and uplifting as a cathedral or the downtown building.

The other thing that I have to offer is the evolutionary paradigm. The two are related, because you need to see everything as part of the tangled bank before you can perceive the need for a database that includes everything. If the tools of science are required to see into a city, then the tools of evolutionary theory are required to reflect on it. It's a Darwinian world out there at two levels: the fast-paced process of behavioral change and the capacity for rapid change that evolved by genetic evolution. If we don't use the tools of evolutionary theory to reflect on a Darwinian world, then *fuhgeddaboudit*. Far from threatening, this should be looked upon as one of the most positive developments in the history of intellectual thought. Think of all that we are on the verge of understanding.

A value system isn't complete until perception and reflection lead to sustaining action. The Binghamton Neighborhood Project was only starting, but white-collar concepts such as prosociality were already beginning to merge with blue-collar concepts such as population density, ethnic heterogeneity, and mixed land use. Even on the basis of our preliminary knowledge, Dan and I were beginning to contemplate how to move.

CHAPTER 10

Street-Smart

AS DAN AND I were reflecting on how to raise Binghamton's valleys of well-being into hills, I acquired a new graduate student named Omar Eldakar, who grew up in the Binghamton area. In many respects, Omar's background was the opposite of Dan's. Dan was an academic star and the consummate good kid. Omar flunked every grade through fourth grade and scored in the sociopathic range on the prosociality and Machiavelli scales. As children, they seemed to be on opposite ends of the bell curve that supposedly measures aptitudes such as intelligence and prosociality. As graduate students, Omar had joined Dan on the high end of the bell curve for academic performance, and his apparent lack of prosociality was not what it seemed. I taught Omar a few things, but Omar taught me a lot about the capacity of people to change their position on the bell curve of life—or the valleys and hills of a GIS map.

Before describing how Omar was raised, let's reflect on what it means to raise anything. When Anne and I raise vegetables every year, we provide the right conditions. The plants do the growing. Raising children requires interacting with active agents. Pity the parents who treat their children like passive lumps of clay. Pity the children even more.

Teachers are seldom said to raise their students, but they provide the same kind of guidance as gardeners and parents. We have all heard stories about transformational teachers. They might be rare, but it is remarkable that they exist at all. If raising children were merely a matter of effort and access, parents would always be more influential than teachers. A transformational teacher has a mysterious key that unlocks the potential of the student. The potential was there before, and turning the key doesn't take much effort or access. If only we knew the key for each and every individual, transformation would not be so rare.

Are transformational teachers real, or do they merely inhabit feel-good Hollywood movies? In his wonderful book *Intelligence and How to Get It,* my colleague Richard Nisbett tells a story about Miss A, a teacher who had a gift for touching the lives of the first-grade students she taught for thirty-four years at a school in a poor neighborhood. Miss A's story is special because it was quantified by scientists, in the same way that Dan and I quantified Halloween. Sixty adults who had attended the school as students over a period of eleven years were contacted. A third of them had Miss A for first grade, and the others were taught by someone else. Miss A had not been assigned the best students, so if hers did better as adults, it could only be because of the difference she made. Comparing Miss A's former students with those of the other teachers was a natural experiment waiting for a scientist to come along. What were the results?

Every adult who had Miss A as a teacher remembered her name, compared with only 69 percent for the other teachers. Three-quarters of Miss A's students rated her as very good or excellent, compared with only a third for the other teachers. Seventy-one percent gave her an A for effort, compared with only 25 percent for the other teachers. If nothing else, Miss A made a positive impression.

The performance of the students in second grade provided another natural experiment. Now nobody was being taught by Miss A, and the students in the first-grade classes were shuffled before being dealt into the second-grade classes. Still, two-thirds of Miss A's students scored in the top third of their second-grade class, compared with less than a third for students of the other teachers.

Like money in the bank earning compound interest, whatever Miss A did for her students kept growing. When the adults in the study were measured for their accomplishments, such as grade of education completed, occupational attainment, and condition of their home, 64 percent of Miss A's students scored in the highest category, compared with only 29 percent for students of the other teachers. It might seem mind-boggling that a first-grade teacher could have such a lasting impact, not only on the *occasional* student but for the *average* student—unless you're a gardener. Then you know that a small difference in how the seedlings are tended can make a huge difference in their yield at the end of the season. Miss A was providing something like good soil conditions during the seedling stage of the human life cycle.

When the other teachers were asked how Miss A taught, one replied, "With a lot of love." The same colleague reported that Miss A expressed confidence in the ability of her children to learn and firmly announced that no student would leave her class without being able to read. She backed up her claim by staying after school to help the slow learners. She also empathetically shared her lunch with students who had forgotten theirs.

For me, this one scientific study is more powerful than all of the feel-good Hollywood movies about transformational teachers combined. Those movies are often said to be based on true life, but who believes Hollywood about anything? This one study is vastly more believable because it is so much more accountable, in part because something was counted. You can inspect it from stem to stern, jump up and down on it as hard as you can, and the conclusions still appear solid. If you don't trust your own judgment, then you can take comfort in the fact that experts also inspected it before it could be published in a peer-reviewed journal. If you're still skeptical, then you can repeat the study on Misses B through Z if you like. If the conclusions still hold after all that scrutiny, then we can stand on them with confidence and build the scaffold of knowledge even higher.

THE STUDY OF MISS A is a tiny fraction of the evidence that Dick Nisbett presents in his book that intelligence can be acquired. This claim is controversial among scientists. In fact, some scientists are convinced that intelligence is almost entirely innate and cannot be altered by experience. You are born somewhere on the bell curve of intelligence, and there is nothing you can do to alter your location. Charles Murray, author (with Richard Herrnstein) of a hugely influential book titled *The Bell Curve,* which was published in 1996, put it this way in 2007: "Even a perfect education system is not going to make much difference in the performance of children in the lower half of the distribution."

Many people who encounter this claim wish that it could be otherwise but think that it is supported by such a great weight of scientific evidence that it must be accepted whether they like it or not. If you are among them, then *Intelligence and How to Get It* should be the next book you read. Dick Nisbett is like a person who detonates large buildings for a living. He attaches dozens of studies comparable to the story of Miss A around the edifice of *The Bell Curve,* like so many sticks of dynamite, pushes the plunger, and down it comes with a roar and a plume of dust that rises high into the air.

That doesn't mean that Dick denies the importance of genes and individual differences that are genetically based. He merely appreciates the fact that all people are like growing plants that interact with their environment, however much their genes influence the rules of engagement. Contrast the growth metaphor with the metaphor of *The Bell Curve,* which is that each person is born a certain way, like a lead soldier, and remains in the same position for the rest of his or her life. Which of these metaphors do you think makes more sense from an evolutionary perspective? As Dick puts it in a discussion of the difference that families can make, "The belief that differences between family environments have little effect on IQ has to be one of the most unusual notions ever accepted by highly intelligent people."

In addition to being wacky in retrospect, the lead-soldier metaphor is a toxic self-fulfilling prophecy. Imagine Miss A walking into class one day and writing the quote from Charles Murray on the blackboard. Since some of her children haven't learned to read yet, she explains it in a way that a first-grader can understand: The slow learners will always be slow, and there is nothing that she or anyone else can do about it. Slow learners are not morally inferior, but a compassionate society must accept the way they are and not waste precious resources trying to change them. A duck can never become a swan, and a slow learner can never become a brilliant scientist.

That lesson would be the very opposite of Miss A's actual teaching style, which was brimming with confidence about the ability of each and every student to learn. If a diabolical version of Miss A could be found who teaches the lesson of the lead soldier to her little seedlings, they would likely become stunted by the experience. We don't need to speculate, because studies along the same lines have been done and are duly reported in *Intelligence and How to Get It*. In one remarkable experiment, children were given a set of problems and told that they had performed very well. Some were praised for being intelligent, and others were praised for working hard. The children were then allowed to choose another set of problems that could be either easy or hard. Most of the children praised for their intelligence chose the easy problems so that they would look smart. Most of the children praised for working hard chose the hard problems so that they could demonstrate their industriousness. Like seedlings reaching toward the light, the children reached toward praise, causing them to go in different directions depending on what was being praised. Praising hard work enhanced growth, and praising intelligence ironically stunted growth.

This was an actual experiment and nicely complements the natural experiment involving Miss A. The actual experiment shows what happens on the basis of a single episode of praise. The natural experiment shows what happens when the right kind of praise is provided to first-grade students for an entire school year. Just as geologists explain major features of the earth on the basis of daily events operating over immense periods of time, major cultural differences in IQ can potentially be explained on the basis of the daily social events documented by the actual and natural experiments. Another study reported by Dick shows that Americans of Asian descent tend to work harder after failure, while Americans of European descent tend to work harder after success. The cultural difference is similar to the different kinds of praise offered in the actual experiment. Perhaps that's why Americans of Asian descent tend to work harder and therefore become smarter over the long term, rather than a genetic difference between the two ethnic groups.

We therefore have two reasons to reject the lead-soldier metaphor in favor of the growth metaphor. First, it isn't true and makes little sense from an evolutionary perspective. Second, it is toxic to the growth of human seedlings. The research synthesized by Dick in *Intelligence and How to Get It* shows how much basic scientific understanding can contribute to improving the human condition. When I finished reading the book, I had more faith than ever that the valleys of my GIS maps can be raised into hills.

That's the good news. The bad news is that it won't be easy. Here is a summary of the growth metaphor: Each of us is a growing organism with complex rules of engagement for interacting with our environment. The rules of engagement are strongly influenced by the particular hand of genes that we are dealt, making us unique coming out of the starting gate. Then the rules of engagement change on the basis of experience, making us doubly unique. We don't even remotely understand our own rules of engagement, which take place largely beneath conscious awareness. Science doesn't understand our rules of engagement, either, or, rather has barely started to ask the right questions. Given such profound ignorance, we often operate far below our potential. For every one of us, a key might exist for unlocking our potential, but the key that works for you might not work for me, and nobody knows where it is hidden in the first place. Only when we stumble upon it or a wise teacher provides it can we come closer to realizing our full potential. It's true that we have an incapacity for change—not because we are lead soldiers but because we've all lost our keys.

SOMEONE LIKE OMAR MAKES MUCH more sense against the background of the growth model than against the lead-soldier model. Omar flunked the first grade so badly that his teacher wondered if he could understand English. He wasn't just on the lower half of the bell curve, he was way out on the tail. He wasn't a defective lead soldier, though, but a growing organism waiting for the right conditions to come along.

By the time I met Omar, he had already transformed into a star athlete on Binghamton University's track team, where some of his school records in short-distance running events stood for years. He took my course on evolution and human behavior and then started taking all of my courses. I didn't notice him until he took my course titled "Model Building in Ecology, Evolution, and Behavior," where he stood out for two reasons. First, this course is primarily for grad students, so the undergrads who dare to take it had better be good. Second, Omar was the only minority student in the class and was a gorgeous physical specimen compared with the out-of-shape Caucasians sitting around him. He is Egyptian, as I was later to learn, and his almond-shaped eyes could have come straight from the illustrations on a pharaoh's tomb. His body was muscular and beautifully proportioned, as one might expect from an elite short-distance runner. Someone had lovingly groomed his hair into corn rows, and a thick shell necklace gave him a tribal look. Women melted when he passed by, which was never one of my problems.

Omar had the verbal banter of a street-smart person, funny and laced with profanity. He was clearly used to holding forth among his friends, which made him speak up frequently in class, a bit too frequently for what he had to say. His way of putting things was often hilariously at odds with genteel scientific culture. Just as Eliza Doolittle shouted out, "Come on, Dover! Move your bloomin' arse!" at the Ascot race course in *My Fair Lady*, I am accustomed to getting e-mails such as this one from Omar, about some articles that he thought slighted my work.

> Dear Dr. Wilson,
>
> I attach the paper and also another one that will really kick you in the nuts. It's on evolutionary suicide and doesn't mention you at all . . . ooooh that cuts deep.
>
> Have fun,
> Omar

The next course of mine that Omar took was an advanced seminar on human social behavior. Whenever possible, I try to do science in my classes, not just talk about it. In this class, the students took a number of surveys to measure their individual differences, including a prosociality survey similar to the DAP and the Machiavelli survey. Then they formed into groups that were given a difficult class assignment, to see if performance could be related to individual differences and the social dynamics of each group. After the assignment was completed, the students switched from being the guinea pigs in the experiment to analyzing the results. I taught this course with Ralph Miller, my friend and colleague in the Psychology Department, and Rick O'Gorman, my graduate student at the time.

Thanks to this course, I know exactly where Omar resides on the bell curves of prosociality and Machiavellianism. At least according to the surveys, he is the least prosocial and most Machiavellian person I have ever measured. In statistical terms, Omar is an outlier. On a graph of the students in my course, the points for the others formed a tidy cluster, and the point for Omar was all by itself, like a runner who has broken away from the pack. Outliers are often omitted from a statistical analysis as aberrant, perhaps caused by a measurement error. Omar might be unusual, but he wasn't a measurement error, and I couldn't rub him out of my analysis when he was a flesh-and-blood person in my class. Moreover, Omar was proud to be an outlier and actually carried a copy of the graph around with him to point out to his friends how different he was.

When Omar graduated from Binghamton University, he asked if he could stay on and get his master's degree with me. Great. I'm the guy who studies the evolution of niceness, and my newest grad student scores in the sociopathic range of the prosociality scale. I'm an out-of-shape, white, middle-aged, nerdy professor, and Mr. Cool has decided that we make a great team. Of course, I said yes. It's hard to dislike Omar when you get to know him. Besides, he was helping me go beyond the simplistic portrayal of nice and nasty, saints and sinners, solid citizens and knaves. Omar was not entirely out for himself and was not entirely unscrupulous. I had a sense that he could be more trustworthy to those who earned his trust than most people who are blandly prosocial. He merely didn't automatically extend his trust to anyone. His response to a survey question such as "I think it is important to help other people" would be "Hell, no!"

Omar wouldn't automatically help other people any more than he would leave his car or apartment unlocked with a sign saying "Help yourself."

ONCE I LEARNED ABOUT OMAR'S family background, his approach to life began to make more sense to me. His parents met as college students at the University of Cairo. His mother's family was filthy rich, and his father's family was dirt poor. It was a case of an uptown girl falling for a downtown boy. Omar's father was so poor as a boy that he and his brother owned only one good shirt between them. One day, when Omar's father was wearing the shirt, he got in a pushing match with another boy and heard something rip. Enraged at the thought that the family shirt had been ruined, Omar's father pummeled the other boy into submission, only to realize that it was his opponent's shirt that had been ripped.

That was Omar's father's approach to life in a nutshell: push hard, and don't give an inch. Such an approach can lead in many directions. In the case of Omar's father, it led to an athletic scholarship at the University of Cairo. Around the world, it seems that the best way for poor men to exercise their minds is by excelling with their bodies. Omar's father was a nationally ranked swimmer who failed to compete in the Olympics only because Egypt didn't send a team during those politically turbulent times. He also had academic ambitions and especially loved zoology. He was good at it, too.

Nevertheless, love trumped academic ambitions. Omar's mother's parents were dead set against the relationship, so they decided to elope to America. Omar's father dropped out of graduate school and went first, arriving in Los Angeles with only $500 to his name. No matter. Omar admiringly says that his father can survive anywhere. He worked at low-paying jobs and stayed in cheap lodging to save money for his beloved's arrival. She completed her PhD in chemistry and applied to UCLA to earn a second degree in chemical engineering. When she arrived in America, it was in the respected and upward-bound role of a graduate student, while Omar's father was trapped in jobs that made little use of his talents. His approach to life served him well in Egypt and gave him a toehold in America, but it put him on a collision course with superiors who expected him to submit meekly to their often stupid and arbitrary demands. Job after job turned into a train wreck when Omar's father started pushing and refused to give an inch, especially when his sense of honor was offended and he became bent on retaliation.

Omar's parents had three children while his mother smoothly ascended her career staircase and his father pummeled his way through life. Their marriage finally came to an end when she got a job at IBM and they moved to Endicott. Omar was wrong about the ability of his father to survive anywhere. He couldn't survive in a little city where there was no way to escape a reputation as a troublemaker. Omar's father returned to LA, where word about one's pugilistic past doesn't travel so fast.

Omar's father did not abandon his children when he returned to LA. He was a good father, eager to see them and impart his life skills by having them visit during their summer vacations. Omar became especially close to his father and attributes his love of biology to the trips they took together to the seashore to collect and identify the tidal organisms. He even received a copy of Darwin's *Origin of Species* as a gift from his father, long before he could appreciate its significance.

In addition to acquiring a love of biology, Omar became schooled in the fine art of pushing and not giving an inch. One day, they were driving together and got involved in a minor fender bender.

"Hand me my pipe," Omar's father said to his son. This was a metal rod wrapped in rubber on one end for a good grip, which his father kept with him most of the time. His father stepped out of the car and brandished his pipe, and the man in the other car drove away.

WHETHER BY NATURE OR BY nurture, Omar had no interest pleasing authority and found school so mind-numbingly dull that he would rather stare out the window, making his first-grade teacher wonder if he could understand English. When he was given an aptitude test in fourth grade, he was told that it wouldn't count for his grade. "Sweet!" he thought, and proceeded to answer the questions at random. When concern was expressed over his dismal score, he protested that he had been told earlier that it wouldn't count—blowing it off made him the smartest kid in class.

Omar's mother was unperturbed and kept reassuring his teachers that he would catch up. Sure enough, he started to earn passing grades, although they were nothing to crow about. Omar's mother also encouraged him to join the track team in high school, which he enjoyed more. He was not a star, however, and wasn't even physically strong until he was diagnosed with diabetes in his junior year. As soon as he started taking insulin shots for his diabetes, his body experienced a surge of growth, and he became the star of his team. He still got in trouble with his coaches, however, by failing to be a team player. For example, if Omar thought that his left leg needed stretching more than his right, he would continue stretching his left leg when the coach said to alternate. Omar also felt pressured to underperform—to win without showing up his own teammates. Eventually, his reputation as a troublemaker with the track coach became so bad that Omar went to the head coach for advice. The head coach was sympathetic and arranged it so that Omar would report only to him, which alienated the track coach even more.

When Omar was admitted to BU on a track scholarship, he wanted to pursue his love of biology but was advised by his athletic advisors that it would be too hard for him. He approached a biology professor for advice, who turned out to be Julian Shepherd, my departmental colleague and one of the most nurturing people you will ever meet.

"Oh, I don't know, I think you can probably handle it," Julian replied to Omar, knowing absolutely nothing about him. Thanks to this kindly nudge at a critical fork in the road, Omar became a biology major. He earned passing grades but didn't catch fire until he took my evolution and human behavior course. Once he started to think about people as part of the tangled bank, everything seemed to fall into place. His grades shot up in every course he took, even nonbiology courses such as history.

For his master's thesis, Omar started to collaborate with Rick O'Gorman, the PhD student who had helped me teach the advanced seminar on human social behavior. Rick is an affable Irishman, and we were studying a hot topic called altruistic punishment, the desire to punish wrongdoing even at one's own expense. In the classic western movie *The Man Who Shot Liberty Valance*, John Wayne and Jimmy Stewart are both solid citizens, but Wayne is willing to stand up to the bad guy (played to perfection by Lee Marvin), while Stewart is a wimp. Keeping the peace is costly for Wayne, and it's Stewart who gets the girl. There you have the concept of altruistic punishment in a nutshell.

In our experiments with college students, a curious result emerged. The students who were most likely to act *selfishly* were also most likely to *punish*. This seemed morally hypocritical but made sense as an evolutionary strategy. Bad guys, no less than good guys, have an interest in getting rid of other bad guys. If you were Lee Marvin, wouldn't *you* want to be the only one tending your herd of wimps like Jimmy Stewart? Omar immediately resonated to the concept of selfish punishment, as he dubbed the phenomenon, and declared that he wanted to study it theoretically.

The ability to create imaginary worlds with mathematical equations or computer-simulation models is commonly regarded as the most demanding kind of science requiring the highest degree of intelligence. I had long since learned not to prejudge Omar and simply stood back to give him elbow room. He quickly found a fellow graduate student named Dene Ferrell in our bioengineering department, who was a Mathematica prodigy. Mathematica is a set of computational tools that can do just about anything, at least after you master its arcane language. Omar had no interest in learning the nuts and bolts of Mathematica, as long as he could pair up with Dene. Why duplicate effort? Instead, Omar concentrated on the plan of the theoretical model, in the same way that an architect designs a building without worrying too much about the materials (Dene was no slouch as an architect, either).

In the imaginary world built by Omar and Dene, every individual has a propensity to be altruistic or selfish and a separate propensity to punish or not punish selfishness in others. If both propensities were measured on a scale from 1 to 10, a nice wimp such as Jimmy Stewart would score 10 for altruism and 1 for punishment. John

Wayne would score 10 for both, and a hypothetical selfish punisher would score 1 for altruism and 10 for punishment. At the beginning of a computer-simulation run, 10,000 individuals were created and randomly assigned a value for each propensity. All combinations of altruism and punishment existed in equal proportions, or at least roughly equal proportions, because Mathematica was doing the equivalent of rolling the dice each time it assigned a value to an individual.

Now that a virtual population of creatures had been created by Mathematica, they started to interact socially with one another. The more altruists in a given group, the better everyone did, but selfish members always received more. The more punishers in a group, the more selfishness was thwarted, but punishment was costly, just as John Wayne didn't get the girl. The fate of any given individual in the model depended on its own propensities and the propensities of other members of its group, which in turn was determined by the luck of the draw.

After social interactions occurred, it was time to reproduce. Each individual accumulated points according to its social interactions and produced offspring in proportion to its points. Omar and Dene also built mutation into their model so that individuals usually replicated themselves exactly but occasionally produced offspring that were different. The virtual world created by Omar and Dene was now complete, including variation, selection, heritability, and mutation. The propensity for altruism and punishment were uncorrelated at the beginning of a simulation run. What would happen as the generations unfolded?

A positive correlation developed between selfishness and punishment, much like the result that we obtained for college students. Altruistic punishers bore a double cost by being good and policing others. Selfish nonpunishers were thwarted by punishers of all sorts. Altruistic nonpunishers bore the cost of being altruistic but did not bear the cost of being punishers and themselves were not punished. Selfish punishers bore the cost of punishment but not the cost of being altruistic. Based on all of these factors, the whole population settled down to an equilibrium consisting of a majority of altruistic nonpunishers and a minority of selfish punishers. A division of labor spontaneously arose in which the punishers were like police who were "paid" for their services by not bearing the cost of altruism.

Omar left the nuts and bolts of the computer-simulation model to Dene, but it was clear that he understood what was going on and was participating in the numerous decisions required to create a computer-simulation model in minute detail. Chess is a popular game among street-smart people, once they encounter it, because it is all about pushing and not giving an inch. We're amazed when an inner-city chess team trounces an Ivy League team because we think that chess requires great intelligence, which Ivy League folks have and inner-city folks don't. Once we appreciate that chess is an abstract representation of a social world where survival hinges on fighting for oneself, we can appreciate why inner-city folks might hold an advantage after they make the connection. In just the same way, Omar became a gifted theoretical biologist by making a connection between his street-smart social skills and the abstract world of a computer-simulation model. The article describing the model was published in the *Proceedings of the National Academy of Sciences* and was picked up by the popular press. That's a feather in anyone's cap, not to speak of someone who flunked his first four grades and was advised not to become a biology major because it was too hard.

OMAR WAS THRIVING AS my master's student but came even more alive when he attended his first major scientific conference. Before, he seemed uncertain about whether to get his PhD, but after, he seemed like a runner impatient for the next race. What happened at the meeting to produce this effect? Was it listening to the distinguished speakers? Was it socializing with grad students from other colleges and universities? When I asked Omar, he said that listening to the talks was like sitting on the sidelines of a track meet knowing that he could crush the opposition. It was wrong for them to be up there and for him to be in the audience. He knew that he could win this race, and now it was merely a matter of training for it.

Even though I used to study the biological tangled bank, I have become so engrossed in the human tangled bank that I expect my graduate students to do the same. Omar had different plans. He loved biology and wanted to study insects. He was good at it, too, learning how to identify them with ease and even serving as a teaching assistant for Julian Shepherd's entomology course. Now that selfish punishment had been demonstrated in humans and the virtual world of a computer-simulation model, he wanted to see if it existed in the biological world. Water striders were one place to look, and who was I to stand in his way?

Earlier, I described male water striders as insensitive jerks, jumping on any female who comes within range, but a closer look reveals that males differ in their aggressiveness. Some qualify as psychopathic sexual predators by human standards, but others act like perfect gentlemen waiting for Mrs. Right to come along. Might these individual differences correspond to selfishness and altruism in some way? Might the psychopathic males limit one another, providing a niche for the gentlemen? Always on the lookout for a win-win situation, Omar teamed up with another graduate student named Mike Dlugos, who was studying water striders under the supervision of pioneering strider researcher Stim Wilcox.

If ever there was a case of using evolutionary theory to study the same subject in utterly different creatures, this was it. There were Dan and I, examining the fates of high-pro and low-pro youth in the city of Binghamton. There were Omar and Mike, examining the fates of high-pro and low-pro striders on the surface of streams. Just as we began by measuring individual differences with our survey, Omar and Mike measured individual differences by placing striders in children's wading pools and scoring the aggressiveness of males when attempting to mate with females. Dots of enamel paint on the back of each strider enabled them to be recognized as individuals. Huge individual differences emerged that were stable. People might have a capacity for change, but strider psychopaths and gentlemen retained their personalities week after week, spring or fall, hungry or full.

In one experiment, Omar and Mike altered the composition of males in the pools, from all psychopaths, to a mix of psychopaths and gentlemen, to all gentlemen. Females did much better in the company of gentlemen, who allowed them to feed and waited to be asked before doing their manly duty. Females were so harassed in the company of psychopaths that they were literally climbing the walls of the wading pools to escape. You can't eat if you're being harassed all the time, and if you can't eat, you won't lay as many eggs. Omar and Mike estimated that females lay three times more eggs in the company of gentlemen than in the company of psychopaths.

What happened in the mixed groups? If selfish punishment takes place, then the psychopaths should target one another, creating a niche for the gentlemen. That didn't happen. Instead, in every pool containing both types, the psychopaths raced around and scored most of the mating. They owned the females, and the gentlemen were just a bunch of pussies, as Omar put it in his street-smart vocabulary.

If this was the whole story, then the hammer blows of natural selection would eliminate the gentlemen, and only psychopaths would be observed. What was creating the niche for gentlemen, if not psychopathic males selfishly punishing one another? In an elegant experiment, Omar and Mike answered this question by allowing the striders to move among groups. Now, when a psychopath entered a group, the females could leave rather than climbing the walls. The psychopaths were free to follow, but the game of perpetual hide-and-seek resulted in most of the females clustering around the gentlemen. That was the advantage of being a gentleman, which balanced the disadvantage of being a wuss in direct competition. The striders formed into high- and low-quality neighborhoods, much like the city of Binghamton. Psychopaths had the local advantage but were stuck interacting with one another. Gentlemen were no match for psychopaths in a direct contest, but they didn't need to be, because they were lounging with the ladies in other groups. With these compensating costs and benefits, the hammer blows of natural selection maintained both types of male in the population.

OMAR WAS BECOMING A RISING STAR. Evolutionary theory enabled him to dance across disciplines, studying any aspect of any organism, and he had great moves. He could speak the lingo of theoretical biology with the fluency of a rapper. He could roll up his sleeves and test his predictions in the real world. He was even a talented field biologist, able to identify the plants and animals encountered on a walk through field and forest. It's little wonder that when he neared the end of his PhD and applied for a postdoctoral position at the University of Arizona's Center for Insect Science, they flew him out for an interview where he was expected to give a seminar on his research to the people who would decide whether he would get the job.

As soon as he could get away after his seminar, Omar called me to say that it had not gone well. Evidently, some of the people in the audience were molecular biologists and neurobiologists who knew little about the evolution of social behavior. They started to bombard him with poorly informed questions only a few minutes into his talk. Omar knew a lot more than they did on this particular subject, but they were the ones who would decide his fate. The situation was eerily similar to Omar's father confronting his superiors and refusing to give an inch. What should Omar do?

Being Omar, he pushed hard and didn't give an inch. He abandoned his prepared talk and launched into a tutorial on the evolution of social behavior, conveying the themes that I have related in my parables of the strider and the wasp. The conversation became much more heated than the average job seminar. Omar was clearly acting as an equal and refused to back down. Every question received a cogent reply, drawing on both theory and data recalled from his general command of the subject. Eventually, Omar returned to his prepared seminar to show how he was using striders to test the foundational ideas with which the others were largely unfamiliar.

Omar returned from Arizona certain that he had blown his interview and was amazed when he was awarded a plush three-year postdoctoral position. I would give anything to have been a fly on the wall when Omar gave his seminar. Science is supposed to be a contest of ideas, so maybe the scientists in the audience were impressed by Omar's pugilistic skills, even if he bruised their own authority. Maybe Omar added enough humor to soften his punches. Maybe he employed some diplomatic skills that Machiavellians are known for. Whatever happened, he is now happily located in Tucson.

Thus ends the story of how one person ascended from a valley to a hill, from one end of the bell curve to the other. What does it tell us about the capacity of an individual to grow, which we might also apply to changing whole neighborhoods? The complex interplay of genes and environment that went into Omar's making will never be known, but it resulted in a person with rules of engagement for interacting with his environment. What counts as smart in some environments can spectacularly fail in others. What works on the streets of Cairo or LA doesn't necessarily work in a job or a school that requires conformity or a world so safe that carrying a pipe and having a hair-trigger sense of honor becomes unnecessary. However one comes by one's rules of engagement, one struggles to grow as best as one can.

When Omar expresses gratitude toward the people in his life who enabled him to grow, it is surprising how little they had to do: his head coach in high school who said that Omar could report directly to him, Julian Shepherd for encouraging him to be a biology major, and me for standing back and giving him as much elbow room as he wanted. He also expresses gratitude to evolutionary theory for making so much sense out of life, another condition for growth. These conditions can be extraordinarily easy to provide, but they make the difference between a life stunted and a life fulfilled.

CHAPTER 11

The Humanist and the CEO

NO MATTER HOW MUCH we try to call our own shots, we are like that silver ball sent hurling in unexpected directions by the pinball machine of life. I thought I knew where I was going with EvoS and the Binghamton Neighborhood Project, but then an e-mail from a total stranger batted me in a new direction with a third project of equal magnitude.

The stranger introduced himself as Jerry Lieberman, a retired professor and president of the Humanists of Florida Association, a branch of the American Humanist Association. He wanted to create a humanist think tank and was persuaded by my book *Evolution for Everyone* that it should be an evolutionary think tank. No other think tank used evolution as a framework for formulating public policy, and Jerry wondered if I could provide some guidance.

Never mind that Jerry was a total stranger, that I knew nothing about think tanks and had no more than a layman's knowledge about humanism. Never mind that the Binghamton Neighborhood Project was only starting and that studying a whole city should be enough to keep anyone busy. Jerry seemed to be offering the prospect of doing for the whole world what I was trying to do for my city of Binghamton, and I found the prospect irresistible.

In no time, we were brainstorming about possibilities. The think tank should be a national organization, not just restricted to the state of Florida. It could be called the Evolution Institute. We decided to select a single policy issue as a proof of concept for how evolutionary theory can be used to address any major policy issue. Our first focus would be childhood education.

Nearly every parent wants to give his or her children a good education, and every citizen wants his or her nation to have a good educational system. Billions of dollars have been spent to achieve it, hundreds of theories have been formulated, and thousands of experiments have been conducted. Entire think tanks are devoted to its study alone. Given so much goodwill, effort, and thought, why do so many children fail to learn, and why does America's educational system rank so low compared with other nations of the world? If Jerry and I could provide new solutions for the subject of education, it would be a major achievement and proof of what the Evolution Institute could do for other policy issues. The new solutions could be implemented in Binghamton along with any other place in the world. It might seem like sheer hubris to expect evolutionary theory to make such a difference, but Jerry and I were committed to finding out.

My job was to assemble the evolutionary expertise on childhood education, and Jerry's job was to raise enough money to bring them together. My job was relatively easy. Thanks to EvoS, I'm like a spider in the middle of a web radiating out in all directions. I already knew some colleagues who were doing outstanding work on child development and education from an evolutionary perspective. They were eager to become involved and knew who else should be invited. In no time, I assembled a dream team of experts who were highly respected within their fields and already stressing the need to approach childhood education from an evolutionary perspective. Had they already accomplished what Jerry and I were trying to achieve? Not in the least. They felt like voices crying in the wilderness and were thrilled to become part of a more concerted effort.

Jerry's job was a whole new world for me. When I need money for my research, I write proposals to agencies such as the National Science Foundation that are set up for the purpose. Jerry was approaching wealthy individuals who wanted to do worthy things with their money. Some had set up their own philanthropic foundations, and others simply had big bank accounts. They had their own ideas about how to save the world and were typically besieged by others eager to tell them how to spend their money. A conservative think tank called the Discovery Institute had raised millions of dollars to spread misinformation about evolution and sneak creationism into public-school education in the guise of so-called intelligent-design theory. A nonprofit organization called the National Center for Science Education was heroically trying to oppose them with a much smaller war chest. Jerry and I wished the best for the NCSE and did not want to duplicate their effort. We had bigger plans for the Evolution Institute. We wanted everyone involved in public policy to consult evolutionary theory for solutions to life problems, starting with childhood education. We didn't just want evolution taught properly in school. We wanted school taught properly with the help of evolutionary theory.

Could Jerry sell this idea to philanthropists? Might he even raise a war chest for the Evolution Institute to rival the Discovery Institute? That would be something. I began to dream of having millions of dollars to change the way the planet thinks about evolution. But who was Jerry Lieberman? Was he up to the task?

JERRY IS A TRIM MAN in his seventies with a penchant for wisecracks. I could well imagine him on the cast of *Seinfeld*, perhaps a friend of Jerry Seinfeld's parents. His ancestors came from Hungary, and his father was the son of a prominent cantor who should have died three times in World War I—first from a bullet in the back, which he kept as a souvenir, then from a bayonet in the intestines, and finally in a Siberian prisoner-of-war camp. After the war, Jerry's father immigrated to the United States and married Jerry's mother, who had grown up on a farm in the same region of Hungary and immigrated to America before the war. Thus it was that Jerry's father found himself in a strange land with a new wife, bursting with potential that was to remain largely unfulfilled.

Even as a boy, Jerry was struck by life's contradictions. At school, he was told about the American melting pot, yet the neighborhoods of Brooklyn were divided into ethnic enclaves. Jerry was happy to jump into the melting pot. He was attracted to peers of all ethnic and racial backgrounds. He also loved the melting-pot joys of rooting for the Brooklyn Dodgers, being one of the guys on the neighborhood baseball team that he organized, and delighting everyone but the teacher as the class clown. His experience was profoundly different from that of his father, who was understandably pessimistic about the ability of nations and ethnic groups to get along with one another.

Jerry was also struck by the contradictions of his religious faith. His mother was not observant, his father was verbally observant but often failed to practice what he preached, and his formal religious education in Hebrew school was dogmatic. Asking questions only got him into trouble. That included the education he received in public school. Jerry was passionate about science, especially big questions about the universe. He read voraciously about astronomy and built his own telescope, but his public-school education was uninspiring. Only when his parents sent him to a private school was his love of learning nurtured rather than thwarted by his formal education.

Jerry earned his PhD in political science, and his academic career was a bit like Omar's father's pugilistic path through life. He was an academic street fighter, fiercely egalitarian and sensitive to social injustice. He was not granted tenure from his first college teaching job for becoming involved in organizing a teachers' union but quickly stepped into a second job at Essex County College in Newark, New Jersey, just after rioting that had occurred there, where he became the coordinator for the social sciences with twenty faculty members under his supervision. His main ambition was to make higher education accessible to everyone and sensitive to their needs. He was apprehensive of professors who behaved like prima donnas, as if they had all the answers when, in fact, they were limited and lacked genuine commitment for getting involved in the community or in real-world matters of importance. His vision was alternately favored and opposed by his superiors in the administration as they came and went. At one point, he served as dean of institutional development, and at another point, he was stripped of everything but his teaching duties for not supporting the decisions of the board of trustees, which he deemed detrimental to the interests of the students and the community.

In his middle fifties, Jerry decided that his college position was too confining and walked away from his tenured full professorship to relocate in Florida. There he gravitated back toward community economic development and became associated with the University of South Florida, where he started the Florida Community Partnership Center. Wherever Jerry went, he stepped into the role of organizer, just as he had organized his baseball team as a kid on the streets of Brooklyn.

THROUGHOUT THE UPS AND DOWNS of his career, Jerry was a humanist. Humanism is a value system that is centered on rational thought in a world without supernatural agents. Humanism shades into liberal and transcendental forms of religion on one end and strident atheism on the other end. If Isaac Bashevis Singer's mother was alive today, she might well be a humanist. At least, her pronouncement "Dead geese don't shriek" could serve as a humanist rallying cry. Humanists are also activists, or at least they aspire to be. They don't just possess humanist values; they want to spread them to make the world a better place.

What are the values that humanists want to spread? Here are some passages from the Humanist Manifesto III, which can be found on the Web site of the American Humanist Association (www.americanhumanist.org):

> Humanism is a progressive philosophy of life that, without supernaturalism, affirms our ability and responsibility to lead ethical lives of personal fulfillment that aspire to the greater good of humanity.The lifestance of Humanism—guided by reason, inspired by compassion, and informed by experience—encourages us to live life well and fully. It evolved through the ages and continues to develop through the efforts of thoughtful people who recognize that values and ideals, however carefully wrought, are subject to change as our knowledge and understandings advance.

In other words, "Work for the benefit of humanity, and good luck with the details. You're probably wrong, so expect to change your mind."

Don't get me wrong. These are also my values, and I'm happy to call myself a humanist, even though I'm not an official member of any humanist organization. Therein lies the problem. Humanists are like the proverbial cats that refuse to be herded. Encouraged to challenge everything, they go off in different directions and can't achieve a consensus. A good fundamentalist religion, in contrast, equips you with a detailed list of do's and don'ts, packaged in a way that makes you feel as if you're fighting a cosmic battle of good against evil. No wonder Christian soldiers line up more easily than humanist soldiers.

Humanists look to science for salvation, but most of them aren't scientists, and they aren't plugged into an organized scientific process for addressing life's problems. That makes them dilettantes, idly speculating based on what they read in books, magazines, and newspapers. They have their hearts in the right place, but when it comes to getting things done, they are like someone showing up at a construction site without a toolbox.

We'll never know how nature and nurture interacted to form Jerry's personality, but his insatiable appetite for organizing caused him to become involved with the Humanists of Florida Association, which he built into one of the most active chapters of the American Humanist Association. His vision for the Evolution Institute was to create the organized scientific process for addressing life's problems that the humanist movement lacks. He realized that not just science but evolutionary science was needed to get the job done. That was the hand that reached out to me in the form of an e-mail message from a total stranger.

Jerry was not at all what I expected from a fund-raiser and director of a think tank. I imagined a silver-tongued diplomat who could ingratiate himself with the rich and powerful without causing offense. Instead, Jerry was more likely to deliver a comedic monologue or give someone a tongue lashing for failing to walk the walk—especially professors sitting in their ivory towers thinking that they knew it all. Still, Jerry was effective in his own way. He had created a large social network of people with humanist values whom fortune had smiled on. Our job was to persuade them that supporting the Evolution Institute was the best way to walk the humanist walk. When I spoke at the annual meeting of the Humanists of Florida Association, I mingled with people who were thrilled by the prospect of a concrete plan for fulfilling the humanist manifesto. I didn't feel the least bit cheapened by asking for their support. Before long, Jerry had raised enough money to hold our workshop on early-childhood education as proof of concept for what the Evolution Institute could do for any major policy issue.

WHILE JERRY WAS DOING his fund-raising magic in Florida, I received an e-mail from Eugenie Scott, who directs the National Center for Science Education in its heroic defense of evolution education in our public schools. One of her board members had become familiar with my work and wondered if he could help out with my efforts to bring evolution and science to the public. His name was Bernard Winograd, he was a businessman who lived in Manhattan, and Eugenie was happy to make the introduction.

Suddenly, I was playing the fund-raising game rather than merely assisting Jerry. Moreover, Bernard Winograd wasn't just a businessman. He was one of the top executives of the Prudential Financial Corporation, in charge of managing between $400 billion and $500 billion worth of assets. When it came to fund-raising, I was like a little boy going fishing for the first time with my uncle Jerry, and on my very first cast I had hooked Moby Dick.

I'll talk about evolution with anyone, from a homeless person to the queen of England. I'm not intimidated by power, money, or prestige. As a bestselling novelist, my dad mingled with the rich and powerful and would frequently tell me hilarious stories about their lives. Everyone was part of the same human comedy for my father, and that rubbed off on me. I was therefore happy to talk with Bernard about evolution for nothing. It was asking for money that made me feel awkward. Even though Bernard had approached me with the generous expectation of helping me out, and despite my recent experience in Florida, a respectful relationship doesn't include asking for money, as far as my own upbringing is concerned. This fund-raising business would take some getting used to.

Bernard and I traded a couple of e-mails and agreed to meet for dinner in Manhattan. I said that I would be arriving by bus, and Bernard chose a restaurant close to the Port Authority bus station, the Bryant Park Grill on 40th Street between Fifth and Sixth Avenues, adjacent to the New York Public Library. I caught the bus in the morning to arrive in plenty of time. The trip passed pleasantly, with me glancing up from my work to see the gorgeous farmland south of Binghamton, the Scranton bus station, the rugged beauty of the Delaware Water Gap, the Manhattan skyline, and then *pow!* The full-frontal assault of Manhattan as the bus emerged from the darkness of the Lincoln Tunnel.

Bryant Park covers only a single city block but pulses with life. The central lawn is surrounded by a tree-lined promenade with tables, chairs, and booths selling food and boutique items. People from all walks of life were enjoying the summer afternoon—old friends talking, lovers embracing, chess players stooping over their games, dog walkers, parents with their children, musicians surrounded by circles of appreciative listeners. It was a joy just to sit and watch the human parade pass by until my appointment with Bernard.

I entered the Bryant Park Grill shortly before Bernard arrived and was shown to our table. I scanned the people entering to see if I could recognize the CEO of a major corporation. Would he be larger than life, like Donald Trump? Instead, an understated man was shown to my table. There was nothing obvious about his clothes or demeanor to indicate his status. Still, the waiter knew to be deferential, not because Bernard was a regular but based on subtle ways that he carried himself. I could well imagine subordinates being terrified by his direct gaze.

Bernard and I fell easily into conversation. I began by describing my campaign to change the way the planet thinks about evolution. When I get going on evolution, I speak in urgent tones, with a direct gaze of my own. Bernard commented that it was amazing how I managed to get so much done. Then he told me what he did. The more I learned about what it takes to run a major corporation, the smaller I felt.

"Bernard!" I exclaimed after he had finished. "How can you be impressed by what I do, given what you do?"

Bernard appeared to be a man of few words. "I have a staff, and you don't," he replied.

"OK, but I still don't understand how you do all that."

"I only do what only I can do."

Whoa! I liked that. I even liked the elegant way it was stated. I had something to learn from this man.

I asked Bernard to tell me his story. His ancestors came from the Ukraine in the late nineteenth century, and Bernard is a third-generation American. His father was a clothing salesman who opened a chain of men's stores in Detroit. Bernard attended the University of Chicago and earned his degree in general studies and social sciences in 1970. His interest in evolution began when he was forced to take a biology course and he became fascinated with the part that dealt with evolution and biological anthropology. Ever since, he maintained an amateur's interest in evolution and saw parallels with the ways that companies compete and financial markets evolve over time.

Bernard's main interest in college was politics. He became active in the Democratic Party as an organizer in the same Hyde Park circle that provided President Barack Obama's base of support. He eventually left politics to join the public-relations department of the Bendix Corporation, a manufacturing and engineering company that was a household name during my youth. From there, he seemed to rise in the corporate world like a cork floating to the surface of the water. He became the executive assistant to the CEO of Bendix, Michael Blumenthal, and accompanied Blumenthal to Washington when he became secretary of the Treasury under Jimmy Carter. Returning to Bendix, he was promoted first to head of public affairs and then to treasurer.

"Does that mean that you became treasurer without any financial training?" I asked.

Bernard gave me a laconic look and replied, "It's not very difficult."

Perhaps, but something about Bernard caused him to float to the top wherever he went, eventually joining the Prudential Corporation and becoming head of all Prudential businesses in the United States in 2008. Not very difficult, indeed.

Smiled on by fortune and his own talents, Bernard started to become involved in projects that interested him, including the National Center for Science Education. He liked to keep a low profile to make it clear that his involvement was as a private citizen, not in the name of Prudential. He was attracted to my work after reading *Evolution for Everyone*. The NCSE was defending the teaching of biological evolution in public-school education. I was gazing at humanity through the crystal ball of evolution. Bernard was interested in helping out with both endeavors.

As our conversation switched back and forth between our respective worlds, I noticed that my world had riches that money couldn't buy in Bernard's world. It's a luxury to ask the big questions about the human condition. Religious sages, philosophers, intellectuals, novelists, and scientists like me get to ask the big questions for a living. Bernard's work must be intellectually demanding and fulfilling in some respects, despite his own modest assessment that it wasn't difficult, but it didn't fully satisfy his desire to ask the big questions. As he told me later in an e-mail, ours was one of his most interesting dinner conversations ever.

Bernard gave me good advice on my projects. He said that I shouldn't treat them as separate. If they went together in my own mind, that is how I should organize and represent them to others. In terms of his own involvement, the Evolution Institute was most worthy of his own attention. He would like to learn more about it by attending the workshop on early-childhood education. He might also like to become directly involved in addition to helping out financially. He wasn't planning on retiring soon, but he was looking for worthy uses of his talents when he did retire.

I needed to catch the last bus back to Binghamton, and Bernard offered to give me a lift to Port Authority. When we left the restaurant, his car was right there at the curb.

"Wow!" I exulted. "How lucky is that, to find a parking spot so close to the restaurant in downtown Manhattan!"

I wasn't looking at Bernard's face, so I don't know if he suppressed a smile as his driver got out and opened the back doors for us. It's a good thing that our emerging relationship didn't depend on my silver-tongued diplomatic skills.

* * *

I WAS FEELING PRETTY GOOD about my big family of projects and especially its newest addition, the Evolution Institute. Bernard was helping me to see them as part of a single enterprise rather than as competing siblings. Jerry's success at funding our first workshop and Bernard's long-term interest made it seem as if the Evolution Institute could become a reality, turning evolution from a pariah concept to an essential survival tool. Watch out, Discovery Institute!

Then the global financial crisis of 2008 hit. Bernard was at the epicenter and e-mailed to say that attending our childhood-education workshop was out of the question. Nothing was said about his long-term involvement. Nothing could be said, with the earth seeming to shake beneath everyone's feet. Before he disappeared into the maelstrom, Bernard sent me this message:

> The question I'd like to pursue is whether we could organize a discussion on the right regulatory framework from an evolutionary perspective, akin to what you will do on child education.... I am curious whether you think you could marshal the right academic resources.

Right. I thought I was being audacious by trying to make a difference in my city of Binghamton. Now I was being asked if I—or, rather, evolutionary theory—could succeed at solving the world financial crisis where the forbidding world of economics had failed. I felt like the gentle hobbit Frodo in *The Lord of the Rings,* being urged by Gandalf the great wizard to leave the shire and travel to the land of Mordor, just before Gandalf disappears to attend to matters of his own. Was I up to the task? Would Bernard reappear to offer his assistance? With a lump in my throat, I began to contemplate what it would be like to rethink economics from an evolutionary perspective.

CHAPTER 12

The Lost Island of Prevention Science

EVEN THOUGH I FELT that I had much to offer my city of Binghamton with my evolutionary perspective, I also knew that I was the new kid on the block when it came to community-based research. Just how new was impressed on me by an e-mail from another total stranger named Tony Biglan, who introduced himself as president of the Society for Prevention Research. He was organizing a symposium on prevention science from an evolutionary perspective and hoped that I could become involved.

If you're unfamiliar with science, you might be shocked by the degree to which it is fragmented into isolated disciplines. Science is like human society before the advent of agriculture. Back then, the largest social units were tribes of only a few thousand individuals. Each tribe had its own customs and frequently its own language, incomprehensible to other tribes. Tribes within reach of one another traded, raided, or simply went their separate ways. At most, they might combine into loose federations, mostly to solve problems associated with warfare.

So it is with science and, more generally, academia, which is why I coined "The Ivory Archipelago" as more apt than "The Ivory Tower." "Ivory Tower of Babel" would also be apt. Members of different tribes use the same words, which makes it *seem* that they are speaking the same language, but the meanings of the words and their associations with one another can be completely different. Say the word "evolution" to a cultural anthropologist, for example, and you are likely to be instantly branded as an enemy. If that's not tribal, what would be?

One reason I am passionate about evolution is that it provides a common language for all scientific and academic disciplines that deal with living processes. Evolution can turn the Ivory Archipelago into the *United* Ivory Archipelago, even though overcoming the xenophobic response of tribes such as cultural anthropology might not be easy. I was therefore pleased to hear from Tony but a bit chagrined that I had never, until then, heard of the field of prevention science. I pride myself on sailing the Ivory Archipelago, but this island was not on my chart.

Tony seemed to anticipate in his message that prevention science was not a household word. He explained that his society was founded fifteen years earlier by behavioral scientists trying to prevent psychological, health, and behavior problems—from cigarette smoking to terrorism, as he put it. He wanted his society to adopt a more holistic perspective provided by evolutionary theory, especially the concept of whole societies as units of cultural evolution that I had developed in my book *Darwin's Cathedral*. My book had floated from my island to his, like a message in a bottle, forming a thin line of communication that he wanted to strengthen.

The pragmatic focus of prevention science piqued my interest. Tony's colleagues might be newly encountering evolutionary theory, but I was newly encountering the challenges of making a difference in my city of Binghamton. At that time, the Binghamton Neighborhood Project was a mere baby about to celebrate its first birthday. I sent Tony a welcoming reply and a brief description of the BNP. His reply was short and sweet: "Oh boy. Could I be of help to you. This is what I do."

Indeed, it was. As I began to learn about the field of prevention science through Tony, I realized that it wasn't just another island in the Ivory Archipelago. Prevention scientists might not use the E-word, but they had developed techniques for accomplishing behavioral and cultural change that actually worked. Moreover, they had high standards for evaluating their techniques. The gold standard for assessment is called a randomized control trial, in which participants are randomly allocated to the experimental and control groups. That way, differences between the two groups can be attributed with confidence to the experimental intervention. Prevention scientists would turn up their noses at the study of Miss A. Since students were not randomly allocated into the classes of Miss A and the other first-grade teachers, the differences coming out might have been the result of differences coming in rather than the difference made by Miss A. Prevention scientists were so snooty about their standards that they mentioned the phrase "randomized control trial" whenever they could, like snooty persons constantly reminding everyone that they are members of Mensa or the Daughters of the American Revolution.

Even though prevention scientists didn't use the E-word, evolution leaped off the pages of the articles that Tony gave me to read. Just as with Pierre Teilhard de Chardin's *Phenomenon of Man,* I felt like a musical instrument resonating to another instrument being played nearby. Evolution is all about variation and selection. Prevention scientists had become highly sophisticated at managing the variation-and-selection process to achieve desirable outcomes. They were applied cultural evolutionists, even if they didn't use the E-word.

I started to ask my evolutionist colleagues if they had heard of the field of prevention science. Almost invariably, the answer was no or a vague maybe without any knowledge of specifics, in the same way that I know exotic names such as "Timbuktu" or "Zanzibar" without knowing anything about them. Prevention science was a lost island of the Ivory Archipelago for them and me. As I learned more about the historical roots of prevention science, I began to understand why it is unknown to evolutionists. I also had to confront xenophobia in my own tribe of evolutionists, similar to the xenophobia exhibited by cultural anthropologists toward evolution.

PREVENTION SCIENCE IS A DIRECT descendant of behaviorism, arguably the dominant tradition in psychology during the middle of the twentieth century. Great behaviorists such as Ivan Pavlov, John B. Watson, and B. F. Skinner made *learning* their focal point, starting with Pavlov's demonstration that dogs salivated at the sound of a bell when it had previously been paired with the presentation of food. Watson made this famous boast in 1930:

> Give me a dozen healthy infants, well-formed, and my own specified world to bring them up in and I'll guarantee to take any one at random and train him to become any type of specialist I might select—doctor, lawyer, artist, merchant-chief and, yes, even beggar-man and thief, regardless of his talents, penchants, tendencies, abilities, vocations, and race of his ancestors. I am going beyond the facts and I admit it, but so have the advocates to the contrary and they have been doing it for many thousands of years.

The ounce of humility at the end of this passage is nothing compared with the pound of hubris that precedes it. *Guarantee*, no less. Watson was clearly selling the brave new theory of behaviorism as a miracle cure for whatever ails us. He would have been a good advertising man. In fact, that's exactly what he became later in his career.

Skinner developed behaviorism into what he called radical behaviorism, which tries to explain as much as possible about how organisms behave, think, and feel on the basis of their previous experiences. He invented the operant conditioning chamber, better known as the Skinner box, which rigorously controlled the environmental input that an animal received so that its behavioral output could be measured. In books for the general public such as *Walden Two* and *Beyond Freedom and Dignity*, Skinner expressed the same confidence as Watson that people are almost infinitely malleable and can be shaped by their experiences to engineer a better society.

In addition to faith in the power of learning, Skinner and his predecessors also had strong ideas about how learning should be studied. Skinner treated the mind as a black box that received input from the environment and delivered behavior as output. He insisted that learning could be understood purely in terms of the input-output relationship without opening the black box. The entire concept of the mind was treated as an unnecessary hypothetical construct.

A school of psychology that avoids the concept of *mind*? This position might seem bizarre today but had a justification early in the twentieth century, when speculation about internal mental events was rampant and the scientific tools for studying them had not yet been developed. If the mind was beyond the reach of science, then the only scientific way to proceed was by studying what went in (experiences) and what came out (behavior). Something was taking place inside the black box to accomplish the transformation, but it had to be inferred from the input-output relationship. For Skinner, the black box was like Pandora's box, which must forever remain closed.

Behaviorism's star began to fade in the 1950s, when other psychologists insisted on opening the black box of the mind. After all, the brain is an enormously complex organ, and it is monstrous to think that it can be understood purely on the basis of environmental inputs and behavioral outputs. Behaviorism had such a strong grip that resisting it became known as "the cognitive revolution." Jerome Bruner, one of the revolutionaries, titled one of his books *Beyond the Information Given,* which elegantly expresses the essence of cognitive psychology. So much information-processing equipment is present inside the brain at birth that the mechanistic details must be studied to understand how environmental input leads to behavioral output.

The cognitive revolution relied heavily on computer science and artificial intelligence, which themselves were new branches of science in the 1950s. Computers must have an elaborate circuitry to function. Brains are more sophisticated than any computer and therefore must have an even more elaborate circuitry. If the message of the cognitive revolution had to be described in a single sentence, "The brain is like a computer" would do.

NOW THERE WERE TWO MAIN islands in the Ivory Archipelago within the Sea of Psychology. A third island arose in the 1990s, as if by a volcanic eruption, which called itself evolutionary psychology. The central insight of evolutionary psychology is that the brain is not a single organ. All species must solve many problems in order to survive and reproduce. Finding an appropriate mate is a different problem from raising a child, which in turn is a different problem from avoiding toxic food. For many problems, it is important to get it right quickly, rather than slowly learning from experience. How many chances do you get to avoid a poisonous mushroom or escape an ambush from a tiger? One lesson from artificial intelligence is that computers and robots must be specialized to be smart. Deep Blue, the computer that can play chess better than any human, can't do anything else.

For the mind to solve many different problems, it must therefore be *modular.* The mind is a collection of many special-purpose computers, not a single all-purpose computer. Furthermore, different species must have different modules to solve the challenges of their respective environments. Birds in northern climes that don't migrate have modules for remembering the locations of thousands of food items that they store over the winter. Migratory birds have modules for memorizing the night sky so they can fly south. To discover human mental modules, we must understand the hammer blows of natural selection that caused them to evolve. That requires understanding the environments that we inhabited during our evolution as a species—our environment of evolutionary adaptedness (EEA), a term that I have already used and which was coined by the pioneering evolutionary psychologist John Bowlby in 1969.

Understanding the EEA might be difficult, but there is no alternative. If I assigned you the task of studying a migratory bird species without telling you anything about its ecology, you could study it for a thousand years without stumbling upon its ability to memorize the night sky. That is exactly how psychologists have been studying the human species, according to evolutionary psychologists. The only way to figure out our modules is to reconstruct the EEA. To be a psychologist, it is mandatory to be a paleoanthropologist.

The job of the evolutionary psychologist is complicated by the fact that our current environment is so different from the EEA. Other species might be slightly mismatched to their current environments, but we are massively mismatched to ours. If we wish to understand our eating behaviors, for example, we must realize that food was scarce in the EEA, resulting in our open-ended craving for fat, sugar, and salt. Those cravings evolved by genetic evolution, and we're stuck with them, even though they're killing us in today's fast-food environment. According to evolutionary psychologists, we should be thinking about all of our behaviors that way.

Evolutionary psychologists portrayed their brave new theory in revolutionary terms, just as the cognitive psychologists and behaviorists had done before them. Their main manifesto was an edited volume, *The Adapted Mind,* which was published in 1992. The past tense of *Adapted* was fraught with meaning. The Marx and Engels of evolutionary psychology were Leda Cosmides and John Tooby, a married couple whose ideas were so entwined that they gave seminars as a pair. Their chapter in *The Adapted Mind,* "The Psychological Foundations of Culture," began with a section titled "The Unity of Science," but in fact it was a masterpiece of polarization. In this corner, the brave new theory of evolutionary psychology (EP). In that corner, the standard social-science model (SSSM), which included both the blank slate of behaviorism and the general-purpose computer of cognitive psychology. Out with the old, in with the new. Leda and John were joined by others, such as Steve Pinker, whose bestsellers *How the Mind Works* and *The Blank Slate* exuded the same kind of bravura as Watson and Skinner.

I HAVE ALWAYS HAD A LOVE-HATE relationship with the ideas that became associated with EP. There's plenty to love, but here's what I hate the most: the appropriation of the term *evolutionary psychology* for a particular school of thought. EP should be defined at face value as "the study of psychology from an evolutionary perspective." Leda and John had a particular vision of EP that might be right or wrong. If major elements of their vision prove to be wrong, that will not be the death of EP. It will be a form of scientific progress in which erroneous ideas are replaced with better ideas. EP will be the better for it.

Virtually all psychologists accept evolution and regard their own ideas as compatible. They don't necessarily know much about evolution, however, or actively use what they know in constructing their own theories. Cognitive psychologists were inspired more by computer science and artificial intelligence than by evolution, as we have seen. The way to make evolution the common language for all psychologists is to hold them accountable for what they already assume. Are their ideas *really* compatible with evolutionary theory? Taking that question seriously can unify all of those islands of thought into a single island.

B. F. Skinner regarded himself as an evolutionary psychologist, as he clearly stated in the summary of one of his classic papers, "Selection by Consequences," published in the journal *Science* in 1981:

> Selection by consequences is a causal mode found only in living things, or in machines made by living things. It was first recognized in natural selection, but it also accounts for the shaping and maintenance of the behavior of the individual and the evolution of cultures. In all three of these fields, it replaces explanations based on the causal modes of classical mechanics. The replacement is strongly resisted. Natural selection has now made its case, but similar delays in recognizing the role of selection in the other fields could deprive us of valuable help in solving the problems which confront us.

Huzzah! That made Skinner my soul mate, alongside Pierre Teilhard de Chardin. In the article, Skinner described learning as both a product of genetic evolution and a process of evolution in its own right, just as I did in my parable of the immune system. In Skinner's terminology, Mother Nature equips us with instincts for selecting behaviors called reinforcers. As we go about our daily lives, our reinforcers are constantly evaluating our behaviors and selecting those that meet their criteria. In this fashion, our behaviors are shaped not directly by the hammer blows of natural selection but by instincts that themselves were shaped by the hammer blows of natural selection.

What are these reinforcers? Pleasure and pain, for sure, but there are others. Smiles and frowns are powerful reinforcers, along with other forms of social recognition. Skinner also distinguished between primary reinforcers, which we are born with, and secondary reinforcers, which themselves are learned.

Some reinforcement takes place consciously. When we make a

deliberate decision, we are employing a variation-and-selection process in which we are fully aware of the selection criteria. Conscious choice is just the tip of the iceberg, however. Beneath the surface of our consciousness, automated variation-and-selection processes are shaping our behaviors every moment of the day, just as our immune systems are churning out and selecting antibodies. There's an old story about a class of students who agreed to smile when the professor walked to one side of the room and frown when he walked to the other side. In no time, the professor was spending all of his time on the smile side. This might be just an academic myth, but the behaviorism literature includes hundreds of comparable examples for people and other creatures alike.

Since Skinner regarded himself as an evolutionary psychologist, it's ironic that Leda and John excluded him as part of the SSSM. Who's right? The parable of the immune system helps us to see that both are partially right. If Skinner and Leda and John were immunologists, Skinner would be describing the adaptive component of the immune system but ignoring the innate component and perversely insisting that we can treat the immune system as a black box without directly studying the mechanisms. Leda and John would be describing the innate component of the immune system but ignoring the adaptive component and its impressive ability to adapt organisms to their current environment using a rapid variation-and-selection process. There's room enough in evolutionary psychology, properly understood, to include both of their visions.

Behaviorism didn't go away with the advent of cognitive psychology and the particular school of thought that became known as evolutionary psychology. Instead, selection by consequences became an essential tool in the branches of psychology that are most involved in actually changing people's behavior, such as psychotherapy. Psychotherapy has come a long way since Freud, who represents yet another island of thought in the Ivory Archipelago. Today, if you seek psychological counseling, the therapy you receive is likely to owe more to Skinner than to Freud—and it stands a good chance of working. As I discovered through Tony, prevention science includes individual therapy and more. Prevention scientists use the principle of selection by consequences to change the behavior of whole groups—even whole states.

I FINALLY MET TONY when he traveled to Binghamton to give an EvoS seminar. Thanks to EvoS, Binghamton is like a busy harbor that receives visitors from all islands of the Ivory Archipelago. Tony has a broad, friendly face and still has a head of thick dark hair and boundless energy for a man in his sixties. His ancestors came from Germany on his mother's side and Ireland on his father's side and settled in the vicinity of Scranton, Pennsylvania. As a boy, Tony lived in a big house in the country until his parents divorced when he was twelve. Suddenly, he found himself living in an apartment in a modest Scranton neighborhood with his mother and two sisters. His father sold the country home and moved in with a wealthy aunt and a bunch of his other unmarried relatives. As Tony explained it, during the potato famine in Ireland, it was customary to be unmarried, and his father's side of the family carried on the tradition in America. They were also devoutly Catholic, which made divorce a grievous sin. For whatever reason, his father provided very little financial support, even though he lived nearby, so Tony's mother had to support the family on her meager nurse's salary. The fatherless family moved to Rochester, New York, when Tony was in the middle of high school so that his mother could find better employment. From Tony's standpoint, he was uprooted from one shabby neighborhood to another, where he was now a total stranger and had to reassemble his social world from scratch.

Tony was smart and did lots of reading on his own but had little interest in school. He even stopped going on some days to earn extra cash bagging groceries. He became absorbed by psychology, in part to deal with his own troubles. His grades were mediocre, but his standardized test scores were fine. That wasn't enough to get him into the most prestigious schools to which he applied, but he was admitted to the University of Rochester, which happens to be my own alma mater. He graduated in 1966, the year before I entered, and went on to earn his PhD in social psychology from the University of Illinois.

Tony worked hard in graduate school but also was swept up in the turbulent 1960s. He became politically active and was president of Students for Robert Kennedy at Illinois at the time of Kennedy's assassination on June 5, 1968. The 1960s were also a turbulent time for psychotherapy, which for the previous half-century had been dominated by Sigmund Freud, Carl Jung, and others of their tradition — more islands of the Ivory Archipelago. Freud also regarded himself as an evolutionary psychologist whose ideas were consistent with Darwin's. His ideas became so prominent during the first half of the twentieth century that the *Encyclopedia Britannica* included him in its pantheon of great books, along with Aristotle, Shakespeare, and Darwin.

Alas, history has not been kind to Freud. His elaborate theory of what goes on inside the head was precisely the kind of speculation that Skinner and other behaviorists were reacting against. Moreover, according to Freud, years were required to delve into your subconscious and emerge healed. My mother underwent Freudian psychotherapy for most of her adult life. She used to drop me off at a candy store, where I would gorge on cotton candy and play pinball machines while she had her weekly therapy session. My stepfather, a no-nonsense orthopedic surgeon whom I loved, used to complain that she spent a fortune and that the only tangible outcome was that she had become less inhibited about saying the word "fuck."

Even though Freudian psychotherapists were required to undergo years of training in addition to their medical degrees, there was nothing even remotely scientific about their methods. The same went for other therapeutic methods that constituted a vast archipelago in their own right. When these methods started to be properly assessed, the results were an embarrassment. Some made matters worse, others had no effect, and even the best were little different from a nurturing relationship with a friend or one's pastor.

Against this background, the arrival of behaviorism on the therapeutic scene was like a meteor striking the earth. Behaviorists brashly claimed that they could change human behavior in a short period of time and prove it. Tony became a part of this movement during its early stages. He helped to expand it beyond psychotherapy to other kinds of behavioral change, such as smoking prevention. The more deeply Tony thought about individual behavior as a product of selection by consequences, the more he appreciated the need for a more explicitly evolutionary perspective.

HERE'S AN EXAMPLE OF THE KIND of research that the emerging field of prevention science became known for. In 1983, some of your tax dollars were used to fund the Oregon Youth Study (OYS). In an Oregon city, more than 200 families from neighborhoods with high juvenile crime rates were studied by a team of scientists headed by two of Tony's colleagues, Gerald Paterson and Deborah Capaldi. All of the families had sons in the fourth grade at the beginning of the study, who were followed in excruciating detail as they grew into young men. In addition to surveys and interviews, they were observed in their homes using the same methods that my evolutionist colleagues use to study lions, chimps, and crows in their natural environments. The observer focused on each member of the family in turn for fifteen minutes, scoring all of the behaviors of that person and the reactions of the other family members. The observers were trained to score the behaviors according to a "Family Process Code" with twenty-five categories for content and six categories for emotional valence. Every observation method was tested for reliability by comparing the agreement among different observers employing the same methods.

In addition to family dynamics, the boys were also observed interacting with their friends starting at age thirteen. Each boy and a friend of his choice were videotaped while being asked a number of questions. First, they were asked to plan an activity together. Then they were asked to discuss four problems that might have occurred during the last month: (1) a problem for the study boy related to getting along with his parents, (2) a problem for the study boy related to getting along with peers, (3) a problem for the friend related to getting along with his parents, and (4) a problem for the friend related to getting along with peers. The order of the questions was counterbalanced; when you're designing a scientific study such as this, each and every detail must be scrutinized. Videotaping the conversations enabled the scientists to analyze them in minute detail, including subtleties such as pauses and laughter, using another coding system that had been validated for inter-rater reliability.

All of this information was related to the presence and absence of deviant behavior, based on parents' reports, teachers' reports, self-reports, and juvenile court records. The OYS team might not use the E-word, but they were clearly studying people *in relation to their environment,* in just the same way that Daphne Fairbairn and Richard Preziosi followed the lives of water striders in such excruciating detail. Both studies were heroic for the amount of work required. Prizes should be given for this kind of work.

What were the results of the OYS? Imagine that you're in a supermarket and overhear a mother and her son arguing over what to buy. The son whines and makes himself as obnoxious as possible. If he's young enough, he might throw a tantrum. The mother is also rude, barking orders, threatening to slap her son, and perhaps even carrying out the threat. It's embarrassing to watch, and you quickly move on, wondering to yourself how people can possibly act that way. If they are ethnically different from you or poorer than you, you might harbor thoughts about "those people."

The OYS shows that these behaviors are *selected by consequences.* In a perverse way, they are adaptive, even though "nobody's having fun," as Tony likes to put it. If the son puts up such a fuss that he gets what he wants, then he is positively reinforced for his behavior. If the mother gives up and her son stops being obnoxious, the cessation of his behavior is momentarily satisfying and therefore another form of reinforcement. In this fashion, what prevention scientists call "obdurate noncompliance" is perfected as a strategy early in life and becomes elaborated later in life as deviant behavior.

The OYS team knows this because they calculated a "nattering and abusive behavior" score for each family, based on observations such as nagging, scolding, humiliating, threatening, and finally physical abuse. The score for parents correlates highly with the antisocial-behavior score of the child, as if they are locked in a tragic coevolutionary race to the bottom. The syndrome also correlates strongly with deviant behaviors in the boys as they mature. Unless they are placed in a different environment, obdurate noncompliance becomes a way of life for them, their rule of engagement with the world, their playbook. Never mind that the strategy is harmful for everyone over the long run. We already know from the parable of the strider that genetic evolution can take us where we don't want to go. Selection by consequences is no different. Our reinforcers, operating largely beneath conscious awareness, shape our behaviors on the basis of their immediate consequences, no matter what the fallout over the long term.

Selection by consequences also takes place among peers. In one of the most striking results of the OYS, the boys were coded for when they laughed while talking with their friends. Some friends laughed while discussing deviant behaviors, while others laughed while discussing normative behavior. This difference was highly predictive of actual deviant behavior, even years later. Earlier in this chapter, I told the apocryphal story of the students who trained their professor to lecture on one side of the room by rewarding his behavior with smiles. The OYS team provided a real-world example. It seems that smiles, laughter, and other forms of social approval can powerfully reinforce almost any behavior, regardless of whether it is regarded as deviant by society as a whole. So it is the immediate social environment that counts, not some larger abstraction. When Johnny Badass acts up in class and gets everyone to laugh, no amount of scolding by the teacher can make up for the positive reinforcement he has received. In fact, the teacher's disapproval can add to the positive reinforcement.

These results can be used in Binghamton or any other city to design interventions that work and avoid those that don't work. In one study conducted by Thomas Dishion and David Andrews, two members of the OYS team, an additional 119 families with at-risk children were randomly assigned to a number of intervention treatments that involved working with the parents and/or their children in groups. Teaching the parents how to reinforce good behavior was somewhat successful at breaking the vicious cycle of nattering and obdurate noncompliance. Meeting with the kids in groups perversely increased the level of deviant behavior measured a year later. You guessed it: the kids positively reinforced one another for deviance, which outweighed the instruction that the adults were trying to provide. Thanks to their insistence on rigorous assessment, prevention scientists can say with authority when something doesn't work, no matter how sensible it seems on the surface. Any intervention program that involves bringing high-risk kids together in groups — and there are thousands of them — is liable to make the problem worse.

What are we to make of colossal problems that refuse to go away, such as dysfunctional families and at-risk youth? Perhaps they will always be with us, like an enormous boulder that can't be moved. Or perhaps they are examples of evolution taking us where we don't want to go. As long as the environment remains unchanged, the evolutionary outcome won't change, as stubbornly as a boulder that can't be moved. But change the environment, and the evolutionary outcome will follow. It will be hard to stop it from changing. Change can be fast when the evolutionary process is the rapid selection by consequences built by genetic evolution. That which defied change can change in a year, a month, perhaps even a day. The key to change is to become wise managers of evolutionary processes.

That was my personal quest since starting the Binghamton Neighborhood Project. Now it seemed that I had a powerful supporting army in the field of prevention science. I couldn't wait to attend the annual meeting in San Francisco, which was preceded by a special symposium on prevention science from an evolutionary perspective. Evidently, others were interested in forging an alliance between evolutionary theory and prevention science, not just myself.

THE MANY TRIBES of the Ivory Archipelago wouldn't even hang together at the scale that they do if it weren't for their annual meetings. Scientific conferences are like the tribal gatherings described by anthropologists. They are also just as rowdy, especially after hours. Prodigious amounts of food and alcohol are consumed, marriages are made and broken, babies are conceived. The Society for Prevention Research conference was held at the Hyatt Regency, a four-diamond hotel on San Francisco's waterfront. The atrium of the Hyatt Regency is seventeen stories high. As I checked into my room on the twelfth floor, I looked down on a lobby with a stunning geometric sculpture, a swank café, boutiques, and enough vegetation to make it seem like an outdoor park. Outside, one of America's greatest cities beckoned me to walk its streets. I was going to have a great time even before having a single conversation.

Just as the field of prevention science was a lost island on my chart until I met Tony, evolution was a lost island on the charts of most of the prevention scientists who were checking into the Hyatt, eager for their own wild rumpus to start. This might seem strange. Didn't B. F. Skinner, one of their deities, regard himself as an evolutionary psychologist? Didn't they subscribe to his notion that selection by consequences is a fast-paced evolutionary process, built by the slow-paced process of genetic evolution? Wasn't Tony one of their chiefs? If Tony appreciated the relevance of evolution enough to reach out to me, didn't the other members of his tribe?

When I posed these questions to Tony, he answered this way: Most scientists aren't that interested in the big picture. They become engrossed in their particular problem, which causes them to become more and more specialized. The entire structure of federal funding doesn't see the big picture, either, and doles out money to solve specific problems, such as smoking, delinquency, or learning disorders. Scientists are selected by consequences, just like everyone else, and before long, they become encapsulated in little groups, with their own specialized languages and concerns. A few remain cosmopolitan (like Tony) but mostly by virtue of their personal preferences and not because they are rewarded for it by the system. The SPR and its annual meeting are needed to keep these tiny groups from drifting apart and becoming separate islands. The people attending the meeting were already stretching their minds by seeing themselves as members of a single tribe. Against this background, my island of evolutionary theory was like a continent on the other side of the world.

Tony's explanation struck me as right. At least, I felt like one of the first Native Americans brought to Europe, as Tony introduced me to his friends during the social mixer on the first night of the meeting. I might be a big chief where I came from, but nobody knew me there. I didn't know most of them, either, since I was only starting to master the prevention-science literature. They were eager to schmooze with their own kind, and I'm a shy person when out of my element, so I was happy to play the role of anthropological observer whenever Tony left me to my own devices.

There was one person I especially wanted to meet, however. Dennis Embry was one of Tony's closest colleagues, and I was wild about his work. Tony had told me a bit about him, and I soon discovered that he is a rarity wherever he goes — a gay man so comfortable with his identity that he puts everyone around him at ease. Because Dennis has shed social conventions like a butterfly emerging from a cocoon, those around him feel encouraged to do the same. They become lighthearted and can't help putting their arms around him, not because of sexual attraction but because of a feeling of liberation.

After I got to know Dennis, he told me his story. His paternal great-grandfather was a freed slave who left Kentucky and settled in Lawrence, Kansas, which had just been established by New Englanders intent on making Kansas a free state. He married a white woman, and their son, Dennis's grandfather, became a doctor and set up a practice in the town of Great Bend. There was plenty of prejudice against black people in Kansas, even though it was a free state, and Dennis's grandfather was sufficiently white in appearance to deny his black ancestry. In fact, he even kept it a secret from his own wife, who thought that black people were evil and so different from white people that they had foam for blood. As someone who has wrestled with the question of whether to conceal his sexual identity, Dennis often looks back on his grandfather's plight with sympathy.

Dennis's father was a socialite as the son of a prominent doctor in a small town. He married a girl who was stunningly beautiful but had a troubled family life. Dennis suspects that she was abused by her father, who departed suddenly when she was young. Dennis was their second child, born eleven years after their first son, and Dennis suspects that he wasn't planned. His mother was so obsessed with appearing young that she would dress like a teenager and pretend that her older son was not her own. Dennis was raised by a nanny named Marnie (Mary) Williams, a black woman whom he adored and became more attached to than his own mother. He used to accompany Marnie and her husband, Pete, to their revivalist church on Sundays. She only used physical punishment on him once, by stripping a twig from a tree and flicking it gently against his ankle, causing Dennis to bawl with remorse.

We'll never know the combination of genes and experience that made Dennis, but from the earliest age, his playbook was to reach out to others. For the most part, he was rewarded in return by good people who must have been amused and impressed by the little entrepreneur. At the age of five, he was selling them watercolor paintings, made with a paint set that had been given to him by an adult neighbor. He stood inside a cardboard box as a pretend television set and charged admission for his neighbors to watch. As far as Dennis was concerned, the world was a safe place, and his efforts would be rewarded.

Except at home, where his parents were sinking into alcoholism and domestic violence. Dennis started living with his grandmother in sixth grade. In seventh grade, a gifted teacher encouraged his artistic talent, leading to a one-boy exhibit at the local hotel. One of his pictures sold for $100, and the woman who purchased it invited him to tea. During their conversation, she leaned over and said, "I know that your mom and dad are in trouble often. Their problems hurt you, but they are not you." Later, Dennis learned that she was the wife of the town's justice of the peace and that her husband had sentenced his father on the day of his exhibit. She became a mentor to Dennis from then on. He would visit her home, the grandest in town, stuffed with thousands of books that he could read. Still later in life, Dennis discovered that she mentored other troubled youth, two or three at a time, without ever introducing them to one another so that each one would feel special.

In this fashion, Dennis was nurtured by the kindness of people beyond his immediate family and became bold in reaching out to them. He charged into college and graduate school at the University of Kansas's well-regarded program in behavioral science with the same outgoing energy that served him so well in childhood. His master's thesis was on how stories could be used to reinforce good behaviors in children. In no time, he was serving as a consultant for *Sesame Street* and as assistant director of a childhood institute at KU. Take someone brimming with energy, talent, and the attitude that initiative will be rewarded, and anything can happen.

DENNIS'S PHD RESEARCH WAS ON accident prevention, in particular, how to prevent children from running into the streets and getting hit by automobiles. This might seem like a blue-collar subject, compared with lofty white-collar subjects such as the meaning of life, but it is deeply interesting when you start to think about it. We are so accustomed to viewing our problems as pathological that it's hard to break the habit. If someone is so beset with problems that they seek therapy, mustn't they be sick in some sense? The idea that they might be perfectly normal but selected by consequences to a place that they didn't want to go is a bit subtle.

This confusion does not exist for accident prevention. When kids dash onto the street in front of oncoming traffic, they are definitely in harm's way, but they are also perfectly normal. It's a matter of selection by consequences. Usually, there are no negative consequences to running onto the street, only immediate positive consequences such as fetching a ball or having extra space to play. The negative consequence of getting hit by a car is too infrequent to shape learning. Thus, normal kids are selected by consequences to put themselves in harm's way. The challenge for Dennis was to design a reinforcement scheme to keep them out of harm's way.

You might think that Dennis shouldn't be needed and that parents should be good at keeping their kids out of traffic, first by telling them and then by smacking them if they don't obey. By now, you should be attuned to the flaw in this argument. Kids who get smacked often respond by becoming obdurately noncompliant. Even kids eager to please their parents can be positively reinforced for misbehavior if it is their only way to get attention. My neighbor has a new dog that he puts out in his backyard at night. The dog gets lonely and starts to howl. After a while, my neighbor can't stand it and bursts through his back door, howling at the dog to stop howling. Reinforced by the attention, the dog starts howling as soon as my neighbor goes back inside. I feel like giving him Dennis's telephone number. Kids can be much like dogs that way.

Evidently, parents aren't doing their jobs well enough, because getting struck by cars is a major cause of injury and death in young children. As a good prevention scientist, Dennis demonstrated that spanking, scolding, reprimanding, and nagging make the problem *worse.* Praising children for safe play is much more effective, especially with the use of tangible rewards, however small, such as stickers. In addition, Dennis drew on his previous research on stories and created a coloring book designed to positively reinforce safe play. A key to its effectiveness was that it made the child the hero of the story. The first page had an outline of a child that could be male or female, which the child colored to be himself or herself. Dennis did the research to show that children were four times more likely to identify with a story about themselves than one about someone else.

Creating a successful intervention program is challenging enough; spreading it adds more challenges. Just as Mother Nature provided us with instincts that shape our learning, we have additional instincts for copying the behaviors of others. As with learning, some of our copying is deliberative. If I see you doing something that makes sense, I might consciously start doing it to see if it works for me. The conscious part of copying is just the tip of the iceberg, however. Beneath the surface, we are bristling with mechanisms that screen the behaviors of others, keeping most out and letting some in based on indirect cues about their effectiveness. A behavior observed in a total stranger will probably be ignored. A behavior observed in a friend or a highly respected person will probably be admitted. Common behaviors will probably be copied more than rare behaviors: when in Rome, do as the Romans do. Stigmatized behaviors will be avoided: don't be a Judas. The behaviors that we copy need not make any functional sense. If I admire you, I might copy your laugh, the jaunty way you position your hat, your brand of cigarette. The behaviors that I copy can be harmful, such as smoking or the lifestyle habits that lead to obesity, especially when the harm accrues over the long term. Copying, like learning, can take us where we don't want to go.

All of this equipment evolved by natural selection, and it worked pretty well on average for our ancestors to sift useful behaviors, like a panner sifting for gold. It doesn't necessarily work in our current world. A multi-billion-dollar consumer industry has been selected by consequences to push our buttons to sell us stuff that we wouldn't otherwise want and isn't necessarily good for us. Conversely, a prevention scientist such as Dennis or Tony might develop an intervention program that works spectacularly, potentially saving thousands of lives in the case of smoking or accident prevention, and the successful program could fail to spread because it doesn't trigger the right copying instincts. The situation is precisely analogous to immune-system dysfunction as discussed in chapter 8. Our instincts have become part of the problem, and we need to alter the environment to make them work better. Successful behaviors need to be marketed to spread in modern environments. Moreover, we need to understand what marketing means from a scientific perspective. Real marketers might be highly effective at what they do, but they don't have an explicit theory of copying behavior.

As an entrepreneur since he was a boy, Dennis took to marketing his accident-prevention program like a duck to water. He was provided a golden opportunity when the New Zealand government invited him to implement his program at a nationwide scale. Pedestrian accidents were a major source of child mortality in New Zealand, and they wanted to mount a national campaign to do something about it. Dennis actually succeeded at doing this, using the methods of prevention science. It would be hard to overstate the importance of this accomplishment. Dennis managed to change a behavioral practice at a nationwide scale, saving hundreds of lives. As a white-collar scientist new to the trenches, I had lofty ideas about becoming a wise manager of evolutionary processes. As a blue-collar scientist and veteran of the trenches, Dennis had done what I merely envisioned as an abstract possibility.

Returning from New Zealand, Dennis found the academic environment too stifling for his entrepreneurial spirit and went into business for himself. One of his clients was the U.S. military, which contacted him with a problem shortly after the onset of the first Gulf War. An epidemic with symptoms resembling posttraumatic stress disorder had broken out among children of military families whose parents were being deployed for duty. Could Dennis help?

When Dennis was confronted with this problem, he had a personal epiphany. A severe mental disorder that develops in thousands of children within the space of a few weeks? This could only be the response of perfectly normal children to an environmental change. Moreover, there was nothing unusual about going to war. The main predator of people has been other people for a long, long time. The response of the children might not be pathological at all. It might be an adaptive response to extreme between-group conflict, the mind working as nature's hammer blows designed it to work, not the mind falling apart. This interpretation did not make the problem go away. There was still an urgent need to alleviate the stress of the children, but thinking of their response as an adaptation rather than as a pathology made all the difference in terms of designing a successful solution.

In a flash, Dennis saw the deep relevance of evolution for understanding psychology. As he worked feverishly to design an intervention for the children of military families, he also started to explore evolution in relation to all things human—my island in the Ivory Archipelago.

Using his story methodology, Dennis produced a book that the military distributed to hundreds of thousands of military families, substantially alleviating the problem. He met with Dick Cheney (then secretary of Defense) and spoke at congressional hearings on the effects of deployment on military families, representing both the American Psychological Association and the Secretary of Defense's Office.

His professional life as a scientific entrepreneur was thriving, but his personal life was a bundle of contradictions, to put it mildly. His marriage had fallen apart while he was in New Zealand. He was coming to terms with the fact that he was gay. He was a scientist to the core but also felt highly spiritual and was attending a church sympathetic to gays and lesbians, where he met a man who would become his lifelong partner. In short, he was a religious gay scientist who worked with children and the military. If that's not a bundle of contradictions, what would be?

Dennis's inner turmoil resolved itself like a bolt of lightning in what he calls a transcendental experience. He didn't know where it came from, but it seemed like the voice of God issuing a commandment that could not be ignored: "This is what you are. I chose you for those things, so go out and bear witness." Or, as Dennis puts it in a less God-like way, "Stop whining, and get on with it." Ever since, Dennis has borne witness for social justice, for being gay, for having deep spiritual beliefs, and for being a scientist. His comfort with himself shines like a beacon that attracts people to him without knowing why. When people glow that way, you just want to copy them.

TONY HAD DONE AS MUCH AS a president of a society can do to expose his tribe to evolutionary theory. In addition to the symposium on prevention science from an evolutionary perspective that he organized for the day preceding the conference, he scheduled me as a keynote speaker during the conference. The symposium was attended by perhaps fifty people, including two program officers from major federal agencies that fund prevention-science research, who were there to see if this was a promising new direction. Some of the speakers were prevention scientists newly interested in evolution, while others were evolutionists like myself newly interested in prevention science. Everyone was a teacher and a student at the same time, which is the ideal climate for respectful and productive intellectual discourse. The audience was fascinated, but it was clear that most of them were encountering evolutionary theory in relation to their subject for the first time.

My talk at the symposium was on the Binghamton Neighborhood Project, then only two years old. I was as nervous as a music student giving my first recital. These guys were the pros. Compared with them, I felt like a baby just learning to stack blocks. Would they tell me to come back when I had grown up? Would they cluck approvingly like indulgent parents? Or did I actually have something to offer, right then and there, with my GIS maps and my way of reflecting on the information based on evolutionary theory? It seemed to me that I did have something to offer, especially the ambitious scope of studying all aspects of a human population the size of a city from a unified theoretical perspective, with an empirical infrastructure to match.

Everyone wanted to know what the two program officers thought about the symposium. Prevention science wouldn't exist without federal support, and many prevention scientists get their salaries from grants in addition to research expenses. If their grants aren't funded, they're out of a job. How likely would it be for a grant proposal based on evolutionary theory to be funded by a federal agency such as the National Institute on Drug Abuse or the National Institute for Mental Health? One of the program officers replied that while the ideas were very interesting, her advice was to avoid using the word "evolution" in a grant proposal. It wasn't that the reviewers of the proposal would be hostile to evolution; it would just be so foreign to them that they wouldn't know what to do with it.

I was stunned. I might expect a member of a local school board in the Bible Belt to say something like this but not a program officer for a major federal scientific funding agency. The Ivory Archipelago was even less united than I thought.

CHAPTER 13

The Lecture That Failed

MY KEYNOTE ADDRESS at the Society for Prevention Research conference was my big chance to present evolutionary theory to prevention scientists. I took a long walk through the streets of San Francisco to plan my strategy. Along the Bay to Fisherman's Wharf and the maritime historical park, then inland up Hyde Street and down Lombard Street, including the block that earns it the title of the curviest street in the world. What a city! I stopped at a café, ordered a beer, and planned my keynote address while watching the human parade. I decided that my best strategy was to discuss three prevention-science success stories from an evolutionary perspective. One involved changing the behavior of individuals, another involved changing the behavior of small groups, and the third involved changing the behavior of entire states. In all three cases, the problem was caused by unmanaged variation-and-selection processes, and the solution involved managing the variation-and-selection process.

As the event approached, I imagined a swelling audience on the edge of their seats, straining to hear what I had to say. Instead, only a few cosmopolitans and people especially loyal to Tony filed into the room. Everyone else was too busy networking with his or her own kind to hear a talk about a far-off land. Oh, well. In science as in vaudeville, the show must go on.

Tony gave me a glowing introduction against the background of my first slide, which boldly proclaimed the title "Evolving the Future:

Evolutionary Theory Meets Prevention Science." The first part of my talk was like a flipbook version of this book, flying through the basics to leave enough time for my three case studies. Right away, I could see that it wasn't working. Too much, too fast. I sounded as if I was trying to sell them something on late-night TV.

Perhaps my three case studies would save the day. The first was a therapeutic method called acceptance and commitment therapy (ACT, pronounced as one word), developed by psychologist Steve Hayes. I had yet to met Steve, and he wasn't at the conference, but Tony regarded him as a modern B. F. Skinner, which is saying something. As a young scientist, Steve was inspired by Skinner's combination of hard science and utopianism. He bounced around a lot before finding his path, like so many other people I have profiled in this book; turned off by school, drawn to humanistic movements and Eastern religions, early marriage and divorce, living on a commune, odd jobs, wrestling with a panic disorder of his own. Eventually, his search turned into a focused academic career that fulfilled his early ambition.

Steve calls his approach post-Skinnerian because he is willing to open the black box of the mind, along with cognitive and evolutionary psychologists. What he finds is an elaborate capacity to create and generalize mental associations. If you learn that A is connected to B and B is connected to C, you will automatically infer a connection between A and C. You will also easily transfer the network of associations to another set of objects such as X, Y, and Z. That's what metaphors are all about. When I say, "My love is a rose," you take a network of associations that you have about roses and use it to think about love. Our ability to relate one frame to another, as Steve puts it, enables us to create whole systems of meaning that organize our perception. It is part of our capacity for culture, which sets us apart from all other species.

This capacity comes at a cost, however. If an antelope has a traumatic experience, such as being ambushed by a lion on its way to the water hole, it will learn to avoid that water hole or to take a different route. If a person has a traumatic experience, it can ramify throughout the network of associations and become crippling. Steve calls this *experiential avoidance*. Suppose that you have a snake phobia and are invited by some friends to visit the zoo. You want to be with your friends and love the zoo, as long as you don't have to visit the snake exhibit. You become consumed with fear that your friends will want to visit the snake exhibit and make an excuse that you can't go. Your immediate sensation is one of profound relief, but that reinforces your decision, which, in fact, caused you to spend a lonely day at home. You try to avoid thinking about snakes, but that only makes you think about them more. As an instructive exercise, for the next minute, try *not* thinking about snakes. The very effort requires thinking about them! Pretty soon, your panic spreads through a network of associations, which become fused around the object of your panic. Your main goal in life is to avoid panic, which severely limits your behavioral options. You are a victim of experiential avoidance.

If something like this describes you, then you can take comfort in the fact that there is nothing wrong with you. You are not broken merchandise. There are no loose screws in your head. You don't need to medicate yourself. You are a perfectly normal evolutionary process that has led to a dysfunctional outcome. Selection by consequences has gotten you into this mess, and it can get you out of it, but only if we manage the evolutionary process.

The goal of ACT is to manage the evolutionary process by first examining one's criteria for selecting behaviors and then expanding one's behavioral repertoire so that there will be sufficient variation. Let's begin with selection. If you are that person with the snake phobia, what are your real goals in life? To avoid panic? Clearly, you only want to avoid panic so that you can do something else. Let's think clearly about your goals and make a commitment to achieve them, regardless of what panic might have to do with it. That is the commitment part of ACT.

Now let's work on your behavioral repertoire. Negative associations are very hard to undo, especially when they are deeply traumatic, such as being abused as a child. Trying to avoid or eliminate them can only make them worse, as we have seen. An alternative is to accept and detach yourself from them, so that they no longer interfere with your life goals. This is where ACT converges with religious practices, especially the concepts of mindfulness and detachment associated with Buddhism.

Metaphors can be powerful tools for changing behavior, which is why they are the stock in trade of religions. Relational frame theory is essentially a scientific theory of metaphorical thinking, and ACT relies heavily on metaphors to cultivate an attitude of acceptance and detachment. Here is one of my favorite ACT metaphors: Suppose that you are a chess player totally absorbed in a game with an archrival. It's a battle to the death that requires all of your wits. Now imagine that you are the chess *board*. The game is going on as before, but the outcome no longer appears as a life-and-death matter to you.

Here's a second ACT metaphor: Imagine that you are standing on a bridge across a slow-moving stream on a beautiful fall day. Colorful leaves are falling onto the water surface and drifting past you from beneath the bridge as you gaze down at them. These leaves are your problems. Look at them carefully, and appreciate them as best you can. They are not you. They are what you are observing from a distance in a cool and relaxed frame of mind. This metaphor is sometimes expanded into an exercise that involves actually writing one's problems on pieces of paper in the shapes of fall leaves.

Here's a third ACT metaphor: Imagine that you are a bus driver trying to get to a certain destination. It's your job to stop and let people onto the bus. You don't necessarily *like* the people who get on. In fact, some of them are downright scary. Nevertheless, your primary job is to get to your destination, managing the people on the bus as best you can.

When I related Dennis Embry's story in the last chapter, I told about the woman who became his guardian angel, comforting him with the words "I know that your mom and dad are in trouble often. Their problems hurt you, but they are not you." This woman was a natural-born ACT therapist.

All very good, you might be thinking, but where's the proof? What distinguishes ACT from all the other feel-good cures that crowd the TV talk shows and the self-help sections of bookstores? In the first place, ACT doesn't need to be different to be effective. Other therapeutic methods that go by different names might have converged on the same principles. Some religions might have converged on the same principles long before psychotherapy came into existence. Steve is happy to acknowledge that ACT is a member of a family of therapeutic methods that combine the best elements of behavior therapy, cognitive therapy, and mindfulness-based therapies. Let's focus on what works, not what it's called or whose name is attached to it.

In the second place, if you want to know if a therapeutic method works, ask a prevention scientist. Nearly fifty controlled studies have been conducted, demonstrating the efficacy of ACT for problems as diverse as depression, stress, psychosis, anxiety, pain, burnout, substance abuse, obsessive/compulsive disorder, epilepsy, weight, prejudice, smoking, and bipolar disorder. Even more amazing, ACT can begin to alleviate problems in a remarkably short period of time. In one study, eighty patients hospitalized for hallucinations and/or delusions were randomly assigned to two groups. One group participated in a single three-hour ACT session that focused on accepting and distancing themselves from their problems. The other group received the normal "treatment as usual" (TAU) that patients received at the hospital. The ACT group was much less likely than the TAU group to seek hospital readmission over a period of 120 days.

This might seem too good to be true until we realize how much our mental associations organize our experience. A hallucination or delusion is not terrifying by itself but only in terms of what it means. If its meaning can be changed in a single session, then the response will follow suit. If I was experiencing hallucinations, I would be very comforted by the bus metaphor. I would still seek help, but I wouldn't be so terrified that I had to check myself into a hospital. That is the degree to which a managed variation-and-selection process can "transform how we live," as Steve puts it.

AFTER I FINISHED THE SEGMENT of my talk on ACT, I glanced around the room to see how I was doing. Some people seemed interested, but others had a stuffed look, as if I had forced them to enter a hotdog-eating contest. Too much, too fast, and two more case studies to go.

My next case study was the Good Behavior Game (GBG), a

program for creating a culture of cooperation in elementary-school classrooms. It was invented by an elementary-school teacher, had been developed and assessed by prevention scientists, and was being marketed to schools around the country by Dennis. Not only would it be useful to implement in Binghamton's classrooms, but it also embodied the same principles that are required for cooperation to succeed as an evolutionary strategy in many social settings.

Some elementary-school classrooms are so dysfunctional that more than 200 disruptions take place every hour. The GBG begins with the teacher asking the students what counts as good behavior. One way to do this is with a game resembling musical chairs, in which the students are seated in a circle and the teacher stands in the center. The teacher goes up to a student, names a good behavior (it might be "not teasing others"), and sits in the student's chair. That student then approaches another student to name another good behavior, and so on, until all students have had their turns. In this fashion, a list of do's and don'ts is established by consensus. This is important. There is a world of difference between being told what to do and agreeing about what to do, even if you're a first-grader. It turns out that most kids know what counts as good behavior, even when they misbehave. The students' list is little different from what the teacher might impose by decree.

The list of do's and don'ts is placed prominently on the classroom wall, perhaps with a big picture of a traffic light to indicate green for the do's and red for the don'ts. Little pictures of traffic lights can also be taped to each desk. Now the class is divided into groups that compete with one another to be good. For example, their challenge might be to do their classwork for ten minutes with fewer than three don'ts. The bar is set low enough so that at least some groups will be able to hop over it. Every group that succeeds gets a small prize, while the groups that fail have to watch. One of the most effective prizes is permission to misbehave. For example, members of winning groups might get to run around the classroom while the members of the losing groups remain in their chairs.

It turns out that group rewards are often more potent motivators than individual rewards. We have already seen how our behaviors are powerfully shaped by social reinforcers such as smiles, laughs, and frowns. When students compete in groups to be good, then good behaviors are reinforced by the students in addition to the teacher. Brilliant!

At first, the game is played for short periods of time, followed immediately by the reward. Gradually, the game is played longer, more often, and with more delayed rewards, such as the end of the day or the week. Sometimes the game is played without telling the students when it begins, simply announcing after a period of time which groups won. Between games, the teacher reinforces the do's and don'ts with abundant positive reinforcement and mild punishment. Rather than calling everyone's attention to someone who is misbehaving, for example, the teacher might approach the student's desk and discreetly tap on the red light. If the student persists, then removal from the group might be the next step. Punishment is sometimes necessary, but it must be mild in comparison with the rewards, and the teacher is bound by the same rules as the students. It's not right to punish by yelling or demeaning, for example. Kids have a sense of fairness just as much as adults.

In this fashion, the norms of good behavior that are agreed on by consensus become the culture of the classroom. This might seem too good to be true, but prevention scientists are proud people who always back their claims with numbers. One of the most comprehensive implementations and assessments took place in Baltimore, Maryland, one of the toughest public-school systems in America. First- and second-grade classrooms were randomly assigned to experimental (implementation of the GBG) and control (normal classroom practices) treatments. Members of the assessment team were present in the classrooms to count misbehaviors before, during, and after the implementation of the game. Before, these were classrooms from hell, with an average of nearly 100 disruptions per hour. When the game was first implemented, the disruptions plummeted to near zero as the kids competed in groups to be good. When the game wasn't being played, the disruptions crept upward but remained lower than before the implementation in the experimental classrooms and the continuing mayhem in the control classrooms. As the game was continued and more bells and whistles were added to the reinforcement scheme, the classes permanently settled into a culture of good behavior.

The GBG was implemented in the first and second grades only, but the students were followed as they matured and in fact are *still* being followed by those tireless prevention scientists. Astonishingly, the GBG has lifelong benefits. At the end of the sixth grade, the GBG kids were less likely to be diagnosed with conduct disorder, to have been suspended from school, or to be judged in need of mental-health services. During grades six through eight, they were less likely to use tobacco or hard drugs such as heroin, crack, and cocaine. In high school, the GBG kids scored higher on standardized achievement tests, had a greater chance of graduating and of attending college, and had a reduced need for special-education services. In college, the GBG kids had a reduced risk for suicide ideation, lower rates of antisocial personality disorder, and lower rates of violent and criminal behavior. The GBG was especially effective at improving the lives of boys. All of the above-cited results are statistically significant and can be attributed to the effect of the GBG because the students were randomly assigned to the two treatment groups. We know that the kids in the treatment groups were similar before the intervention, giving us confidence that the intervention *caused* the differences.

This cornucopia of benefits might seem inconceivable until we remember that kids are like plants, and providing the right growth conditions can make a huge difference to their development, like money in the bank earning compound interest. A gifted teacher such as Miss A had a lifelong effect on her students. The GBG has the same kind of effect, probably for some of the same reasons, but in a way that can be understood, replicated, and improved in a systematic fashion.

The GBG is one of the greatest success stories in the annals of prevention science. I used it as my second case study to stress its similarity with ACT as a managed variation-and-selection process. Just as ACT begins with an examination of personal goals, the GBG begins with examination of do's and don'ts for the class. Both establish the behaviors that are to be selected by a conscious and deliberative process. Then, like the martial art of jujitsu, which harnesses the energy of one's opponent, the learning and copying instincts that caused the problem when unmanaged are harnessed to provide the solution. The GBG is ACT for the classroom.

I also pointed out that the culture of cooperation established by the GBG is more normal than the typical classroom environment. We are designed as a species to cooperate in small groups that are coordinated and policed by norms established by consensus. We hate being bossed around and distrust the assurance that it is for our own good. Any group of adults attempts to establish this kind of guarded egalitarianism unless circumstances dictate a more hierarchical structure. Even then, "leaders" are held accountable by their "followers" to a remarkable degree. The GBG creates a social environment for children that adults spontaneously create for themselves whenever they have the chance.

GLANCING AT MY WATCH, I was shocked to see that my time was running out. I'd need to talk like an auctioneer to finish on time.

My third case study was about underage smoking. In America, federal agencies regulating tobacco sales employ underage kids as secret agents who enter retail stores and attempt to purchase cigarettes. When they are successful more than 20 percent of the time in a given state, the state is put on notice that it stands to lose millions of dollars provided by the federal government in the form of block grants. Wyoming and Wisconsin were in this dilemma, with cigarette sales to minors hovering above 30 percent. They came to Dennis and Tony for help. Since the states had millions of dollars to lose, they were happy to throw money at Dennis and Tony to fix the problem. But how do you change a cultural practice at the scale of an entire state? Even if you succeed, what impact will a reduction in tobacco sales to minors have on reducing their use of tobacco, which is the ultimate goal? After all, kids are resourceful and can get access to tobacco in many ways—by borrowing from an older person, getting an older person to buy for them, or stealing from parents and stores. Restricting one source of cigarettes might merely increase reliance on other sources without affecting use.

Not only did Tony and Dennis succeed, but they accomplished the job in sixty days and have the numbers to prove it. Here's what they did.

Their first step was to build a consensus against illegal sales. Dennis and Tony made the rounds of key legislators, state department heads, and other important people to stress the need for action. Even though most of these people had a genuine interest in the long-term welfare of their constituents, the immediate danger of losing millions of dollars in federal support was a more powerful incentive. Antitobacco organizations and other stakeholders were also brought into the process, resulting in a declaration endorsed by leaders at the state level that could then be endorsed by leaders at each locality within the state.

The declaration was publicized by an advertising campaign using the same techniques that are so effective at marketing cigarettes—social branding, rather than product branding. TV and radio commercials portrayed a convenience-store clerk being rewarded for doing the right thing. Slogans were invented, such as "Wyoming Wins!" Political figures and celebrities endorsed the cause. Owners of retail outlets were informed of the consensus and provided with materials to distribute to their clerks.

All of this was required to establish the criteria for selecting behaviors, just as with ACT at the individual level and the GBG in a single classroom. Much more effort was required to establish a meaningful consensus at the scale of an entire state, but it could still be done, as Tony and Dennis were able to demonstrate.

Now that "the right thing" was clear in everyone's mind, the next task was to reinforce the right thing by making our instincts for learning and copying work for us rather than against us. Tony and Dennis created task forces with their own underage secret agents who attempted to buy cigarettes. Clerks who turned them away were richly rewarded with coupons from local businesses, articles in the local newspaper, and their picture on the wall of the store. Clerks who obliged were mildly punished with a reminder to uphold the law. In an especially brilliant move, Tony and Dennis held a contest among the Wisconsin clerks for the most clever thing to say when faced with a minor wanting to buy cigarettes. After all, it's a difficult situation. Imagine that you're a convenience-store clerk facing a group of tough-looking sixteen-year-olds who might vandalize your car if you refuse them. The winning entries were printed in the form of cards that could be handed to the underage customers, which were provided to all the clerks. One of the cards reads: "I don't think so. Folks like me make about $7 an hour. If I sold tobacco to you, which is illegal, I could get fined $500. I'd have to work 107 hours to pay for that. That's about 2½ weeks full-time. How many shifts will you work to help me?" Hooray for variation-and-selection processes!

This "reward and reminder" procedure was designed to be implemented in sixty days. The scale of the operation is difficult to fathom. Wyoming has about 550 tobacco outlets, and Wisconsin has about 13,000. Every one was visited by underage secret agents, who recorded the data electronically with handheld devices, resulting in the creation of a massive database. This information was combined with data collected by other federal and state agencies on the incidence of tobacco use in underage youth and how they get their tobacco. The monitoring, rewarding, and reminding continues to this day. I am reminded of the specialized cell types of the immune system, continuously circulating in the body to detect and stamp out disease organisms.

Here's what the numbers say. Before the intervention, federally reported average illegal sales of tobacco to minors were 43 percent in Wyoming and 35 percent in Wisconsin. After the intervention, those numbers declined to 10.8 percent and 8.1 percent, and they have remained stable to the present. At the very least, Tony and Dennis saved millions of dollars of revenue for the two states in the form of block grants that would have been withheld otherwise, making the cost of their program a huge bargain.

Did their intervention merely cause underage kids to get their tobacco from elsewhere? According to self-report surveys given to underage kids every year by federal agencies (not by Tony and Dennis), the intervention decreased access to tobacco from all sources, not just direct purchases. Underage kids were finding it harder to borrow from adults or have adults buy for them. They were even stealing tobacco less. All that publicity about doing the right thing, coupled with the inconvenience and added expense of borrowing and buying from adults, decreased the actual incidence of daily smoking in underage youth from 14.5 percent to 9.2 percent in Wyoming and from 13.3 percent to 7.6 percent in Wisconsin. As scientists to the core, Tony and Dennis compared these declines with nationwide averages, analyzed the intercepts and slopes of the Wyoming and Wisconsin data to confirm a causal effect of their intervention, and concluded their analysis by lamenting that they were unable to choose the states and years randomly for the intervention!

In their final tally, Tony and Dennis estimate that their intervention resulted in millions of packs of cigarettes not being sold to minors, resulting in a gross sales loss of more than $4 million in Wyoming and $69 million in Wisconsin. That's a big financial hit for the store owners, yet they gladly accepted it, not only because they were doing the right thing but also because they were socially rewarded for doing the right thing in a way that made full use of their natural-born instincts for learning and copying. Money isn't everything, but the other things must be communicated through our senses.

What Tony and Dennis accomplished at a statewide scale takes place naturally at a small scale. For our hunter-gatherer ancestors, most challenges to survival were obvious, a consensus was established around the campfire, and social rewards and punishment took place through the spontaneous expression of emotions. What comes naturally at a small scale does not happen automatically at a large scale. Something must be constructed at a large scale that interfaces with our genetically evolved instincts for learning and copying. If that something isn't added, then large-scale society won't work well. We will be like the solitary wasp whose ring of pinecones has been displaced, like the wasp colony whose offspring can't feed the adults, like the immune system destroying its own body. Tony and Dennis knew how to make a large group function like a small group, preventing hundreds of smoking-related deaths. How many other problems faced by large-scale society might be solved in the same way?

* * *

AT LAST, MY LECTURE WAS OVER, and my dazed audience filed out of the room. A single lecture was not enough to unite evolutionary theory and prevention science. Even more discouraging, prevention science is only one of thousands of islands in the Ivory Archipelago. Science is a large-scale society, no less than everyday life. Left to their own devices, scientists are no more likely to become united at a large scale than anyone else. They will be selected by consequences, rewarded for getting grants and writing papers, becoming really good at some things and phasing everything else out. Their groups will drift apart, and their isolation will inevitably result in a divergence of what they mean when they use the same words. It will be a stretch for them to form tribes of even a few thousand. Something must be constructed at a larger scale to turn the Ivory Archipelago into the United Ivory Archipelago. That something requires a common language, in the form of a common theoretical framework, and evolutionary theory is the only framework that applies to all living processes.

At least, Tony, Dennis, and I were speaking the same language. We had become the Three Amigos, ready to take on neighborhoods, cities, and the world.

CHAPTER 14

Learning from Mother Nature about Teaching Our Children

ONLY A YEAR AFTER the Evolution Institute was conceived, Jerry Lieberman and I held our inaugural workshop at the University of Miami. I was amazed by how fast it came together. We quickly decided to focus on the topic of childhood education, an issue highly relevant to Binghamton and all other cities, as a proof of concept for how any public-policy issue can be approached from an evolutionary perspective. Jerry quickly raised the funds. I quickly located the evolutionary expertise. Now the dream of an evolutionary think tank was about to become a reality.

We were meeting at the University of Miami thanks to William Scott Green, UM's newly appointed senior vice provost. Bill is a religious scholar by training who spent most of his career at the University of Rochester. He's such a bundle of energy that he became dean of undergraduate studies at UR while maintaining his very active academic career. We met when he read my book *Darwin's Cathedral* and invited me to talk to one of his religious-studies classes. I was flattered. In *Darwin's Cathedral,* I was brashly sailing to yet another region of the Ivory Archipelago — the study of religion — and claiming it for evolutionary theory. You'd think that the natives would grab me and put me in a stew pot, but Bill was a friendly native. Moreover, he was the perfect guide to this part of the Ivory Archipelago, with an expansive knowledge of the great world religions in addition to his detailed work on Judaic studies. Bill and I became good friends and started to collaborate. In his new capacity as senior vice provost at UM, he loved the idea of an evolutionary think tank and offered to host our first workshop. Knowing Bill, I was confident that he would do it in style.

I felt like a person who had been granted a wish by a genie. I asked for an evolutionary think tank, and *poof!* There it was. I left my city of Binghamton on a cold cloudy morning, and *poof!* Here I was in warm, sunny Florida. Last night, I was eating leftovers in my rumpled clothes, and *poof!* Tonight I was in my best jacket and tie, drinking cocktails and eating fine food at the reception that Bill had organized on the eve of the workshop.

As we know from so many folktales, people often end up regretting what they wished for. Now that my wish had come true, I realized how much I had taken on faith and how spectacularly the workshop might fail. We all desire a quality education for our children and our nation as a whole. Billions of dollars are spent to achieve it, including millions of research dollars for scientists to study it. There are more theories of education than flavors of ice cream. Despite all of this goodwill, expense, and expertise, many schools flunk on the basis of their performance, and few earn an A+. Wasn't it presumptuous to claim that evolutionary theory could succeed when so many previous efforts had failed? Wasn't it absurd to think that David and Jerry could fell this Goliath of a problem with a single workshop, like a single stone hurled from a sling?

To make matters less certain, I had studied religion long enough to have a strong sense of how it could be approached from an evolutionary perspective, but I had not studied education with the same thoroughness and had to rely on the experts I had convened. They shared a common interest in evolution but came from different islands of the Ivory Archipelago with respect to their disciplinary training. Some of them knew one another, but others would be newly meeting and encountering one another's ideas. Tribal warfare or incomprehension could break out at any time.

To make matters even less certain, Bernard was still away fighting his own battles, but other important observers would be attending the workshop. Jerry had invited other potential donors. Bill had invited the dean of UM's well-regarded School of Education and several of its most distinguished faculty to sit in. These folks would be encountering evolutionary theory in relation to their subject for the first time and might well feel territorial. Bill's reputation was on the line with his new colleagues at UM. He would look pretty foolish if the workshop turned out to be a dud. I could tell that Bill was feeling the pressure when he pulled me aside just before the workshop was about to begin on the first morning and told me with uncharacteristic sternness that *this had better work.*

The seven experts at the workshop included four who knew one another well and whom I also knew either personally or by reputation: Daniel Berch, David Bjorklund, Bruce Ellis, and David Geary. They were interdisciplinary but spent lots of their time sailing in the vicinity of evolutionary psychology. I was eager for them to meet my two amigos from prevention science, Tony and Dennis, who were equally eager to join the fray. The seventh participant was Peter Gray, a psychologist who wrote the first introductory psychology textbook that prominently featured evolution. The other participants might have known about Peter's textbook, but they didn't know about his interest in childhood education. For most of his career, Peter conducted research on the neuroendocrine system of rats. He didn't start thinking about childhood education until his son Scott began to experience problems in elementary school. At first, Peter was merely trying to help his son, but the solution that he found caused him to appreciate just how much we can learn from Mother Nature about teaching our children. It was only by chance that I heard Peter tell his story at a conference a few months before the workshop. I begged him to join us, and he graciously accepted my invitation at short notice. As far as everyone else at the workshop was concerned, Peter was a wild card.

PETER IS SLIM AND BOYISH-LOOKING, like a white-haired Peter Pan. His ancestors were among the earliest settlers of New England, but Peter grew up with his mother and stepfather in Minnesota and Wisconsin. His stepfather was a printer at a time when this was a skilled occupation. His mother had a restless personality that caused them to move often. She dreamed of running a local newspaper, so when

Peter was ten, they moved to a village with a population of 300 optimistically named Hill City, Minnesota. The local newspaper was for sale, and Peter's mother saw this as a great opportunity. The business was doomed from the start, forcing them to leave two years later, but Peter has only fond memories of his short time there. He helped his stepfather run the printing press and even wrote a column on Boy Scout news. Peter enjoyed moving frequently. In every town, the children seemed to engage in a different activity—playing baseball, making kites, shooting marbles. Whenever Peter moved to a new town, he immersed himself in what the other kids were doing until he got good at it, which caused him to be quickly accepted by them. Fishing was popular wherever he went, and Peter remembers all-day adventures with his friends during the summer, catching crappie, northern pike, and the coveted walleye pike. At that time, a lake was likely to have a communal boat that anyone was welcome to use.

Peter moved with his family to the tiny village of Cabot, Vermont, when he was midway through high school. This was during the Sputnik era, and kids were being encouraged to become scientists and engineers. Peter wanted to take physics, but the school's single science teacher, who had gone to Vermont Agricultural School, wasn't up to the task. Fortunately, he wasn't threatened by bright students, and Peter ended up teaching the class to the other students. Peter discovered that he learned much better by mastering the material himself and explaining it to the other students than by passive learning.

Peter knew that he wanted to attend college but that his parents couldn't afford it. He therefore worked hard and became the prodigy of Cabot—editor of the school paper, president of the senior class, captain of the basketball and debating teams. Being the biggest fish in a small pond proved to be a successful strategy, and Peter was admitted to Columbia University in New York City with a full scholarship. A freshman psychology course that taught Skinner's radical behaviorism appealed to Peter as a "physics of the mind," combining the scientific rigor that made physics appealing with the big questions about understanding and improving the human condition to which he was also drawn. He soon realized the limitations of radical behaviorism, however, and began to think of psychology as a biological science, despite what his professors might say.

When Peter graduated from Columbia, he was tired of taking courses and applied to join the Peace Corps. Then he heard about a new graduate program at the elite Rockefeller University. Rockefeller is more of a scientific institute than a university. The professors earn huge salaries to do research and have no teaching obligations at all. The new program would enable twenty graduate students, covering all fields of science, to learn in the presence of these scientific gods. There would be no distinction between biology and psychology and no formal courses. Evidently, the students would just shoot science directly into their veins.

Peter was admitted to this ultra-elite program and spent the next few years learning science the same way he learned baseball, kite making, and marble shooting as a boy. The scientists at Rockefeller were at the forefront of the cognitive revolution and animal-behavior research, stretching Peter's mind beyond behaviorism. He started doing research immediately, learning the necessary techniques as they became needed. He had plenty of time for reading and could choose what to read rather than being given mandatory assignments. He especially loved tracing current ideas to their sources by reading the classic scientific literature, such as Charles Darwin, William James, and Walter Cannon.

Peter landed a good job at Boston College but found himself drawn more to teaching than to research. Research questions were too specialized to engage his interest for long. He decided that if you can't describe your work to a layperson in a way that he or she finds interesting, perhaps it isn't interesting. Unlike most professors, who prefer research and teaching advanced courses, Peter preferred teaching the introductory psychology course and came to think that he had an "introductory mind." His niche was to explain the fundamentals to students encountering them for the first time. He would write an introductory psychology textbook.

In the sixth edition of Peter's textbook (*Psychology*), a section titled "The Idea That the Machinery of the Mind Evolved through Natural Selection" begins on page 8. Placing evolution at the forefront was so radical in the late 1980s, when he was writing the first edition, that he had to fight with his publisher to retain it. Peter's version of evolutionary psychology was largely independent of the version that Leda Cosmides and John Tooby were formulating at the same time. Since Peter was merely the author of an intro psych text and Leda and John were announcing a new revolution, Peter's version was less influential, but in retrospect, it was more balanced.

Peter was sailing through life on an even keel until his son Scott began to develop problems in elementary school. Scott had a voracious mind, as one might expect with his father and equally intellectual mother. Yet when Scott started school, he began to rebel almost immediately. He didn't necessarily do things that the other children were doing or the way the teacher said to do them. He would solve math problems in a different way from how the teacher instructed, which would be marked wrong even though he got the right answer. He wouldn't do an assignment at all if he thought it was silly. He began to write in all lower-case letters. like poet e. e. cummings, just to rebel. The school started testing him for psychopathology, and the other students started to peg him as weird.

The problem culminated in a meeting when Scott was in fourth grade. One little boy was surrounded by a circle of adults — his parents, the principal, his teacher, the school guidance counselor, an external guidance counselor — all telling him to behave.

"Go to hell!" the defiant boy replied.

Peter and his wife burst into tears. At that moment, Peter realized that the school, not his son, might be the problem and that he needed to take his son's side. But what was the alternative? Homeschooling had not yet become common, and private school didn't seem much different from public school.

BY SHEER CHANCE, Peter lived close to an alternative school called the Sudbury Valley School, which was founded in 1968 by physicist Daniel A. Greenberg. In an odd intertwining of fate, at about the same time that Peter was becoming tired of classes at Columbia University, Greenberg was becoming disillusioned with academic life as a professor. His students were interested primarily in their grades, and his colleagues seemed like rats pressing a bar in their incessant quest for grants and publications. Aristotle said that "the human being is by nature curious," but this motive seemed to have been beaten out of students by the time they arrived at college. Greenberg felt so strongly about this that he walked away from his job at Columbia and moved to Sudbury, a relative wilderness at the time, to indulge his own curiosity about human nature, human history, democratic values, and education. He started the Sudbury Valley School with like-minded associates when his own children reached school age.

The Sudbury Valley School is located on an old farmstead with a big house, a barn, and lots of land. Students make their own choices and have the liberty (and responsibility) to do what they wish with their time, as long as it is not disruptive to others or to the school community. The rules of the community arise from the democratic institutions of the school, described in this way on its Web site:

> The school is governed on the model of a traditional New England Town Meeting. The daily affairs of the school are managed by the weekly School Meeting, at which each student and staff member has one vote. Rules of behavior, use of facilities, expenditures, staff hiring, and all the routines of running an institution are determined by debate and vote at the School Meeting. At Sudbury Valley, students share fully the responsibility for effective operation of the school and for the quality of life at school.
>
> Infractions of the rules are dealt with through the School Meeting's judicial system, in which all members of the school community participate. The fair administration of justice is a key feature of Sudbury Valley and contributes much to the students' confidence in the school. Parents participate in setting school policies. Legally, the school is a non-profit corporation, and every parent becomes a voting member of the Assembly, as the corporate membership is called. The Assembly also includes students, staff, and other elected members. It meets at least once a year to decide all questions of broad operational and fiscal policy.

That's right, everyone in the school has an equal vote, from a four-year-old on up, and every major decision is voted on, including the hiring and firing of the teachers. Rules must be obeyed, but they are agreed on by consensus, and the judicial process is as fair as the decision-making process. The governance of the Sudbury Valley School is like an extreme version of the Good Behavior Game.

Apart from the rules of conduct passed democratically by the school community, the school does not structure the lives of the students in any way. Instead, the school relies on students to structure their own lives in their own time and to develop self-discipline. Students are left free to develop as they will, as full members of the community, living, working, and playing according to their own wishes — as if on a summer day in Minnesota. The adult staff is primarily responsible for caring for the institution, like the professionals hired by a town to care for the town. They manage the school's building and plant, respond to emergencies, tend to scraped knees, and oversee the purchase of equipment that the School Meeting has decided on. To the degree that they teach, they do so by responding to students' questions and requests. Partly because the students learn from one another in this age-mixed setting and partly because the School Meeting is very careful about expenditures, the school can operate at a fraction of the cost of traditional schools, with a tuition in 2011 of less than $7000 per student. That's less than half the cost of a public-school education, not to speak of the $40,000 price tag of an elite boarding school.

Scott was wild to attend this school, and Peter reluctantly agreed. Despite his own liberal leanings, it sounded too good to be true. What would prevent the students from goofing off? They might learn the fun stuff, such as computer games, but how about the tough stuff, such as math? How could they get into college without grades? What kinds of careers would they have? Peter began to study the school out of concern for his son. He did a comprehensive survey of the alumni and discovered that their college admission rate and adult careers compared very favorably with those of other schools. As his concern for his son subsided, his intellectual interest in the school grew.

He began to think more deeply about the nature of education. We are a cultural species. Our capacity for culture evolved by genetic evolution. Given that capacity, most of our behaviors come from our cultures, not directly from our genes. We are such a cultural species that our life cycle is stretched out. We remain immature until our late teens and live into our eighties to give children extra time to learn and adults extra time to teach. How does education take place in other cultures? How does it take place in hunter-gatherer cultures, which was the lifestyle of all humans before the advent of agriculture?

When Peter consulted the anthropological literature, what he found looked a lot like the Sudbury Valley School. There was very little that resembled formal education. Kids ran around in mixed-age groups. The older kids were strongly motivated to become adults, and the younger kids were strongly motivated to be like the older kids. Most learning took place in the context of play and practice. Adults provided instruction when asked, just like the adult staff of the Sudbury Valley School. Against the background of the anthropological literature, it was *our* system of formal education that seemed weird.

Mixed-age interactions seemed especially important for this kind of spontaneous education. Learning is a step-by-step process. Learning the next step is easy, but learning ten steps ahead is impossible. Adults are many steps ahead of small children, but slightly older children are a single step ahead. Kids love to be in the company of slightly older kids, whereas adults can be threatening. And every teacher knows that the most effective way to learn something is by teaching it, so older kids gain by teaching younger kids.

You might think that bullying would be a problem, but it turns out that mixed-age interactions tend to moderate bullying. A fourteen-year-old might try to bully other fourteen-year-olds to take their stuff, establish dominance, or impress the girls, but he's not going to bully an eighteen-year-old. Eighteen-year-olds aren't going to bully other eighteen-year-olds when eight-year-olds are in the vicinity. Same-age interactions bring out the competition, and mixed-age interactions bring out the nurturance in kids. Mixed-age interactions also accommodate individual differences. Slow learners can proceed at their own pace without being stigmatized. There is elbowroom to make the most of one's strengths and avoid exposing one's weaknesses.

THE INSIGHT THAT MOST CULTURALLY acquired information is transmitted across generations spontaneously, without requiring a formal system of education, stunned me when I first heard Peter tell his story. My mind flooded with questions that never would have occurred to me otherwise. I found it difficult to shake the conviction that it was too good to be true. Yet the more I thought about it, the more it accorded with my own experience. I hated formal classes, and everything I did well had been learned in the context of play or practice at something I really wanted to do. Oddly enough, while K-12 and college education departed from the hunter-gatherer mode, graduate education returned to it. Most of what I learned in graduate school was acquired not from lectures but from older graduate students and other people who knew the next thing I needed to learn. I was always guiding the process. No one knew my next step better than I.

This was true even for the most difficult subjects, such as mathematics. I took only one math course in college — freshman calculus — and it almost killed me. In graduate school, I had a strong reason to learn math, so I did. I purchased *Calculus for Dummies,* practiced hard, and pestered more knowledgeable graduate students when I got stuck. It wasn't exactly fun, but every time I figured something out, I had a feeling of triumph that motivated me to take the next step. I published my first theoretical paper while still a graduate student, and now I'm a well-known theoretical biologist. Peter received the most elite graduate education possible, and it came closer to the hunter-gatherer mode than his less advanced education. Whenever my evolutionist colleagues and I decide to study a new subject or organism, which is often, we learn in hunter-gatherer mode, not by taking formal courses.

When I learned more about the Sudbury Valley School, I realized that it was essentially making graduate-style education available to younger kids. Here's a story that appeared in a 2006 article on the Sudbury Valley School published in *Psychology Today:*

> Outsiders commonly choke upon hearing that no one even teaches reading. Sometimes insiders get a bit antsy, too. When Ben was in the second or third grade, anxiety temporarily overtook his well-read father, who offered the boy a dime for every 15 minutes he'd spend reading at home. Ben accepted the bribe long enough to prove that he could do it. But true to the Sudbury spirit, his reading proficiency took a huge leap forward only after he began playing with airplanes and then an electronic flight simulator—because that led him to read the flight manual. And that led to the discovery of flight simulator communities on the internet, which led to mock airplane battles, which led to communicating with squadron leaders, which led to spelling and writing, which ultimately got Ben into Swarthmore, where he is now finishing his freshman year.

Is this any different from the way I learned math or Peter learned advanced experimental techniques?

Once I began to accept the possibility that education at all levels can be this spontaneous, another mystery confronted me: How can our formal educational system go so wrong? For example, why do we segregate kids by age when mixed-age interactions are so beneficial? The answer, I quickly realized, is that age segregation has a surface logic, and the negative consequences are not easily traced to the cause. Let's say that you're a school administrator who decides to segregate kids into classes by age because it makes the bookkeeping easier. That's an obvious benefit. After several months, you learn to your dismay that academic performance has declined and an epidemic of bullying has broken out. The connection between those problems and your age-segregation policy is not obvious. You might well decide to continue the policy on the strength of the obvious benefit, oblivious to its hidden costs.

Policies are like wishes. You make one, and *poof!* It comes true, and the consequences might not be what you had in mind. Suppose that your policy for increasing academic performance is to cut back on recess so that the kids spend more time in class. *Poof!* Your wish has come true, but in fact you made things worse, because kids need to move, and learning takes place best in the context of play. Now that some kids are squirming uncontrollably in their chairs, suppose your policy is to medicate them. *Poof!* Your wish has come true, but now you are meddling with extremely complicated brain processes to solve a problem that could have been solved environmentally. Suppose that you implement a no-touch rule to avoid sexual harassment. *Poof!* Your wish has come true, but you didn't reckon on the fact that people, like other social primates, become physiologically stressed when they aren't touched. Suppose you decide that student progress must be monitored and implement a system of standardized tests. *Poof!* Your wish has come true, but the entire educational system becomes selected by consequences to focus on test scores, to the detriment of other forms of learning. *Poof! Poof! Poof!* Our educational system is like a folktale about wishes gone horribly wrong.

How can we find our way out of this maze of unforeseen consequences? It might seem that the best practices would prevail over the long run. Isn't that what cultural evolution is all about? Perhaps, but cultural evolution must be managed to function properly in modern society, as we have seen. The Sudbury Valley School has become a model for about thirty-five other schools worldwide since its inception in 1968. That's pretty good, but centuries will be required for it to become truly widespread at that rate. Montessori schools feature mixed-aged interactions, child-directed learning, and learning in the context of movement and play. That's pretty good, but why aren't these practices more widespread since Maria Montessori developed them more than 100 years ago? Experienced public-school teachers might arrange to have fifth-grade students read to the first-graders. That's pretty good, but why aren't child-mentoring programs implemented more widely in public schools or the problems of same-age interactions more generally appreciated? Once again, we are faced with the paradox of practices that work but don't spread.

This is where a theory becomes useful. Peter's story demonstrates the usefulness of a theory that shows *how* good practices work, rather than just knowing *that* they work. When his son was experiencing problems, Peter was lost in a maze of unforeseen consequences. He didn't know if the problem was caused by his son, his own parenting skills, the school, something in the water, or an infinitude of other potential causes. The Sudbury Valley School provided a solution. Peter demonstrated that the school worked when he carefully studied the alumni, but he still didn't know how it worked. Daniel Greenberg, the school's founder and principal architect, designed the school and therefore had strong ideas about how it worked. Greenberg had a theory that he developed on the basis of his deep reading and reflection, which emphasized the principle of democracy and Aristotle's belief that "the human being is by nature curious."

Greenberg's theory was good enough to design a great school, but Peter took theorizing to a new level. Philosophers such as Aristotle, political theorists such as the American forefathers, and educational theorists such as John Dewey based their ideas on their understanding of human nature. Peter was providing an updated conception of human nature, based on current scientific knowledge. Peter could go beyond Aristotle's observation that human beings are by nature curious. Peter could say that we are designed in a more complex way to learn as individuals and to acquire information from others. Political theorists traced democracy to the Greeks. Peter traced democracy to fiercely egalitarian instincts that evolved by genetic evolution. The wise people of the past would have made use of this information had it been available to them. So should we.

Peter was able to show that our instincts for learning and teaching functioned spontaneously for most of human existence, without requiring a formal system of education. How they work today depends on the environment that we construct. The Sudbury Valley School constructs a benign environment, enabling us to do what comes naturally to us. Our current educational system constructs a hostile environment in many respects, resulting in the dumbfounding of our instincts, like the immune system attacking its own body. Intervention programs such as the Good Behavior Game vastly improve the formal educational environment using some of the same principles as the Sudbury Valley School. They can be regarded as a renovation of a previous construction, rather than a new construction from the ground up.

Once we think of evolutionary theory as an updating of our conception of human nature, something that the great philosophers and social theorists of the past would have been eager to do themselves, then it becomes an essential tool for diagnosing and solving the problems of modern education. Moreover, it begins working immediately. Once you view the world through an evolutionary lens, it's difficult to look away. When I heard Peter tell his story at the conference preceding my workshop, he was the banquet speaker. Even the waiters clearing the tables stopped to listen, nodding their heads in agreement, not only because Peter is a fine speaker but also because his message made so much sense. It even seemed like common sense, except that there was nothing common about the Sudbury Valley School.

AT THE MIAMI WORKSHOP, seven experts had two days to provide a more comprehensive update on human nature and how it can be used to improve childhood education. The twenty observers were welcome to join the conversation on the basis of their own disciplinary expertise. We met in a room with a large central table and additional chairs against the walls. A projector screen was at one end, and a cameraman was on hand to videotape the presentations that would begin the workshop. Coffee, juice, pastries, and freshly cut fruit were available on a side table. Bill knew how to do things right and was sitting next to the dean of UM's School of Education, waiting for me to hold up my end of the bargain. Jerry looked relaxed as he circulated around the room, but he must have been as keen with anticipation as I was to see how the Evolution Institute would fare on its maiden voyage.

I set the stage with a short presentation on evolution in relation to human affairs and the need for an evolutionary think tank. It seemed to serve as a good pep talk, judging by the smile on Bill's face. Next up was David Geary, who holds an endowed chair at the University of Missouri and whose books include *The Origin of Mind: Evolution of Brain, Cognition, and General Intelligence*; *Male, Female: The Evolution of Human Sex Differences*; and *Children's Mathematical Development*. David was not an outsider like me but a highly distinguished insider in the fields of child development and education. He had served on the President's National Mathematics Panel and the National Board of Directors of the U.S. Department of Education's Institute of Education Sciences. When he said that educators can learn from evolutionary theory, educators had good reason to listen. Yet despite his distinguished credentials, David's perspective was so new and the field of education is so large that most of the observers in the room were hearing it for the first time, including faculty from UM's School of Education who had distinguished reputations in their own right.

David provided a broad overview of human brain evolution and what it means for modern education. In addition to covering some of the points that I have already provided through Peter's story, David made a distinction between primary and secondary knowledge. Primary knowledge is learned easily and spontaneously because it was present in our ancestral environment. Secondary knowledge is learned slowly and arduously because it didn't exist in our ancestral environment and we are not genetically adapted to learn it. Spoken language is a form of primary knowledge that does not require formal education. Written language and mathematics are examples of secondary knowledge that didn't exist until only a few thousand years ago. David conjectured that it will always be difficult for people to learn reading and mathematics, just as it will always be difficult for a bear to learn to ride a bicycle. The reason formal schooling is necessary today, in contrast to hunter-gatherer societies, is that modern societies require so much secondary knowledge.

The next speaker was Bruce Ellis, who holds an endowed chair at the University of Arizona and with David Bjorklund wrote the pioneering book *Origins of the Social Mind: Evolutionary Psychology and Child Development*. Bruce added a new dimension to David Geary's presentation by focusing on individual differences, including differences in resilience. Why are risk factors such as poverty and parental neglect so damaging to some kids, while others emerge unscathed? What can we do to make the vulnerable kids more resilient? Bruce surprised his audience by showing that this is the wrong question. The essential individual difference is reactivity to one's environment. Less reactive kids don't pay much attention to their environment, whether it is good or bad. More reactive kids pay a lot of attention to their environment, with variable outcomes. They do worse than less reactive kids in harsh environments, but they do better in benign environments.

Bruce used the terms "dandelion child" (an actual phase in Swedish) and "orchid child" to describe these differences in reactivity. Dandelions thrive even when you run over them with a lawn mower. Orchids require special conditions to grow, but then they become objects of special beauty. Just as dandelions and orchids can both be found in nature, dandelion children and orchid children can both be found in human life. Moreover, the human individual differences are in part genetically based. Modern molecular genetics makes it easy to know when we differ at specific genetic loci. Some of these loci influence how our brains work, such as the DRD4 locus, which alters the sensitivity of our nerve cells to the neurotransmitter dopamine. Individual differences in reactivity can be partially traced to genetic differences at this locus, not only for humans but throughout the animal kingdom. There are dandelion and orchid rhesus monkeys, even dandelion and orchid birds. Genetically based individual differences in reactivity have been maintained on the basis of their respective costs and benefits since long, long before our species existed on earth.

Bruce's talk showed how transformative the evolutionary perspective can be, even on the basis of a twenty-minute presentation. When highly reactive children are seen as vulnerable to harsh environments as if this is the whole story, their reactivity seems pathological, and trying to make them more resilient seems like a sensible goal. The surface logic is compelling, but evolutionary thinking immediately makes it suspect as the whole story, because traits that are just plain costly don't stick around very long on the evolutionary stage. Might there be some benefits to reactivity that balance the costs? This is merely an intelligent guess that need not prove to be correct, but in this case, it leads to the discovery that reactivity is a strategy, not a sickness. Evolutionary thinking also encourages comparisons across species, something that the average expert in education would never think of doing. Once again, this is merely a fertile line of inquiry that can lead in many directions. In this case, it leads to the discovery that individual differences in reactivity are found throughout the animal kingdom and even have a common genetic basis.

This new information transforms, or should transform, educational practice. Orchid children don't need to be cured; they need to be placed in environments that enable them to bloom. Dandelion children won't necessarily respond to the enrichments that orchid children thirst for. These individual differences are not entirely genetically based. Perhaps it's possible for orchid children to become dandelion children and vice versa, but it might not be as simple as exchanging a red shirt for a blue shirt. Moreover, children shouldn't necessarily be expected to change their strategies, any more than a tortoise should be expected to change into a hare. Our educational system needs to accommodate and capitalize on individual differences, not eliminate them.

The third expert to speak was David Bjorklund, a professor of psychology at Florida Atlantic University whose books include *Why Youth Is Not Wasted on the Young: Immaturity in Human Development*; *Children's Thinking: Cognitive Development and Individual Differences*; and *The Origins of Human Nature: Evolutionary Developmental Psychology* (with Anthony Pellegrini), in addition to his book coauthored with Bruce. David's talk at the workshop included a cautionary tale about birds. Birds hear sounds while they are still developing in the egg, but they don't see objects until they hatch. All sensory systems require environmental inputs to develop, so the auditory system of birds is designed to develop before the visual system. This normal course of events can be circumvented by cutting a window into the egg and prematurely exposing the developing bird to light. Premature visual development interferes with auditory development, and the chicks can't discriminate their mother's call from other sounds in their environment.

What do birds in eggs with a view have to do with human development? Consider a product called BabyPlus, a "fetal enrichment technology" that fits around a pregnant mother's belly and plays sounds that are supposed to enhance brain growth before birth. How about "lapware," computer programs that are supposed to be played to babies as young as six months to develop their language and conceptual skills? How about preschool programs that attempt to accelerate skills such as reading and math as much as possible? Might these be the equivalent of cutting a window into a developing bird's egg?

Trying to accelerate child development is a classic case of a wish based on surface logic that can go horribly wrong based on unforeseen consequences. Most parents are eager to do everything possible for their children. They burst with pride when their children precociously develop adult skills such as reading and math. They rush to buy products or to enroll their children in programs that promise to develop adult skills. They are seduced by a surface logic that states, "If you want your child to develop adult skills, teach adult skills as early as possible." They ignore the basic biological fact that development is a sequential process. If we portray the passage from infant to adult as A→B→C→D→**E,** then trying to teach E to someone at stage B can be futile and even downright harmful, like cutting a window into a bird's egg.

Very few products and programs that attempt to accelerate child development have been scientifically assessed. The slim evidence that exists is alarming. In one study, every hour that babies watched a video claiming to teach language skills resulted in six to eight *fewer* words in their receptive vocabularies, which is measured by asking them to point to pictures representing the words. Another study tracked children who had received age-appropriate preschool education compared with a preschool program that provided formal instruction. The kids who received formal instruction in preschool were earning lower grades in fourth grade. They were also more stressed, were less creative, and had more test anxiety than the kids who were simply allowed to be kids in preschool. The parents who wished the best for their children and attempted to carry out their wishes in this particular way were just like the people in folktales who end up regretting what they wished for. Were it not for scientific assessment, they would be lost in a maze of unforeseen consequences, unable to trace the malaise that had descended on their children to the decision that they had made six years earlier.

Peter Gray was next. He was the only speaker who didn't use PowerPoint slides. Instead, he simply stood in front of the blue illuminated screen and eloquently told the story that I have already recounted. I was struck by how the spontaneous educational process described by Peter solved some of the problems with formal education identified by the previous speakers. Dandelion children and orchid children, four-year-olds and sixteen-year-olds, early and late bloomers could all thrive because they got to choose their next step, which nobody knew better than they did. Adults assisted but did not attempt to micromanage their education. A strong system of governance managed conflicts of interest and made the values of the community part of the daily discourse. Virtually everyone else at the workshop was steeped in the formal educational environment and was amazed by Peter's story, as I was the first time I heard it. I knew what many of them were thinking: this is too good to be true.

THUS ENDED THE FIRST MORNING of the first day. Jerry and Bill were beaming. Already, the workshop was having an electric effect on the observers who were encountering the evolutionary perspective for the first time, including the faculty of UM's school of education, who could have been the greatest skeptics. The speakers had provided a glimpse of an updated conception of human nature, including the nature of the mind, development, and sociality. It was hardly the final word, but it was vastly more detailed and scientifically grounded than previous conceptions of human nature that informed current educational theory and practice. The speakers were like prospectors announcing the discovery of gold. Something of inestimable value was waiting to be mined, even if hard work might be required.

The updated conception was based on hard science, but it was also fully intuitive. You didn't need a PhD to wrap your head around it. It was more like a lightbulb turning on in your head once you are led to see things in the right way. An updated conception of human nature that is also fully intuitive? Who wouldn't get excited?

The excitement built during the afternoon session and erupted in a volcano of conversation at the dinner Bill organized at a posh Italian restaurant in downtown Miami. Get a bunch of intellectuals intoxicated with ideas, then get them intoxicated with alcohol, and stand back. We were talking so boisterously and probably sounded so incomprehensible to the average American that the diners at nearby tables gave us sidelong glances, as if they had suddenly been transported to a foreign country.

The first day of the workshop had succeeded in generating tremendous energy. Our challenge for the second day was to harness that energy to produce useful outcomes. The participants and observers broke into brainstorming groups. What do we recommend for future research? How can we inform the vast education community? How can we inform the general public? Most important, how can we move these important ideas off the drawing board and into the real world and actually improve childhood education?

At this point, disagreements began to surface among the experts. One problem concerned David Geary's distinction between primary and secondary knowledge. According to David, secondary knowledge such as reading and math is so alien to the evolved mind that it can never be learned spontaneously. It will always be laborious and unpleasant for children, requiring a formal educational system constructed by adults. Here's how David puts it in one of his articles:

> The gist is that the cognitive and motivational complexities of the processes involved in the generation of secondary knowledge and the ever widening gap between this knowledge and folk knowledge leads me to conclude that most children will not be sufficiently motivated nor cognitively able to learn all of secondary knowledge needed for functioning in modern societies without well organized, explicit, and direct teacher instruction.

Peter Gray begged to differ. He agreed with the distinction between primary and secondary knowledge but claimed that secondary knowledge exists in all cultures, even those without formal instruction. Consider the differences among walking, playing soccer, and tracking game. We are genetically adapted to walk, which is why children from all cultures learn to do it spontaneously at an early age. We are not genetically adapted to play soccer or track game, which is why countless hours of practice are required to perfect both skills. Children in hunter-gatherer cultures learn to track game without formal instruction because it is a valued activity. If you tried to teach them soccer without making it a valued activity, they would resist your efforts. You would have to force them to attend soccer school, and they would resentfully ask you, "Why do I need to know this?" They would quickly drop out and forget what they learned. Conversely, if you tried to teach animal tracking to an inner-city kid without making it a valued activity, you would get the same response.

According to Peter, we flatter ourselves by thinking that reading and math are such difficult skills, compared with the secondary knowledge that other cultures must master. Is it really more difficult for an American boy to learn his times tables, for example, than it is for an Australian Aborigine boy to learn the song lines of his ancestors, which are songs that function like maps and must be passed down from generation to generation? Math is hard to learn, Peter suggested, not because it is intrinsically difficult but because it is not a valued activity in our culture. Learning math is like a hunter-gatherer kid learning soccer or an Italian kid learning to track game.

Peter felt that the Sudbury Valley School provided abundant support for his position. Everyone there learns to read without formal instruction, because the older kids are doing it and the younger kids want to get in on the action. Math is not learned as readily, because it is not as valued. Most kids learn what it takes to get by in everyday life, which is all that most other people remember after all those years of formal schooling. Whenever math becomes important to learn, the kids learn it, just as they learn any valued activity. This commonly happens when they decide that they want to attend a college that requires an admission test, such as the Standard Aptitude Test (SAT). They purchase an SAT study guide, just as I purchased *Calculus for Dummies* in graduate school, and learn in a year or so what most other kids spend twelve years learning against their will. A few kids fall in love with math for its own sake. It acquires a personal value for them, and they get very good it at. For Peter, all secondary knowledge can be learned without formal education as long as it is valued. Formal instruction just plain isn't necessary.

Another controversy arose over a teaching method called direct instruction (DI), which involves breaking a complex subject such as math into building blocks that are taught one by one, with lots of teacher involvement, drill, and testing. It seems like the ultimate in adult-managed education, the polar opposite of the Sudbury Valley School. Many teachers and educators reject it as the worst kind of rote learning, in favor of other teaching methods that cultivate self-esteem, a natural love of learning, and a deeper understanding of the subject. These other methods seem closer to the Sudbury Valley School's "back to nature" philosophy, but it turns out that DI works much better than the other methods. Tony was an authority on the subject and pointed out all of the randomized control trials that proved, beyond a shadow of a doubt, the efficacy of DI compared with "back to nature" methods. DI also had impressive side benefits. It even increased student self-esteem better than methods that attempt to increase self-esteem directly.

Despite its proven success, DI was not being implemented in schools as fast as it should be. Educators who found it philosophically repugnant simply chose to ignore the scientific evidence and continued using "back to nature" methods that didn't work. Tony declared that he couldn't endorse a set of recommendations from the Evolution Institute that didn't showcase the efficacy of DI, whatever else it might also say about the Sudbury Valley School model.

I could appreciate the merits of both. They reminded me of the concept of adaptive peaks in evolutionary theory. Tortoises and hares are both well adapted to their environments, but if you could succeed at hybridizing them, the result would be a disaster. DI and the Sudbury Valley School model were like the tortoise and the hare, two very different educational systems that could each work well but couldn't necessarily be mixed. If education is going to become spontaneous and learner-motivated, then it might be necessary to go all the way, remembering that the Sudbury Valley School model includes a strong system of governance where anything doesn't go. If childhood education is going to be carefully managed by adults, then it needs to be constructed in the right way.

The Good Behavior Game provides a well-constructed educational environment by implementing a system of governance resembling the Sudbury Valley School model. The GBG creates a culture of cooperation but says nothing about how a complex subject such as math should be taught. DI takes what David Bjorklund said about development seriously. Learning a complex subject such as math is a sequential process, which can be represented by A→B→C→D→E. Teaching stage E to someone at stage B is as futile as trying to teach language skills to a fetus, however well meaning. DI takes children through the sequence and richly rewards them every step of the way. There's nothing wrong with drill and testing when appropriately motivated. All forms of secondary knowledge require hours of practice. Kids at stage B feel great when they have mastered C and are eager to move on to D. It's no wonder that DI increases their self-esteem. In contrast, kids at stage B can feel defeated when they are expected to appreciate the beauty of E directly. If we are going to micro-manage the educational environment for our children, let it be the GBG and DI.

A single day was not nearly enough to sort through these issues, much less to derive an action plan for research, communication, and implementation. Clearly, the workshop was not like a single stone hurled from a sling, felling the Goliath of problems associated with childhood education. Yet as the intense two-day event drew to a close, it still seemed like a spectacular success. We had discovered a tremendous source of insight for educating our children, however long or hard it might be to implement. The Evolution Institute had survived its inaugural voyage, and I was now totally committed to its mission.

TWO DAYS LATER, AND *POOF!* I was back in my cold, cloudy city of Binghamton, furiously trying to catch up on my other work. I often ask myself why I take on so many projects, each one a universe in its own right. Bruce Ellis was so impressed by the Miami workshop that he approached Jerry and me with the idea of a second workshop on the subject of risky adolescent behavior. Of course, we said yes. Bernard had set me on the path of rethinking economics. Jerry and I had plans of our own—war and peace, obesity and other health problems, quality of life, the workplace environment. How could I possibly expand in all of these directions without compromising my other projects?

Then I realized how much the projects contribute to one another. The BNP was an outgrowth of EvoS and will always rely on its network of faculty as the "whole university" part of our "whole-university/whole-city" approach. The EvoS consortium provides a worldwide network of evolutionists for me to draw on for EI projects. Thanks to the EI workshop on childhood education, I'm in a much better position to improve education in my city of Binghamton, which in turn can provide a testing ground for solutions that can be implemented anywhere in the world.

CHAPTER 15

The World with Us

IN HIS BOOK *The World without Us,* journalist Alan Weisman conducts a fascinating thought experiment. How fast would the earth recover from humanity's impact if we were suddenly to vanish? As Pierre Teilhard de Chardin might put it, how fast would the noosphere turn back into the biosphere? When would plants and animals reoccupy our cities? What structures would last the longest? Would the wanton shortsightedness of our civilization cause our structures to crumble faster than the structures of previous civilizations? Would aliens visiting from another planet a thousand years after our disappearance see pyramids and cathedrals but not skyscrapers?

Part of the appeal of this speculation for me is the righteous feeling that humanity got what it deserved. We neglected our duties as stewards of the earth, and now the earth has exacted its revenge. We thought we were everything, and now we are nothing. Oh, God, the pride of man, broken in the dust again!

For those of us who feel contrite, the best way to atone for our sins is to begin thinking about the world *with* us. Plants and animals are already reentering our cities without waiting for us to vanish. We should welcome them, in part because we are animals who feel more comfortable in a state of nature than in cities of our own making. Just because we made cities doesn't mean that they're good for us. Cities didn't even exist until a few thousand years ago, and the vast majority of people didn't live in them until the last few generations. We are mismatched to our cities in terms of both genetic and cultural evolution. Well-meaning urban-planning attempts have a surface logic but frequently go wrong, like wishes in folktales. We are lost in a maze of unforeseen consequences, just as with childhood education, and the only way out is by approaching urban planning from an evolutionary perspective.

There's a strip of nature in the heart of Binghamton that runs along the bank of the Chenango River, from where it joins the Susquehanna at Confluence Park to just north of Court Street, the main street of the downtown district. It occupies the space between the river's edge and the River Walk, a path along the river that was created only a few years ago. Previously, the river was separated from the city by a concrete floodwall as high as a fortress, a monument to urban planning gone wrong. Of course, it is necessary to protect a city from flooding, but the floodwall made it impossible to enjoy the river at all. Once the River Walk was created on the other side of the floodwall, the river drew people to its banks like a magnet. Walkers, runners, bikers, lovers young and old, friends in earnest conversation, parents with their children feeding the ducks — all came to the river, as if to be healed. In a single stroke, the quality of life in the city of Binghamton had been improved.

Huge stones were brought in to stabilize the strip of land between the river's edge and the River Walk. The width of the strip varies with the level of the water. At times, it is completely submerged, along with the River Walk itself. It would be impossible to landscape the strip, even if the money were available. Therefore, nature did the landscaping. Plants have 1001 ways of dispersing their seeds. Some seeds are wafted through the air on tiny parasols. Others are adapted to be eaten by animals and to survive the passage through their digestive systems. Still others are designed to stick like Velcro on passing animals. In fact, the person who invented Velcro was inspired by seeds that had stuck to his clothing and the fur of his dog after a hunting trip.

All of these modes of dispersal have been perfected by the hammer blows of natural selection over countless generations. As a result, millions of seeds are deposited along the strip of land between the river's edge and the River Walk every year. Of these, many can't survive the tough conditions, including competition with one another. The survivors grow into a garden landscaped entirely by nature. I love walking along this strip whenever I am in town. The summer is especially glorious, with a profusion of yellow, purple, and white flowers amid luxuriant green foliage. I was so impressed that I asked Mark Blumler, a friend and colleague in BU's geography department, to help me identify them. Mark has a Talmudic knowledge of plants and was happy to spend a summer afternoon with me cataloging nature's garden in the heart of Binghamton.

Mark told me the story of how he became a scientist over lunch at a new sushi restaurant on Court Street before we began our adventure. Unlike most of the scientists I have profiled, he did not form a deep connection to nature in childhood and did not develop an interest in plants until relatively late in life. His ancestry is English and German. He had an intellectual upbringing and entered Columbia University intending to major in math, then switched to anthropology, then dropped out after two years. Something about college wasn't providing what he needed. He knocked about for a couple of years, aimless and depressed, ending up in Los Angeles in 1971. There he got involved in primal-scream therapy, which involved bawling your eyes out to get rid of repressed pain. Mark described it as like a cult. At first, you learned to bawl as an individual, and then you did so in groups. The bawling sessions were held in a padded room so that no one would get hurt as they flung themselves about. Evidently, they would get themselves going with movies such as *Old Yeller* and *Bambi*. Watching Bambi's mother get shot was enough to send the room full of adults into paroxysms of weeping. Mark said that at first, it was totally bizarre, but after a while, it became second nature. When it comes to the cultures that we create for ourselves, you can't parody reality.

As Mark was groping his way through life, he chanced to read a book titled *Stalking the Wild Asparagus*, published in 1962 by Euell Gibbons. Gibbons grew up in the American Southwest during the Dust Bowl era and was taught by his mother to forage for wild foods, like our hunter-gatherer ancestors. *Stalking the Wild Asparagus* got people in touch with their inner gatherer and was an instant best-seller. I loved my copy and have many happy memories of boiling acorns to get rid of their bitterness, peeling the roots of burdock to get at their tender core, and harvesting cattail flowers just before they mature, which are delicious steamed and eaten with salt and butter, like diminutive corn on the cob.

Mark was also captivated by *Stalking the Wild Asparagus* and began to make a hobby out of learning to identify and gather edible plants. Until then, his only exposure to nature was the annual summer camping trip as a kid with his mother and stepfather. Soon he became interested in identifying all of the plants that he encountered and concerned about the rare wildflowers that were becoming endangered. He returned to college, this time in plant ecology, and became an expert on how crop plants, such as wheat, millet, and corn, became domesticated by our ancestors, a coevolutionary relationship that eventually led us to abandon our hunter-gatherer ways. Getting in touch with his inner gatherer was more therapeutic for Mark than primal-scream therapy.

After lunch, Mark and I walked to the Court Street Bridge to begin our adventure. You've never seen anyone more contented than Mark, naming the plants as if they were old friends. Some were obvious from a distance, such as Japanese knotweed and purple loosestrife. Others required a quick check of diagnostic body parts. He felt and smelled them in addition to inspecting them visually. So great was his knowledge that he could tell where they came from in addition to their names. Only about half were native to North America. The others came primarily from Europe and the Mediterranean, with a few from Asia. Some of the plants bore names that told of their uses during earlier times when people made their own medicines, such as boneset (taken for rheumatism and many other ailments) and fleabane (when burned, the smoke was reputed to drive away fleas). I acted as his scribe, writing down the names in my notebook as passersby looked at us quizzically.

In less than an hour, Mark had identified seventy-four species of plants, assuring me that he could easily surpass one hundred with a little more effort. Hundreds of species of insects were visiting their flowers and munching on their leaves, stems, and roots. Yellow jackets were in abundance, commuting from their own underground cities. We often hear about the wonderful diversity of coral reefs and tropical forests, but Darwin's tangled bank had assembled itself right there in downtown Binghamton.

I WAS FASCINATED BY THE FACT that roughly half of the plant species were immigrants. Where did their ancestors come from, and why did they arrive in North America? That's my first question whenever I ask a person to tell me his or her story, and now I wanted to know the same for the plants. The human and plant stories are intertwined, of course. Some were brought over for food, animal forage, or medicine. Others were stowaways, contained in soil or stuck like Velcro on the surfaces of animals and other objects that made the passage. Many were ornamentals, which means that we brought them to North America for the sole purpose of making ourselves feel good.

The very concept of an ornamental plant is remarkable, when you think about it. Why do we need plants to make us feel good? The answer is straightforward from an evolutionary perspective. We are animals, and our animal natures didn't vanish when we became human or when we moved into cities. Unlike plant seeds, which must try to grow wherever they land, animals can move and have instincts for seeking out elements of their environments that will help them to survive and reproduce. Put a honeybee or a wasp in a cardboard box, and it will batter itself to death trying to get out. Their sensory systems are elaborately designed to seek out flowers, the source of their food. If there are no flowers in the vicinity, the insects are powerfully motivated to leave. If they can't, they get frantic.

Putting a person in a room with four bare walls is like putting an insect in a cardboard box. Our mental equipment is starving for input from our environment so we can decide where to move and what to do. Remove the environmental input, and we become stressed to the point of insanity. In sensory-deprivation experiments, people wear masks that admit only diffuse light, listen to white noise, and sometimes are placed in warm-water baths with their arms outstretched to eliminate tactile stimulation. People who volunteer for the experiments often think that it will be restful and that they can pass the time contemplating their lives, but they become desperate within a few hours and often start hallucinating. Solitary confinement in a bare room is one of the cruelest forms of punishment for the same reason. It's not just the absence of other people; it's also the absence of the environmental input that our mental equipment craves.

That's why we decorate our indoor and outdoor environments. If we can't move to better locations, then at least we can provide the stimuli that our mental equipment craves. Decorations are anything but superfluous. They are literally required to keep us from going crazy.

How we decorate our environment speaks volumes about what our mental equipment craves. Many of our decorations are personal and social. If we can't be with our loved ones, we can gaze at their faces in photographs. If we can't indulge in sex with nubile women or hunky men, we can drool over their images. Other decorations attest to our cultural natures, such as religious iconography or modern art. Recent immigrants are famous for trying to bring the old country with them, and their descendants often use decorations to remind themselves of their cultural roots.

Our decorations also reveal our deep craving for nature. There is truly a hunter-gatherer within all of us, scanning the horizon for game, lush vegetation, a source of water, and a safe shelter with a view. Even if we don't hunt, fish, or camp, we decorate our homes, yards, and parks in a way that would have made our hunter-gatherer ancestors feel at home. We even delight in a crackling fire.

All of those species of plants and animals that came with us didn't stay in our homes, yards, parks, and farms. They spread out into the countryside and often thrived in competition with the native species. Regardless of why or how they came, some of them started to alter the course of nature in ways that were never intended, like a wish gone horribly wrong.

A former graduate student in my department named John Maerz discovered this for himself when he began to study a native species of salamander that lives in the woodlands of eastern North America.

John's ancestors came from the British Isles, Germany, and France. His family moved several times during his childhood, eventually settling in the Washington, D.C., area, but John satisfied his inner hunter-gatherer wherever he went—fishing with his grandfather, camping with his parents, and crabbing on the eastern shore of Maryland. If it was outdoors, John wanted to do it. He attended the University of Maryland expecting to be a physician like his father and his grandfather, but the biology courses geared to premed students didn't interest him. As soon as he started taking ecology and evolution courses, he was hooked. Until then, he didn't know that it was possible to be a scientist studying the outdoor world.

As a graduate student at Binghamton advised by my colleague Dale Madison, John focused on one of the most abundant species of salamander in our local woodlands, the redback salamander, which can be found by turning over rocks or logs wherever the soil is suitably moist. It's easy to identify by the dark red stripe along the back of its narrow body. A large redback is only two inches long, but they are so abundant that according to Dale, their combined biomass exceeds the biomass of deer. John thought that he was getting away from it all by studying a native species in its native habitat, the eastern deciduous forest, but the more he learned, the less natural everything appeared. Part of his research involved pumping the stomachs of the salamanders to examine what they ate. This delicate operation is performed by inserting a slender tube down their throats and irrigating their stomachs with water from a syringe. Their prey consisted of tiny invertebrates that inhabit the soil and leaf litter, such as mites, springtails, and beetles. When John identified these species with the help of experts, much as Mark helped me identify plants, he was shocked to discover that about 50 percent of the prey species were nonnative, primarily from Europe. The immigrants were occupying the so-called natural habitat as easily as humans were occupying farms, towns, and cities.

ONE OF THE INTRODUCED SPECIES is so familiar that you probably don't recognize it as a foreigner: earthworms. Native earthworms couldn't survive the ice age 10,000 years ago, which pushed the northern edge of their distribution several hundreds of miles south of Binghamton. When the glaciers retreated, the native earthworms moved north much more slowly than other species, so slowly that they still haven't made it. The northern forests therefore existed for thousands of years without earthworms until they were colonized by European earthworms, which arrived as stowaways in soil.

Earthworms might seem like humble creatures, but collectively, they are a mighty force, as Darwin showed when he wrote his final book about them. They are called "ecosystem engineers" in the modern scientific literature because of the degree to which they can alter the physical environment and the flow of energy and nutrients. In the absence of worms, the forest floor becomes a thick carpet of fallen leaves, which might require two or three years to decompose. Earthworms gobble them up in less than a year, leaving the ground bare. This changes the physical microclimate of the forest floor, making it less moist and therefore less suitable for redbacks and their invertebrate prey. Adult redbacks can eat small worms, but even the smallest worms are too big for baby redbacks. Thus, when worms are abundant, they decrease the amount of habitat for all redbacks, create a famine for the smaller redbacks by reducing their prey, and provide a feast for the few adult redbacks that survive. The net result is a few very fat adult redbacks in the presence of earthworms. John worked out this web of interactions as part of his PhD thesis.

Earthworms also disrupt a symbiosis between trees and fungi called mycorrhiza. The trees provide the fungus with energy-rich carbohydrates and receive minerals from the soil in return. This symbiosis developed over thousands of years in the absence of earthworms and many species of northern trees can't live without their fungal partners. Earthworms disrupt the symbiosis by plowing up the soil and eating the fungus. They even contribute to global warming by causing the carbon that would otherwise be stored as plant biomass to be released into the atmosphere as carbon dioxide.

It might seem strange that introduced species of plants and animals can so easily invade so-called natural habitats, but they have two advantages over the native species. First, the longer a species exists in an area, the more it serves as lunch for diseases, parasites, and predators that evolve to exploit it. Many introduced species are uncommon in their native habitat because they are limited by their specialized enemies. No one would guess that they might pose a problem when transplanted elsewhere. They become common in their new habitat because they left their own enemies behind and not enough time has elapsed to acquire new enemies.

The second advantage follows from the first. Introduced species that have shed their enemies no longer need to maintain defenses against them. They can concentrate their resources on outcompeting the native species, like an enemy with the luxury of fighting on only one front instead of two. This kind of reallocation of resources can evolve rapidly by genetic evolution, compared with the novel adaptations required for native diseases, parasites, and predators to attack the introduced species.

An elegant study of purple loosestrife (*Lythrum salicaria*) shows how its success as an introduced species is based in part on genetic evolution. This attractive wetland plant is native to European wetlands, where it is not particularly common, but it has aggressively invaded North American wetlands, displacing many native plants and animals. It arrived in North America in the early 1800s, probably as seeds in soil that was used by the ships for ballast. A team of scientists headed by Bernd Blossey at Cornell University grew seeds of purple loosestrife from North America and Europe under common environmental conditions. The North American plants grew larger. They had evolved into super-competitors that could even outcompete their own ancestral stock. Yet the North American plants were also more vulnerable to an insect species that specializes in purple loosestrife in Europe and didn't cross the Atlantic. Evolution is a matter of trade-offs. The North American population was free to evolve into a super-competitor because it didn't need to defend itself against the herbivore. It was like an army that could fight on one front instead of two.

Another example of rapid genetic evolution comes from Australia, where the cane toad (*Bufo marinus*) was deliberately introduced in 1935 to control pests of sugar cane. It was another example of a wish gone horribly wrong. This monstrous toad, which can grow to a weight of more than four pounds and has toxic skin that protects it from predators, proceeded to march across the tropical and subtropical regions of Australia, eating everything in sight. Today it occupies more than a million square miles.

Initially, the front of the invasion advanced at a rate of about 10 kilometers per year. That was how far a cane toad could hop in a year in the 1940s. Try thinking about the invasion front as an evolutionist. Cane toads vary in how far they hop in a year. Those that hop the farthest form the invasion front. Their offspring also vary in how far they hop in a year. The fastest of those form the new invasion front, and so on, generation after generation. By definition, the invasion front consists of toads that hop the farthest. Today the front is advancing at a rate of 50 kilometers a year, five times faster than only seventy years ago. Toads at the invasion front have longer legs than where the invasion originated and can hop as far as 1.8 kilometers in a single night. They're so passionate about running that they even develop leg injuries, like athletes who push themselves too hard. Biologists who tried to predict the rate of advance without taking rapid genetic evolution into account were completely wrong.

Wishes can go horribly wrong, and rapid genetic evolution can take place even without introducing new species. Remember Bambi's mother, whose death at the hands of a hunter induced paroxysms of weeping in the primal-screamers? Deer are native to North America, but they were never as abundant as they are today. In an odd alliance, both deer lovers and deer hunters want to see as many deer as possible, and their wish has come true. Deer have become like the cane toad, devouring everything in sight. Their status as native rather than introduced is beside the point. In some forests, the only places to find wildflowers are on the tops of boulders where the deer can't reach them. Understory shrubs are gone, along with the bird species that nest in them. In their place is a continuous stand of ferns, one of the few plants that deer don't eat. The ferns look restful to the eye, but they're an ecological disaster. They form a closed canopy, preventing light from reaching the ground and preventing owls from seeing their prey. Mouse populations explode, eating the seedlings of other plant species that struggle to grow above the fern canopy. Tick populations also explode, because baby ticks feed on mice, and adult ticks feed on deer. Ticks give people Lyme disease, a very serious illness if it isn't treated immediately. Thus, a major human health problem is part of the maze of unforeseen consequences that result from a well-meaning deer-management policy gone horribly wrong.

Remember the magnificent stag that fathered Bambi? Big males fathered most offspring before deer-management rules made them the primary targets of hunters. Now genetic evolution favors males that mature at a small size with unimpressive antlers, just as it favors cane toads with long legs. Genetic evolution also favors fish that mature at a small size whenever we harvest the largest fish. One reason the lobster industry remains healthy while the fishing industry is collapsing is that the largest lobsters can't enter the traps. Bigger is still better if you're a lobster, but smaller is better if you're a fish or a deer.

STORIES OF ECOLOGICAL DISLOCATION and rapid genetic evolution can be recited without end. Countless species have become dumbfounded by human-altered environments and are being pounded into new shapes by the hammer blows of natural selection. One recent scientific article calls humans "the world's greatest evolutionary force." Species introductions are taking place at a faster pace than ever before. The most recent introductions are likely to come from Asia rather than Europe, such as the Japanese tiger mosquito and species of earthworms from Asia that are muscling their way into our soils, like the European species before them. According to John Maerz, every time someone builds a dream home in the country and landscapes with plants purchased from a big-box store such as Lowe's or Home Depot, they're carrying the newest wave of foreign invaders into our so-called natural environments. In one clever study performed in Germany, plant seeds were collected in one-way highway tunnels in which the traffic was either leaving or entering the city of Berlin. The seeds were being carried on the surfaces of the vehicles, and some blew off and fell into the traps as the vehicles were passing through the tunnels. More seeds were leaving the city of Berlin than were entering. The city had become a net exporter of seeds, especially the seeds of introduced species.

All of these horror stories can induce a kind of despair and lethargy, as if nothing can be done. It's true that the problems are enormous, but this can become a source of courage and determination. As the proverb says, when the going gets tough, the tough get going. We have no choice but to become stewards of the earth. The fact that so much has gone wrong means that some things can be put right almost immediately, if we use science and evolutionary theory to listen and reflect in the right way.

The first step is to get back in touch with our own animal natures. One of the saddest cases of wishes gone wrong is the cultural practice of formula-feeding our babies. We are mammals, and nothing can be more natural than breastfeeding. Yet that has not prevented most women from abandoning the practice in favor of formula-feeding, which had a surface logic when it first became popular during the Industrial Revolution. Wishes always do. Back then, people didn't know about such things as germs, that human breast milk is nutritionally highly complex and different from cow's milk or artificial formula, that it includes antibodies in addition to nutrients, that breastfeeding has important hormonal effects on the mother, and so on. Later, the surface logic of bottle-feeding was supplemented by powerful commercial interests that promoted the practice in developing countries. Among its other surface virtues, bottle-feeding was perceived as the modern and upwardly mobile thing to do. The result was and remains an unmitigated disaster, in terms of infant mortality and a maze of nonlethal unintended consequences that we are still lost within. All of the necessary science has been done to determine that breastfeeding is a better practice than bottle-feeding. It is also pleasurable for the most part, my wife, Anne, assures me (she breastfed both of our kids). If we could get all women to breastfeed their babies, the world would become better in a single stroke. Yet so great is the power of our cultural constructions that breastfeeding can be regarded as painful, disgusting, and even unnatural. Work is required to do what comes naturally and benefits us in every way, as bizarre as that might seem. We have the tools to accomplish the cultural transformation, as we saw in the case of selling cigarettes to minors. Once established, the healthful practice of breastfeeding could become as entrenched as the current unhealthful practice of formula feeding.

We can also get back in touch with our animal natures by taking the concept of decorations seriously. What we view in our daily lives is a matter of utmost importance, as I have already shown. We'd all go crazy without decorations in our homes, yet we often fail to decorate our public spaces. Most of us would leap at the opportunity to live next to the banks of a river, yet we build a floodwall that prevents access to a river that flows through the heart of a city. We fall prey to the surface logic that aesthetics is a frill that can't be afforded on a tight city budget, rather than a necessity for our physical and mental well-being.

To illustrate the powerful benefits of managing our visual environment, consider a recent study of hospital-room decorations. Eighty female patients who underwent thyroid surgery were randomly assigned to rooms that were identical except for the presence or absence of plants. Patients with foliage and flowering plants in their rooms required less painkilling medication; had lower ratings of pain, anxiety, and fatigue; had more positive feelings about their rooms and their hospital stays; and were discharged from the hospital sooner than patients in rooms without plants. The conclusion? We can improve our health and save millions of dollars in health-care costs merely by decorating hospital rooms with plants.

Similar experiments have been conducted in other indoor settings, such as schools and offices, and outdoor settings, such as parks and neighborhoods. In one classic study published in 1991 by Roger Ulrich and colleagues, subjects were first stressed by watching a ten-minute video on accidents that can happen in a woodworking shop. Then they watched a second video depicting either scenes of urban life or scenes of nature. The nature scenes caused them to recover from the stress much faster than the urban scenes.

Dozens of subsequent studies have confirmed and extended this result. In one exceptionally well-controlled study published in 2006, Finnish psychologist Jari Hietanen and colleagues digitally manipulated the amount of vegetation in five versions of a photograph. At one extreme, the photograph consisted of a building with no trees.

Trees were progressively added until they entirely obscured the building at the other extreme. A second set of photographs showed a person with a facial expression that was either happy or disgusted.

In the experiment, subjects were told that they would see a sequence of pictures in pairs. The first picture would be of the environment, and the second would be of a facial expression. Their task was to categorize the facial expression as happy or disgusted by pressing one of two buttons. This might seem like a strange task, since it is based only on the second picture, and all of the expressions are classified correctly because they are unambiguous. However, the *speed* with which the second picture is classified depends on the emotion elicited by the first picture. Subjects who like the picture of the environment will classify a happy face faster than a disgusted face because it corresponds to their own mood. Similarly, subjects who dislike the picture of the environment will classify a disgusted face faster than a happy face. The difference is only a fraction of a second but can be measured by the time that elapses between the presentation of the second picture and the pressing of the appropriate button. This method is called affective priming, and it reveals the immediate emotional response to a visual stimulus.

Trees made the subjects happier on an immediate basis—or did they? Perhaps the subjects were responding to the color green or to the abstract shapes of the trees. To investigate this possibility, the pictures of the environment were digitally manipulated in a second experiment to preserve their color and shape profiles without anything resembling either a building or a tree. These altered images lost their emotional impact. It was trees that made the subjects happy, not the amount of green or certain abstract shapes.

Thanks to these and other studies, we already have the knowledge to improve our lives by welcoming nature back into our cities. It's just a matter of doing it. Evolutionary theory helps us to see that it's not a frill but is necessary for our psychological and physical health.

We can enjoy nature even more by understanding it more deeply and becoming more involved in its stewardship. Earlier in this chapter, I said that people flocked to the banks of the Chenango River, as if to be healed. Now I can say that they really were being healed, merely by the sight of the river and the glorious display of flowers and foliage landscaped by nature. Yet if they were to get to know the tangled bank the way that John Maerz, Mark Blumler, and I do, they would be happier and healthier still. Not only are the lives of plants and animals fascinating in their own right, but we also can begin working together to find our way out of the maze of unintended consequences within which we are currently lost. That is part of the mission of the Binghamton Neighborhood Project, along with more exclusively human-oriented goals such as childhood education. Recently, the BNP was joined by an ally from an unexpected source: an urban seed named Latisha Williams who blew in from New York City.

THE SCIENTISTS OF TODAY are statistically likely to come from backgrounds that are close to science, such as having highly educated parents or a strong connection to nature in the case of evolutionary science. As we step down the generations, however, the association rapidly diminishes. Jim Hunt's father was an educator, just a short hop from being a scientist, but his grandfather was a sharecropper. Dan O'Brien's parents were also educators, but his ancestors came from the average ranks of Irish and Italian society. People are like seeds. Most of them fall close to the tree, but a few are carried far away, where they take root as best they can. If they succeed, then their own seeds easily sprout in their vicinity.

Latisha Williams came from a background that seldom directly produces scientists, which makes her story especially interesting and relevant to those who think about expanding access to scientific careers. Her ancestors came from India by way of the West Indies. Her mother immigrated to the United States from Guyana with her sisters to seek a better life. Her father grew up on the island of Grenada and became politically active as a teenager. Politics is a violent game in Grenada, and his mother began to fear for his life. He immigrated first to England and then to the United States, where he met Latisha's mother.

Latisha was the third of five sisters and one brother and grew up in Queens, one of the five boroughs of the City. Her father had settled down as a carpenter and jack-of-all-trades who was happy to work long hours so that her mother could stay at home with the kids. He was gone before the kids arose and returned home after they went to bed, making weekends a special time. They owned their home but rented out the upstairs, so Latisha and three of her sisters occupied a single room and fought constantly. The neighborhood was close to some of the housing projects and wasn't very safe, although the guys in the neighborhood kept a protective eye out for "the Indian sisters," as they were called. Latisha's parents felt that it was better for her mother to stay at home than to earn a second paycheck.

Even though the house was crowded, there was room for plants. Latisha's mother's parents worked on a commercial farm in Guyana, and her father's parents raised their own food. Gardening was a way of life down there. If you wanted something to eat, you went outside and picked it. Both of Latisha's parents brought their love of plants to Queens. Even in their crowded home, the furniture was jammed into one half of the living room so that the other half could be devoted to plants. Her mother could grow anything. If she saw an unfamiliar plant on someone's porch, she would boldly knock on the door to ask about it and request a leaf or a stem, which she artfully rooted for her own. The front yard was devoted to flowers, and the backyard was devoted to vegetables, herbs, and a barbecue grill. Her dad grew the vegetables in compound buckets that he got from work. Green beans, tomatoes, sweet peppers, hot peppers, and lots of cucumbers burst forth from the buckets.

Two doors down, someone else from the West Indies had an even bigger garden and became part of the extended family. What Latisha's parents didn't grow, Mamma A did. She even grew the flowers that Latisha and her sisters used as corsages when they started to date. Everyone else in the neighborhood had standard lawns and gardens. Latisha's friends flocked to her house, adopted her parents, and loved eating their food.

When it was time for Latisha to start high school, her mother made sure that she attended the only agricultural high school in New York City. John Bowne High School in Flushing, New York, actually has a working farm that dates back to 1917. During World War I, young men from New York City were recruited to work on farms in Upstate New York to replace soldiers who were fighting overseas. Many of them enjoyed the experience and asked to learn more about agriculture when they returned to the City, so a reform school in Flushing with a working farm that had just closed was converted to an agricultural school. Today students who enroll in the year-round agriculture program at John Bowne get a second education in plant or animal science in addition to their normal high school education. The 4-acre farm includes a poultry house, a large animal barn, an exotic animal laboratory, a greenhouse, an orchard, and field crops. Each freshman in the program is given a 15-square-foot plot of land to raise a vegetable and flower garden. In subsequent summers, they are placed on family farms or become interns in veterinary hospitals, florist shops, nurseries, garden centers, zoos, and aquariums. I never dreamed that such a school existed in the City until Latisha told me about it.

Latisha knew that her family couldn't afford to send her to college or even provide much pocket change, so she was always working jobs in addition to her double high school curriculum. She decided to join the Marines as a way to fund her college education but was talked out of it by a cousin who had joined the Marines and was deployed to Iraq just after September 11, 2001. Her cousin put it bluntly by saying that he would break both of her legs if she tried to join the Marines. Latisha therefore attended Queens College, one of the City University of New York campuses right next to John Bowne in Flushing. Latisha hadn't yet left the City, but she could still obtain a quality education in environmental science, thanks to inspiring professors such as Gillian Stewart, who provided a model of a scientist who is motivated primarily by the love of her work. The federal work-study program enabled Latisha to support herself by doing meaningful work in the laboratories of her professors. She also worked part-time for the New York City Parks and Recreation Department, where her extensive training and experience were quickly noticed and put to use.

One of Latisha's most memorable experiences was participating with Gillian Stewart in the 2007 Jamaica Bay BioBlitz. A BioBlitz is a marathon effort to inventory all of the species in a given area, with the help of both scientific experts and the general public. Everyone, from young children to the elderly, assists in collecting and identifying all creatures great and small, from tiny invertebrates to plants, fish, birds, and mammals. The first BioBlitz took place in 1996, and they are now held around the world as a celebration of biodiversity and a way to get people back in touch with the natural world. For many children, it is their first experience of looking through a microscope or binoculars. The Jamaica Bay BioBlitz involved hundreds of people and took place around the clock for forty-eight hours so that nocturnal species would not be missed. It had the same kind of intensity as a religious ritual or a rock festival.

By the time Latisha was a junior in college, she knew that she wanted to attend graduate school to study urban biodiversity. As someone who watched every penny, it irked her that every application to a graduate program would cost almost a hundred dollars. She discovered a program that waived application fees for some colleges and universities, including Binghamton. Long accustomed to planning her future carefully, she visited Binghamton to check us out. She discovered a group of faculty members who were thrilled by her interests and wanted to develop their own research in the same direction. The only problem was that we hadn't started yet. If Latisha entered our program, she would be helping to lay the groundwork.

That suited Latisha just fine. Freedom to create her own research program was much more attractive to her than being plugged into someone else's. Thanks to her, the Binghamton Neighborhood Project is adding the study of biodiversity and the human bond with nature to its portfolio of projects. Latisha has created a GIS map of the green spaces in Binghamton and has conducted detailed plant inventories for a sample of them. With the help of Dan, she has conducted a door-to-door survey of Binghamton residents to measure their attitudes about nature, which we can add to our ever-growing database.

We're not at the forefront of the movement to welcome nature back into our cities. As with the field of prevention science, I'm happy to acknowledge the outstanding work taking place elsewhere. We can rapidly move to the forefront, however, with our combination of a comprehensive database and unified evolutionary perspective. I never expected to receive so much help from such an unexpected source. In addition to a rose that grows in Spanish Harlem, there's a kickass environmental scientist who grew in Queens, thanks to a family who brought their love of plants with them from the West Indies, a high school with a farm, and enough support from a state-funded university and federally funded programs to help a kid from the poor side of the City receive a quality higher education.

I don't claim to have a fix for every problem, but some solutions, such as breastfeeding our babies and welcoming nature back into our cities, are no-brainers when you view them from an evolutionary perspective. We merely need to do what is manifestly good for us. Even that is not easy, as we have seen, but can still be accomplished by wisely managing the cultural evolutionary process. The world with us can be a Garden of Eden, as soon as we decide to end our self-imposed exile.

CHAPTER 16

The Parable of the Crow

PART OF OUR SELF-IMPOSED exile from nature is that we imagine ourselves as set apart from the rest of creation. We might not be angels, but we are closer to them than to the beasts. We are more intelligent, moral, aesthetic. We make tools, speak, transmit culture. If we bear comparison with any other species, it must be with chimps and bonobos, our closest relatives, which are barely hanging on in the last remnants of African forests.

This reasoning ignores commonalities that result not from common ancestry but from being pounded into the same shape by the hammer blows of selection. As far as convergent evolution is concerned, we share a deep affinity even with crows, one of the newest species to invade our cities, which phylogenetically are more closely related to dinosaurs than to us.

If you're up the road in Ithaca, New York, and see a sixty-year-old woman high up in a tree, it's probably my wife. Anne B. Clark, an evolutionist who works as hard as I do, studies crows instead of people. To identify crows as individuals, it is necessary to mark them. To mark them, it is necessary to catch them. The easiest way to catch them is when they are still nestlings unable to fly. Thus, every spring, Anne and other members of the Crow Research Group, as they call themselves, climb trees to mark the next generation of crows. The average crow nest is about 70 feet high, and some reach as high as 120 feet. You'd think that Anne would leave this job entirely to the younger members of the Crow Research Group, but she is never happier than when climbing trees. Small and light, she can squirrel up the most slender trees with the most tightly spaced branches.

The American crow (*Corvus brachyrynchos;* there are many other crow species worldwide) is one of the species that has reentered our cities without waiting for us to vanish. Fifty years ago, they inhabited only farmland and areas uninhabited by people. Today they mingle with us on our busiest city streets, along with pigeons, a species that adapted to urban life much earlier. How did crows become urbanized in only a few decades? Was it genetic evolution? Could it have been cultural evolution?

Anne is far more interested in studying crows than people. In fact, I get a lot of grief at home for switching my own interests from the biological tangled bank to the human tangled bank. Anne liked me better when I was studying beetles and fish. To make matters worse, she's the one who's more genuinely empathetic about helping people. She finds it ironic that after decades of being a local slacker, I have mounted my high horse and gone galloping off to save my city of Binghamton and the whole planet. Yet Anne is possibly discovering more about people by studying crows than I am by studying people, by breaking down the conceit that we are unique. Crows can be like us in ways that we never imagined, until animal behaviorists such as Anne started to unlock their secrets.

ANNE'S STORY WILL HELP TO explain why a highly empathetic person might find crows more interesting than people. Her mother's ancestors were among the first British colonists whose descendants moved to Minnesota when it was the frontier. Her great-grandfather and her grandfather were wealthy bankers, and one of her great-grandfather's banks had the distinction of being robbed by the Jesse James gang. Anne's mother, Barbara, grew up in luxury and developed a love of nature by accompanying her father on hunting and fishing trips among Minnesota's many lakes. She became a pediatric physician, which was unusual for a woman at that time.

Anne's paternal grandfather was an orphan of Scottish descent who earned a living as an insurance agent, but his passions were natural history and photography. Anne's father, David, grew up outside of Chicago and shared his father's love of nature. The family kept dogs, horses (which were still a major means of transportation back then), and even a pet magpie. David became a doctor and a pioneer in the field of pediatric neurology at Johns Hopkins University. When Anne was four, the family moved to a farm outside Baltimore that allowed all of them to indulge their love of animals and nature. Anne and her sister, Cindy, had an idyllic youth surrounded by domestic animals of all sorts and a mother who loved to take them on collecting expeditions. They raised tadpoles, baby birds, baby turtles, baby mice. They had a pet raccoon and several dozen barn cats in addition to their house cat. Both Anne and her sister had ponies and were allowed to ride them, unsupervised, even at the tender age of eight.

When Anne started to think about what she might do for a living, she looked long and hard at her father. David had an international reputation as a neurologist specializing in young children. He was in part a research scientist trying to answer fundamental questions about the developing brain. At the same time, he was a clinical neurologist working with adult patients in addition to children. His patients' needs were overwhelming. David was constantly being called during the evenings and weekends to deal with their crises. To make matters worse, many of their problems were self-inflicted, and they were unlikely to change their ways. David was a responsible person being swallowed up by irresponsible people, or so it seemed to Anne, who did not want the same fate for herself. She would be a research scientist.

Anne's decision was reinforced when the family moved from their idyllic farm to Lexington, Kentucky, where David was recruited to chair the neurology department at the University of Kentucky's medical school. It was a move that David couldn't resist, but it was hard for Anne, who spent the last two years of high school in what she regarded as an alien culture. At the time, Kentucky's public-school system was ranked near the bottom for the nation, and girls were expected to become housewives, secretaries, and schoolteachers, not research scientists. She survived only by finding a tiny group of fellow misfits and fled the state to attend the University of Chicago, her father's alma mater.

Like so many other people I have profiled, Anne quickly discovered in college that being a research scientist didn't necessarily require donning a white coat and spending your life inside a laboratory. She could be an ethologist, the branch of science that focuses on the behavior of animals in their natural environments. A faculty mentor invited her to spend the summer helping him redesign a small zoo in Indiana according to ethological principles. At that time, zoos were not designed to emulate the natural environments of their inhabitants. They were more like animal prisons, resulting in pathological behaviors such as obsessive pacing. This zoo was in especially bad shape, a sorry collection of animals both underfunded and understaffed. Anne's job was to create a reptile collection, but the work was so captivating and so much needed to be done that she stayed on for a year as the zoo's only full-time keeper. She was put in charge of a veritable Noah's Ark, including two lions, two tigers, two pumas, an ocelot, two full-grown chimps, macaques, spider monkeys, bushbabies (a nocturnal primate), kinkajous (a raccoon-like mammal from South America), a few birds, and her reptile collection. She was on the board of directors, cleaned the cages, fed the animals, opened the zoo up to the public, and ran its educational program. She was nineteen years old and looked sixteen.

Anne eventually earned her PhD at Chicago studying bushbabies, which had piqued her interest as a zookeeper. We met and fell in love in Costa Rica during a tropical biology course that we both attended as graduate students. We married just after she received her PhD in 1975, and I accompanied her to South Africa to study bushbabies in the wild. I was proud when she became the president of the Animal Behavior Society in 2002. My wife bore our two children, became president of a major scientific society, and still climbs trees at the age of sixty!

When I ask Anne why she finds animals so much more fascinating than people, she replies that people are too familiar to be interesting, especially when they come from our own culture. When we discover something about ourselves scientifically, it usually affirms what we already knew in some sense. Anthropologists get a bigger thrill when they discover how people from other cultures behave, but she gets an even bigger thrill when she discovers how other species behave. Always up for a challenge, she gravitates toward species that have the longest life spans and are reputed to be the most intelligent, such as primates, elephants, dolphins, parrots, and crows. Primates are a bit boring, because they're too much like humans. It's a hassle to study animals in far-off lands. Crows can be studied at home, even in a city.

ANNE STUDIES CROWS IN ITHACA rather than Binghamton because she joined forces with a colleague named Kevin McGowan, who began a long-term study of crows in 1989. Kevin's ancestors were also among the first British colonists, whose descendants moved to Ohio rather than Minnesota, but Kevin's upbringing was completely different from Anne's. His parents were schoolteachers with no interest in nature, and they lived in a residential neighborhood of Springfield, Ohio. Kevin's love of nature came from within, and he searched for outlets even when they were hard to find. As he puts it, "Other kids were into dinosaurs, but I was *really* into dinosaurs." He treasured Sunday visits to his grandmother's farm. He discovered that his library had a whole series of books about nature geared to children, the Golden Guide books, edited by a naturalist and public-school teacher named Herbert Zim, who is also credited with introducing laboratory instruction into public-school education. Kevin was so enthralled with the Golden Guide books that he says he knew the facts of Mendelian genetics before he knew the facts of life.

Kevin began to focus on birds when he joined the local chapter of the Audubon Society at the age of thirteen. Restricting his attention to a single group of organisms caused him to see more, not less. With his own binoculars and field guide, he began to appreciate the diversity that was all around him. Seventy-four species could be identified without even leaving his neighborhood. More species beckoned to him on field trips organized by the local chapter of the Audubon Society to various habitats in the area. Still more species beckoned from a farther distance—914 species in North America, nearly 1600 species in Ecuador alone, more than 10,000 species worldwide. Kevin couldn't visit these places yet, but he could dream about it by poring over field guides.

Kevin's passion for birds led him to expand his horizons more than most kids who grow up in a Midwestern town. Most of his classmates with more "normal" interests remained in the area and became "normal" workaday people, for better or for worse. A "weird" interest in birding caused Kevin and the small number of other children who became involved in the Springfield chapter of the Audubon Society eventually to travel the world and develop careers that they never would have contemplated otherwise. Kevin currently directs Cornell's world-famous Laboratory of Ornithology's home-study course in bird biology. He is such an accomplished birder that he was on the team that won the 2002 World Series of Birding, a competitive fund-raising event that is held annually in New Jersey. Kevin and his team from the "Lab of O" identified 224 species in a twenty-four-hour period.

Today the study of crows that Kevin initiated is entering its twenty-first year. The Crow Research Group consists of Kevin, Anne, and an assortment of graduate and undergraduate students. The group monitors the population year-round and manages to locate between 80 and 100 nests in their core study area every spring. Of these, some are abandoned, and others are lost to predators. The group manages to mark nestlings from about thirty surviving broods every year. A surviving brood can have between one and six nestlings. To mark the nestlings, someone climbs the tree using climbing equipment, often supported by a belayer on the ground for safety. The nestlings are placed in a bucket and lowered to the ground. In a flurry of activity, each nestling is measured with calipers, a sample of blood is taken, a numbered metal band and a unique combination of plastic colored bands are placed on the legs, and a larger alphanumeric tag is placed on each wing. Then the nestlings are returned to the nest, requiring a second climb. The procedure must be precisely calibrated to the age of the nestlings. If they are too young, they can't be outfitted with the leg bands and wing tags. If they are too old, they attempt to escape by bailing out of the nest, like someone jumping from a burning building, and must be caught by members of the group standing underneath. Throughout the procedure, the parents and other adult crows might be angrily protesting and dive-bombing the climber, although only very occasionally coming so close that they graze one's head.

AFTER THE YEAR'S YOUNG have been marked, the Crow Research Group settles into a routine of monitoring the saga of the crows and focusing on specific aspects of their lives. This involves cruising the streets of Ithaca at a slow speed, stopping at frequent intervals to make observations, and getting permission from residents to walk on their property. The Ithaca police department knows about the Crow Research Group, so they can allay the fears of concerned citizens reporting odd behavior. Kevin and Anne might be known around town as the "crow man" and the "crow woman," but their stories pale in comparison with actual crow men and women. As the human denizens of Ithaca grapple with issues such as which café features the best lattes, the crows face monstrous predators, deadly and crippling diseases, the bitter elements, the daily problem of what to eat, and battles among themselves for prime real estate. Most of these challenges are faced not alone but with the help of kith and kin.

Meet AP HART93, a real crow man. "AP" are the letters on his wing tags, "HART" refers to the location of his natal nest, and 1993 is the year he was hatched and banded. AP was born into a crow dynasty. Crows maintain territories as families, and AP's parents had been blessed with abundant offspring during previous years. Three brothers born in 1990 (OL, PK, and RI), one brother born in 1991 (ET), and one sister born in 1992 (SA) were still present on their family's territory to welcome AP into the world. Offspring stay at home with their parents in part to help raise their younger siblings but mostly as the best way eventually to get their own territories. Crow women tend to leave home earlier than crow men because there are more vacancies available on other territories. If you're a crow woman, you place yourself in mortal danger when you incubate your eggs. A red-tailed hawk or a great horned owl can swoop down at any time of the day or night. Crows post sentry during the day to protect themselves from hawks and mob them to drive them away from the area whenever possible, but there isn't much they can do about a great horned owl in the dead of night. It's a fairly common occurrence for Anne to check out a nest and find that the mother has vanished or even to see her bones and feathers on the ground underneath. Imagine being a crow husband or child roosting on your territory at night and witnessing a monster three times your size swoop down and devour your wife or mother in front of your eyes without being able to do anything about it, except to remain motionless, since you are incapable of flying at night and don't want to be the next victim.

Every time a mother crow is killed, there is a vacancy to be filled by a woman from another territory. Crow men are less vulnerable to predators, but that makes territories harder to find. One of the best ways is by extending the boundaries of a family's territory. All three brothers born in 1990 established territories of their own in this way between 1994 and 1996. Two of them (PK and OL) maintained amicable relationships with their parents and siblings, but a falling-out occurred with RI. Evidently, RI's parents didn't want to part with some of the real estate that RI claimed for himself, so he had to battle with his own family to keep it. AP, now two years old, sided with his parents in this dispute. The next year, however, he moved onto RI's territory and sided with him against their parents. Even though crows are physiologically capable of mating as young as two, AP patiently played the role of his brother's keeper until 2001, when RI disappeared and AP became master of the territory at the age of eight. By this time, his own parents had died, and their territory had been inherited by his younger siblings. The extended family covered six territories — until the great plague of 2002–03.

Just as the Black Death swept through Europe in the 1300s and we worry about similar human pandemics today, a crow pandemic occurred when the West Nile virus arrived from somewhere in the Old World, possibly the Mediterranean area, in the early 2000s. This was also a human health scare, because the virus is transmitted by mosquitoes and can cause encephalitis in people, but for crows it is 100-percent fatal. If you get it, you're dead. The pandemic started in the New York City area in 1999, reached the Ithaca area in 2002, and spread throughout the Ithaca crow population in 2003. The Crow Research Group had not bargained on studying a plague but had no choice. They took on the grim task of collecting the dead bodies of crows that were reported by people calling into the county health departments, creating one of the best databases on the microgeography of the disease that exists. The deadliest year was 2003, when almost 40 percent of the crow population died from the disease, a proportion comparable to the Black Death's toll on Europeans in the 1300s.

It was heart-wrenching to see the plague's toll on the marked population of crows. AP's wife and one of his daughters who had remained as a helper succumbed, leaving him alone. His two older brothers, PK and OL, both disappeared from their territories. The family dynasty was in tatters. Elsewhere, young orphaned crows who lost their entire families were trying to attach themselves to surviving families, which received them with varying degrees of toleration.

Life goes on. In the winter of 2004, a marked female named 0Y BRTN02, who had been born near Barton Hall on Cornell University's campus, appeared in AP's territory and made her availability known. The courtship appeared to be going well until a four-year-old marked female named N1 HANS00, who had been biding her time as a helper two territories away, showed up and forced the young 0Y girl to leave. AP didn't appear to have much say in the matter. He might have been chosen by N1 more than doing the choosing.

PEOPLE PREACH MONOGAMY, but they're not very good at it. Crows are much more monogamous than we are. With very few exceptions, they really do stay together until death parts them. The typical crow couple builds a nest together, and the man feeds the woman while she is incubating the eggs. When the nestlings hatch, the parents take turns feeding them. When the man brings food, he usually gives it to his wife to distribute to the kids. But N1 proved to be as bossy with AP as she was with her young rival, and she made AP do most of the work. When she left to feed, she often returned without anything for the kids. She even made AP distribute the food in addition to bringing it. At first, the Crow Research Group attributed her selfish ways to inexperience, but as the years passed, she didn't get much better. The young 0Y girl would have been a better choice. She became established a few territories away and proved to be a much more attentive and successful mother.

As of this writing, AP is seventeen years old and still being bossed around by N1 within just a few hundred meters of where he was born. The oldest known crow in captivity reached the ripe old age of fifty-seven, but the Crow Research Group was the first to show that crows can survive as long as seventeen years in their natural environment. Yet not a single one of AP's children has survived, as far as the Crow Research Group knows. AP has been lucky to survive so long as an individual crow, but he will leave no genes to future generations unless his luck in the reproductive lottery improves.

For the perspective of a crow woman, meet H8 JSUP98, who happens to be 0Y's mother and who was born in a patch of woods amid Cornell University's athletic fields. She remained on her parents' territory until the age of three, when she left with her sister to check out breeding opportunities in the area. She found one eager male who was trying to establish a new territory on the Cornell campus. They built a nest together, but then H8 abruptly left him to fill a new vacancy on an established territory nearby. Who needs a struggling apprentice when you can have Donald Trump? When you're a crow woman, fidelity begins after marriage and not a minute before.

In 2002, the Crow Research Group was lucky to catch H8 and most of her family by using a rocket net launcher. A dummy version of this cumbersome piece of equipment was first placed on the family territory long enough for them to get used to it. Then the real thing was put in place, and the area in front was baited with food. When the family gathered to eat the food, the device was detonated by remote control, and rockets carried a huge net over the surprised birds before they could fly away. Then the Crow Research Group burst out of their cars to gather up the crows, mark those that had not already been marked as nestlings, and outfit some of them with radio transmitters. The group rarely uses this method because it requires so much work and has such a high failure rate. Usually, on the trapping day, the crows remain up in the trees as if to say, "Are you *kidding?*" This time, they got lucky.

The radio transmitters enabled the Crow Research Group to find the outfitted crows at any time, at least in principle. Crows do not spend all of their time in their territories. They also fly off to forage in the surrounding fields, especially during winter, and can move beyond the range of the radio receiver. Some crows even leave for extended periods, making the Crow Research Group think they are dead, until they show up months later and rejoin the family as if they had never left. Even when a crow outfitted with a radio transmitter is within the range of the receiver, the signal can bounce off the sides of buildings and be difficult to locate. It is therefore an art to walk around with the receiver issuing the signature beep of a given crow in one hand and waving a big antenna around with the other hand to determine the direction of the signal. If members of the Crow Research Group didn't look suspicious to the residents of Ithaca before, they did now.

Thanks to the radio transmitters, the Crow Research Group learned how much crow families stay together and coordinate their activities, even when they leave their territories. Wherever you find one, you are likely to find at least some of the others. During the winter, thousands of crows from farther north migrate south to escape weather even more frigid than our own New York winters. They roost in huge aggregations, and for a period during the winter of 2002, they chose to roost in the territory of H8 and her family. This was a bit like a Southern farm family hosting General Sherman's army during the Civil War. Thanks to the radio transmitters, the Crow Research Group could discover where the family chose to roost during this period. They did not join the ravenous horde but stuck to themselves in a little patch of trees on the edge of their territory.

Crow families stick together, but meltdowns can also occur. In the winter of 2003–04, Anne witnessed a titanic battle between 3L (H8's husband and master of his territory) and 2K, a male helper. Both were caught and banded as part of the rocket-launch operation, so their age and exact relationship are not known. Evidently, 2K couldn't stand his subordinate status anymore and thrashed 3L, who signaled defeat by adopting a submissive posture on the ground. 2K did not become the new dominant male but took over land adjoining 3L's territory, taking one of the other helpers with him. 0Y chose to strike off on her own at this time, leaving H8 and 3L with a single three-year-old daughter, 0Y's older sister. This remnant of a family pulled up stakes and tried to establish a new territory in one of the gorges for which Ithaca is famous. Even their last remaining daughter didn't help but found a male with whom to build her own nest nearby, although she frequently came home to visit. H8 and 3L disappeared in the summer of 2004, immediately after all activity ceased around their nest. H8 was most likely taken by a predator, although the pair's emigrating to a distant location can never be ruled out. She was only six years old at the time of her disappearance.

IF YOU ARE UNFAMILIAR with this kind of detailed research on the lives of animals in their natural environments, you are probably surprised at how human the crows seem. Anne contemplates writing a book of her own about the lives of her crows, omitting only one fact: that they are crows. The saga would make perfect sense as a human saga that took place during earlier times, when people lived in clans and fearsome predators roamed the earth. How is this possible, when crows are more closely related to dinosaurs than to us? Because what we call the human condition is based on our relationships with one another and our environments. Insofar as other species experience the same relationships, they will arrive at roughly the same solutions by the hammer blows of natural selection. The solutions will not be exactly the same, and at some point, the fact that crows and people come from vastly different lineages will make a difference in how they think, feel, and act. We shouldn't second-guess the differences, however, and above all, we should avoid our bad habit of regarding ourselves as more intelligent, cultural, and moral in every way, as if crows are simpletons compared with us.

Take tool use, for example. We are highly adapted to use tools and tend to regard it as a hallmark of general intelligence. Since we are the smartest, we assume that our closest relatives must be the next smartest and swoon with delight whenever a chimp inserts a stick into a termite mound, uses a branch as a walking stick for support while crossing a river, or uses a rock to crack a nut. Against this background, the discovery of a tool-using crow species found only on the Pacific island of New Caledonia became an international sensation. The New Caledonian crow subsists largely on insect grubs that live inside wood. If woodpeckers lived on New Caledonia, they would get at the grubs by hammering away at the wood with their chisel-like beaks. Crows don't have that kind of beak, so they arrived at the solution of fishing for the grubs with a hooked stick. Moreover, they don't just find hooked sticks to use; they *make* them out of the leaf of a common plant called *Pandanus*. They seem to have a sense of foresight and planning that was previously thought to be uniquely human or restricted to the great apes. In laboratory experiments, they will even use a tool to get access to another tool that provides access to food.

Why is the New Caledonian crow the only species of crow known to make tools in the wild? Does that make it a genius among crows? Not in the least. Tool use is not a sign of general intelligence. It is a particular skill that was important on the island of New Caledonia and not elsewhere. Other crows are roughly as smart as the New Caledonian crow, just in different ways. The very concept of intelligence as a single thing that can be ranked on a linear scale needs to be challenged.

American crows are very, very smart when it comes to the mental skills important for their survival and reproduction. They seem to weigh their social options in the same way that we do, which is why their soap operas seem like ours. They are exquisitely aware of their surroundings, perhaps even more than we are. The next time you pass some crows in an urban setting, such as a park or a city street, try a simple experiment. Stop walking, and look directly at them. Within seconds, they will realize that they are being watched and will beat a hasty retreat, even though they were mingling with hundreds of passersby before. How long would it take for you to notice that you were being watched in a crowd?

Crows recognize one another as individuals and keep careful

track of their relationships, as we have seen. They also recognize us as individuals much better than we recognize them. The Ithaca crows know the members of the Crow Research Group, their cars, and the typical behavior of crow watchers. Some fly immediately at the raising of binoculars in a car a block away. Birds disappearing over houses didn't make observing crows easier, so the Crow Research Group made the somewhat uncomfortable decision to feed peanuts, irregularly but often enough to encourage staying rather than fleeing. Since then, a Subaru wagon — the preferred vehicle of several crowers — will typically turn crow heads if it moves slowly along a roadway. If it stops, they might fly over it and wait. A few will give sharp calls as if to encourage peanut delivery. But it isn't just the cars they learn. Cars change, and the crows appear to look directly into new cars and spot familiar Crow Research Group members. They also pay attention to what the researchers are doing. They readily approach Kevin's car for a handout of peanuts, but they get enraged when there's a ladder on top, because that signals a tree-climbing event. John Marzluff of the University of Washington conducted a clever experiment in which he caught and banded crows while wearing a distinctive mask. Later, the crows that had observed this gave alarm calls when they saw him wearing that mask but not other masks.

Crows are clearly sensitive to the context of human behavior. Carolee Caffrey, who studied a population of crows in Los Angeles that lived on a golf course, decided to see what would happen if she showed up with golf clubs rather than her customary binoculars. The crows did their version of a double take. They knew Carolee as the person who was studying them, and they saw hundreds of golfers every day, but the sight of Carolee with golf clubs just didn't compute.

Crows can detect the smallest change in their environment. In one clever experiment that Anne and her students devised, a rope was threaded through six lengths of plastic pipe and knotted to form a loop. The loop was placed on the ground in a certain shape, and peanuts were placed in the center. After the crows became accustomed to eating the peanuts, the shape of the loop was changed. The crows noticed this difference and had to reacclimate to the new shape. This is why they are so difficult to catch with the rocket launcher. First, they must acclimate to a dummy version, and even then, they are likely to notice tiny differences between the dummy version and the real version on the actual day of the capture attempt.

Crows are a fascinating combination of bold and timid at the same time. They are fascinated by novelty but approach it carefully. After they decide that it's safe, they incorporate it into their repertoire if they can get something out of it. Thus, when you see crows tearing open plastic garbage bags, Dumpster diving, sticking their heads into fast-food containers, or artfully dodging traffic on a highway to get at some tasty roadkill, this is only after a trial period. The ultimate urban skill has been perfected by yet another species of crow that has moved into the city of Tokyo, which has learned to drop nuts onto the pavement at an intersection when the light turns green, so that the cars will crack the shells, which are then retrieved when the light turns red.

WHENEVER LEARNING IS DANGEROUS and time-consuming, learning from others provides a powerful advantage. If a trusted associate does something, that's a pretty good indication that you can safely do it, too. The idea that we are uniquely cultural makes little sense from an evolutionary perspective. The juvenile period of crows and many other species is a time when skills are acquired from others. If this period is disrupted, as when wild animals are raised in captivity, then they can become incapable of surviving in their native habitat. Old crows become set in their ways, just like old people. In 1997, Cornell University started composting the food waste from its dining halls, providing a bonanza for crows. The composting site is a regular stop for the younger members of the Crow Research Group's study population, but AP, who was already adult and set in his ways by 1997, is conspicuously absent.

So many skills of city and country life are learned that crows hardly ever switch from one to the other. The Ithaca crows will fly into the surrounding countryside to forage. During the winter, they will range even more widely, migrating as far south as Maryland, just as the crows farther north migrate to central New York. When it comes to setting up a breeding territory, however, not a single marked urban crow has moved even a few miles into the country, and there is no indication that country crows are moving into the city. The Ithaca crows appear to have developed a separate culture that isolates them from their country cousins. The cultural difference is so great that one crow marked in Ithaca has set up a breeding territory in the city of Cortland, forty-two miles away. This crow skipped all of the countryside in between to find another city environment.

Just because urban crows choose to live in cities doesn't mean that city life is good for them. One of Anne's graduate students, Becky Heiss, compared the growth rate of nestling crows in Ithaca and the surrounding countryside. The country crows grew larger and had more protein circulating in their blood than the city crows. Becky was able to increase the growth rate of city crows by providing more nutritious food in the vicinity of their nests. The human foods that form a large fraction of the city crows' diet are as bad for crows as for humans.

Where cultural evolution leads, genetic evolution follows. Many generations ago, a few New Caledonian crows learned to fish for grubs with sticks. They might not have been very good at it, but they did well enough to have access to a rich new food source. A cultural tradition formed that altered the course of genetic evolution. From then on, the hammer blows of natural selection favored bill shapes that could hold the stick more securely and mental abilities better suited to this particular task. Today New Caledonian crows have a bill shape unlike any other crow species, and young crows raised in captivity know how to probe holes with sticks without any learning at all. That which was once learned has become instinctive.

Urban American crows in Ithaca, Binghamton, and other cities are now undergoing the same process. Their inquisitive natures and cultural traditions enabled them to enter our cities approximately fifty years ago, where they encountered a rich new food source and a place to live. Now that they are here, they are being genetically reshaped by the hammer blows of natural selection. The Crow Research Group has only begun to investigate genetic differences between the city crows and their country cousins only a few miles outside of town, and I'll bet that they exist. The first crows to enter might have been genetically predisposed to begin with, and the differences would only become greater with the passage of generations. Even crows are a product of gene-culture coevolution, not just genetic evolution.

Anne admires my crusade to understand and improve the human condition, even if she enjoys razzing me about it. As for me, I wouldn't have her passionate interest in crows any other way. Seeing the world through the eyes of crows forever changes the way we see ourselves.

CHAPTER 17

Our Lives, Our Genes

JUST AS THE FIRST New Caledonian crow that learned to fish for grubs with tools altered the course of genetic evolution for its species, we have been altering the course of our genetic evolution with our cultural innovations. When our hunter-gatherer ancestors spread first throughout and then out of Africa, they encountered vastly different climates and environments, from rain forests, to arid grasslands, to coastal environments, to the frigid tundra. The exodus from Africa took place about 60,000 years ago, agriculture was first developed about 10,000 years ago, and the first cities were formed about 6000 years ago. A human generation is approximately 25 years, so that works out to 2400, 400, and 240 generations, respectively. Those numbers seemed small when evolution was regarded as a slow process, but they allow plenty of time for genetic evolution when the hammer blows of natural selection are sufficiently strong, as they have been throughout human history. Whenever a famine, a plague, or a war strikes, the genetic composition of the population is likely to be altered. Even opportunity can be a strong evolutionary force. A new food source, such as grains or milk for adults, selects for the ability to digest the new food source. The first people to reach North America across the Bering Strait were like the first cane toads to reach Australia, with a vast continent stretching before them. It is fully plausible that Native Americans became genetically adapted for wanderlust, just as cane toads evolved to move five times faster in only a few generations.

The concept of rapid genetic evolution was new for evolutionary biologists in the 1970s, and it is even more new when applied to our own species. Leda Cosmides and John Tooby, who spearheaded the brave new theory of evolutionary psychology in the 1990s, stressed the psychic unity of humankind, as they called it. According to Leda and John, all of us have the same tool kit of mental modules, in the same way that we have the same physical organs. Everything varies, usually as a bell-shaped curve around an average value, and part of this variation is usually genetically heritable, but it isn't very interesting. Who cares if your liver is a bit bigger than mine or if I get a little more jealous? The most important and interesting differences among people arise when the mental modules that we share in common are triggered by different environmental stimuli. Leda and John disagreed with behaviorists such as B. F. Skinner in their conception of the mind, but they agreed that individual differences are based primarily on environmental differences, not genetic differences.

Many people find it more comforting to think that we are a product of our environment rather than our genes, for the simple reason that our environment seems more changeable. Genetic influences seem to deny the capacity for improvement, and environmental influences seem to affirm it. Many people reject evolution or deny its relevance to human affairs, not because it threatens their religious beliefs but because it threatens their belief in human potential.

By now, I hope you can see that this tired formulation is wrong and that knowledge of evolutionary theory is required to realize human potential, in Binghamton or anyplace else on earth. Changing for the better is not as simple as it seems. It requires sophisticated mechanisms. If we don't understand the mechanisms, then change becomes impossible or takes a turn for the worse. We need to understand our capacity for change in the same way that we need to understand automobiles to drive them and especially to fix them when they break down.

Individual differences are like different kinds of automobiles. If you're a car owner or a mechanic, general knowledge about cars is useful, but you also need specific knowledge about the particular model that you are driving or fixing. If some of the differences among us are based on genetic differences, then so be it; that will be part of the knowledge that's required to know how to operate and fix ourselves. Understanding our genetic differences becomes an exciting challenge rather than a threat.

WE DON'T NEED TO SPECULATE about our genetic differences, because they are increasingly becoming an open book, thanks to the techniques of molecular genetics. If you swirl Scope mouthwash around in your mouth for forty seconds and deposit it into a sterile test tube, that will provide all of the cheek cells needed to analyze your DNA. Your entire genome could be sequenced — it's technologically possible now and will become affordable in the near future. Already, it is affordable to focus on specific regions of your chromosomes that are known to be polymorphic, which means that different individuals carry different DNA sequences. In some cases, differences at the genetic level are inconsequential, just as the words *gray* and *grey* mean the same thing. In other cases, the genetic differences make a huge difference in how we fight diseases, digest our food, or think and feel as we interact with one another. In many cases, we understand the chain of events that leads from gene expression to consequences for survival and reproduction in considerable detail. Thanks to this knowledge, we are discovering that your genetic ancestry influences how you respond to medication, what you should eat, and even how you get along with others and make your way through the world. Denying this information is like throwing away the owner's manual to your car.

Our DNA also contains an amazingly detailed historical record of genetic evolution. We know that the cultural practice of using milk as an adult resource arose separately in Europe and Africa, for example, because the specific genetic changes that evolved to enable adults to digest milk are different in each case. We can even estimate when a beneficial mutation arose and how fast it spread through the population. Imagine a deck of cards in which the queen of spades is preceded by the two of hearts and followed by the ten of diamonds. Now imagine cutting the deck. Unless you separated the deck precisely at the queen of spades, she will retain her two neighbors. You will need to cut the deck many times to randomize the neighborhood in the immediate vicinity of the queen.

In just the same way, when a beneficial mutation originates, it is surrounded by a particular sequence of DNA on each side. Many generations of sexual recombination are required to randomize the neighborhood around the mutation. The degree of randomization therefore provides a signature of when the mutation arose and how quickly it spread through the population. In this fashion, we can estimate that the ability to digest lactose as an adult arose approximately 3000 to 5000 years ago in Europe and even more recently in Africa. It was so advantageous that it became the most common gene in cultures where milk was readily available but not elsewhere. Thus, your ability to digest milk depends on where your ancestors came from. A similar story is emerging for the ability to digest starch and other components of our agricultural diet. Depending on your ancestry, gluten can be easily digested or a major health hazard.

When a beneficial mutation originates and begins to spread, it exists primarily as a heterozygote, paired with the most common genes at the same locus that it is in the process of replacing. As it becomes more common, it increasingly becomes paired with itself as a homozygote. Sometimes a double dose is even better than a single dose, and the new gene sweeps through the population, completely replacing the previous genes at the same locus. In other cases, the double dose produces a very different effect from a single dose. The most famous example is sickle-cell anemia, which protects against malaria as a single dose but produces anemia as a double dose. Natural selection has no foresight and doesn't know that a mutant gene conferring a benefit as a single dose will create its own problem as a double dose. The problem simply arises as the gene becomes more common. Eventually, the frequency of the gene reaches a stable equilibrium when the benefits enjoyed by the heterozygotes are balanced by the costs suffered by the homozygotes. If you suffer from sickle-cell anemia, then you are a victim of this cruel balance sheet. In fact, most of us carry at least some genes that harm us rather than helping us. They persist in the population because they are beneficial in other individuals.

OF THE MANY GENETIC POLYMORPHISMS discovered by molecular geneticists, some influence the brain and therefore the ways we think and feel. Let me introduce you to two of my own genes that influence how I respond to the neurotransmitter dopamine. Each nerve cell is connected to many other cells by projections known as axons (a single long projection) and dendrites (many short projections). When an electrical signal reaches the end of an axon or dendrite, it causes chemicals known as neurotransmitters to be released into the tiny gaps between nerve cells called synapses. The neurotransmitters diffuse across the gap and bind to receptor sites on the adjacent nerve cells. If a sufficient number of neurotransmitter molecules bind to the receptor sites, then the adjacent nerve cell fires, and the signal is propagated electrically down its neuron and dendrites. The neurotransmitter molecules are then released from the receptor sites and diffuse back across the synapse to be reabsorbed by the signaling nerve cell. In this fashion, information is transmitted by electrical signals within nerve cells alternating with chemical signals between nerve cells.

A long list of chemicals serve as neurotransmitters, of which dopamine, serotonin, epinephrine (adrenaline), and oxytocin are among the most familiar. The density of receptor sites sensitive to each neurotransmitter also varies. The variety of neurotransmitters and the density of receptor sites give our nervous system the flexibility it needs to carry out the immensely complicated task of surviving and reproducing. Dopamine is especially important in orchestrating our reward system. When dopamine is chemically blocked in rats, they cease to want things and won't even bother to eat. They appear to derive pleasure from eating when they are force-fed, but they still have lost their desire. Thus, there appears to be a difference between wanting and liking as far as the organization of the brain is concerned. Among its many functions, dopamine increases our desire.

A gene on our eleventh chromosome called DRD4 is polymorphic and influences the density of receptor sites sensitive to dopamine. A sequence of forty-eight base pairs (the "letters" of the genetic language) is repeated a variable number of times, from two to eleven, which are often classified as "short" (two to six, with four being most common) and "long" (seven to eleven, with seven being most common). The more repeats, the fewer the number of receptors sensitive to dopamine. The fewer the number of receptors, the more dopamine is required to get a nerve cell to fire. Stronger experiences are required to release more dopamine. Thus, if this long chain of inference is correct, people with the "long" version of the DRD4 gene desire stronger experiences, which they need in order to get the same feeling of reward as people with the "short" version.

I happen to be a double seven, which makes me the envy of my friends. We're all curious about our genes and have swished with Scope to have our cheek cells analyzed. It seems glamorous to have strong desires, so my friends all hope to have the long allele and are crestfallen when they come up short. I like to strut around as if I am James Bond but only in jest. The actual behavioral differences between people with the long and short alleles are much more complex, as we shall see. I have strong desires in some respects but not others. I'm a crusader for evolution and boldly travel to remote corners of the Ivory Archipelago, but I'm shy around women, you'll never get me on a roller coaster, and I won't even go to a movie if there's too much violence.

What does scientific research say about the DRD4 polymorphism? The four-repeat (4R) allele is ancient, stretching back 500,000 years or even earlier. The DNA sequences on each side have been thoroughly randomized by sexual recombination. The seven-repeat (7R) allele is more recent, with highly nonrandom DNA sequences on each side. Yet 7R is found throughout Africa in addition to rest of the world. It probably originated shortly before our ancestors left Africa. The 2R allele is even more recent and evolved as a mutant version of the 7R in Asia.

Even though 4R and 7R are found throughout the world, they are very unequally distributed. The 7R allele is rare in Asia and especially common in South America and the Pacific Islands. If it's true that people with 7R desire stronger experiences, then perhaps they were less likely to settle down and more likely to explore beyond the horizons. If so, they would be among the first people to reach South America and the Pacific Islands, just like those fast-moving cane toads. In the meantime, Asians have been living in densely populated agricultural societies for thousands of years, allowing plenty of time for people who thrive under settled conditions to be favored by natural selection.

A lovely study of an African tribe that inhabits northern Kenya demonstrates these global trends in miniature. The Ariaal are traditionally nomadic cattle herders, but some of them have settled into villages. Benjamin Campbell, an anthropologist at Boston University, took blood samples and detailed measurements of eighty-seven men living in settlements and sixty-five men who were still nomadic. Back in Boston, the blood samples were used to genotype the men by Daniel Eisenberg, one of my own former students. The frequency of the 7R allele was about 20 percent in both groups, but men with 7R had a higher body mass index (BMI) in the nomadic group and a lower BMI in the settled group. The Ariaal are an undernourished people, so a high BMI is a good thing. The 7R allele was providing a benefit for the nomadic men but became a liability when they settled down. One can well imagine the worldwide trends resulting from the differences in fitness revealed by this study, operating over many generations.

HUNDREDS OF SCIENTIFIC STUDIES have attempted to link the 7R gene to clinical disorders and to personality traits such as novelty seeking, impulsivity, and extroversion. The results are more complicated than you might think. One study will show an effect that is statistically significant, but subsequent studies will fail to replicate the effect. The number of studies is so great that a method called meta-analysis has been developed to study the studies. Three recent meta-analyses have examined the linkage between the 7R allele and attention-deficit hyperactivity disorder (ADHD), addiction, and personality traits. There is indeed an association with ADHD, although a lot of studies were required to draw that conclusion. The symptoms of ADHD are inattention, hyperactivity, and impulsivity. Kids with

ADHD have a big problem sitting still and focusing on their classwork at school. It's easy to imagine that kids with the 7R gene would be poorly suited for the classroom environment, in the same way that Ariaal men with the 7R gene were poorly suited for settled life. They're not sick but merely mismatched to their environment. Should they be medicated, or should we attempt to change the classroom environment to accommodate their particular temperaments? Even if ADHD is a disorder that calls for medication, it's easy to imagine that kids with the 7R gene would be disproportionately represented, because their intolerance for routine would combine with other risk factors.

The other two meta-analyses conclude that there is no association between the 7R allele and addiction or personality traits such as novelty seeking, impulsivity, and extroversion. To illustrate the scale of this kind of research, the meta-analysis of personality was conducted by a team headed by Marcus R. Munafo at the University of Bristol in England and included a review of forty-eight published studies, along with a consideration of publication bias, since studies that fail to show a statistically significant effect often go unreported. Just as history can be inferred from our DNA sequences, a publication bias can be inferred from the pattern of published results. After reviewing the literature, the team then conducted its own study based on a sample of more than 40,000 men and women from the southwest of England who had taken one of the standard personality tests. The 1000 highest-scoring individuals and the 1000 lowest-scoring individuals for the trait of extroversion were contacted to obtain DNA samples. After all of this prodigious work, the authors concluded that there was no association between the 7R gene and any approach-related personality trait. Another nearby polymorphism called C-521T *might* be associated with novelty and impulsivity but not extroversion.

The problem with this herculean and seemingly authoritative approach is that it fails to distinguish among contexts. Suppose that Ben Campbell and Dan Eisenberg had conducted their study of Ariaal men without distinguishing between the nomadic and settled groups. They would have found no association between the 7R gene and BMI, because the strong positive association in the nomadic group and the strong negative association in the settled group would have canceled each other out. Many strong associations can look like no association when you fail to distinguish among contexts.

The meta-analysis of addiction, authored by John McGeary at Brown University, does a better job of distinguishing among contexts. When the voluminous literature relating the 7R gene to clinically diagnosed addiction is evaluated, no consistent picture emerges, just as with personality traits. McGeary then went further and analyzed component traits that might lead to addiction in some contexts but not others. The 7R gene appears to increase desire, but many additional factors must come into play before desire leads to addiction. When McGeary reviewed the evidence linking the 7R to desire, he found a more consistent picture than the linkage with addiction.

James MacKillop, who joined McGeary's laboratory after obtaining his PhD at Binghamton, has provided some of the best evidence for the link between the 7R gene and desire, as well as the looser link between desire and addiction. A study that he performed at Binghamton illustrates the care with which this kind of research in clinical psychology is conducted. He began by recruiting Binghamton University students age twenty-one or older who were heavy drinkers. Only men who drank more than twenty and women who drank more than fourteen alcoholic beverages per week need apply. They also had to list beer among their favorite forms of alcohol and to rate their enjoyment of beer at seven or greater on a ten-point scale. Thirty-five students responded to the ads placed around campus, which promised both money and beer for participating. They exceeded the minimal requirements by a long shot, with an average of thirty-six drinks per week for the men and twenty-one for the women, placing them at very high risk for alcohol-related problems.

Each student participated individually in a session that lasted ninety minutes. First, they were taken to a room without any alcohol-related cues to fill out a consent form and take a breath test to make sure that they weren't already hitting the bottle. They completed some questionnaires about their alcohol use, their craving for alcohol, and their current mood. They were told that they would receive a minimum of ten dollars and possibly more. After these preliminaries, the main experiment began. Half of the participants were randomly assigned to enter a room filled with beer paraphernalia, including posters, barroom trifolds, and empty beer bottles. A bottle of their favorite beer was poured into a glass, and they watched a video that provided more cues about the joys of drinking beer. The video also instructed them to hold the beer to their nose for a good sniff. The rest of the participants were taken to another room filled with cues about the joys of drinking water.

After these two different cuing experiences, all of the participants completed the craving and mood questionnaires a second time and were given a choice between a sip of beer (1.5 ounces) or one dollar. If they chose the beer, the amount of money offered was increased to as much as five dollars until they chose the money. If they chose the money, the amount was decreased to as little as ten cents until they chose the beer. The amount of money that they were willing to forgo for a sip of beer provided a measure of how much they *desired* the beer. After they made their choice, they were given either the money or the sip and led back to the first room to be debriefed and relax until the session was over. Finally, cheek cells were obtained with a cotton swab. Their genotypes were determined only after the experiment was over.

What were the results of this fascinating experiment? The participants varied widely in their subjective craving for alcohol when they began the session. Those who received cues for alcohol craved it even more. The more they craved, the more money they were willing to pay for that one sip—but individuals with the 7R gene were willing to pay more money than individuals with the 4R gene. Their desire for the sip was stronger.

This experiment illustrates why the literature as a whole might be inconsistent. Talk is cheap, making self-reported craving a relatively crude measure of desire. Finding the break-even point between a sip of beer and money provided a much more refined measure. Thus, dozens of studies in the literature that measure only self-reported desire might fail to show a result because the measurement of desire is too crude. The linkage between desire and addiction is also complex. Desire can be directed at many things and only leads to addiction under special circumstances. Some people become addicted not because they crave the pleasure of intoxication but because they are avoiding problems in their lives. Avoidance is different from attraction and relies on other neurotransmitters, such as serotonin, more than dopamine. A strong connection might exist between the 7R gene and certain kinds of addiction that result under certain circumstances but not with addiction as a whole. Life is complex. If you don't take context into account, strong patterns cancel themselves out and seem like meaningless noise.

THE STORY OF MY 7R gene is only one of many that can be told for genetic polymorphisms at other loci. If genes are the "words" and single nucleotide base pairs are the "letters" of the genetic language, the human genome has approximately 25,000 protein-coding word genes and 4 billion base letter pairs. Of the 4 billion base pairs, about 4 million are polymorphic and are called single nucleotide polymorphisms (SNP). That's a very small fraction (4 million in 4 billion equals one in a thousand) but still a very large number of polymorphisms. Enough human genomes have been sequenced around the world to calculate the degree of geographical variation for each of these SNPs using an index called Fst. Let's say that a particular SNP consists of two alternative forms at a frequency of 50 percent worldwide. If the frequency is exactly 50 percent at each geographical location, then the value of Fst would equal 0. If the frequency is either 0 percent or 100 percent at any particular location, then the value of Fst would equal 1. Some SNPs have no effect on survival and reproduction, such as the alternative spellings of *gray* and *grey*. Their frequency will vary geographically based on chance processes, resulting in intermediate values of Fst. Especially low Fst values signify stabilizing selection, which means that the hammer blows of natural selection are keeping the polymorphism at the same frequency and not allowing it to drift. Especially high Fst values signify disruptive selection, which means that the hammer blows of natural selection are favoring different forms at different geographical locations.

In a display of scientific brawn, a team headed by Roberto Amato at the University of Naples in Italy used a massive database called the

HapMap project to calculate the Fst values for all 4 million SNPs in the human genome, or 3,917,301 SNPs, to be precise. Using another massive database called the Genetic Association Database (GAD), they assigned the SNPs to categories based on gene action, such as cardiovascular, metabolic, or psychological processes. SNPs that influence psychological processes had higher Fst values, on average, than any other category. The bottom line: recent genetic evolution has been hammering our psychological traits into different shapes at different geographical locations even more strongly than it has been hammering our physical, metabolic, or even immunological traits. We might not know the details for any particular trait, but how we think and feel as we make our way through life depends in part on our genetic ancestry, even more than how we digest our food or respond to disease organisms.

REGARDLESS OF THE GENETIC EVOLUTION that took place during the last 10, 100, and 1000 generations, modern life is a whole new ball game. The ancestors of the people I pass on the streets of Binghamton could have come from anywhere on the face of the globe. They are carrying genes that suited their ancestors well enough to survive and reproduce, but how well those genes function in the city of Binghamton is anyone's guess. The only way to find out is by paying close attention to context.

As soon as I began to appreciate the importance of context, I realized that the Binghamton Neighborhood Project could uniquely contribute to the age-old nature-nurture debate and the gargantuan field of modern human genetics. All of those sophisticated studies were only sophisticated on the *genetic* end. When it came to an appreciation of context, most of them were grossly inadequate. The BNP was gathering richly contextual information about people in relation to one another and to their environment in the city. By combining this information with genetic information, we could become sophisticated with respect to both genes and context. Koji Lum, a molecular biologist in our anthropology department, could handle the genetics end, and EvoS supplied an enthusiastic cadre of students.

We decided to call the project "Our Lives, Our Genes" (OLOG) and to concentrate on the elderly population of the Binghamton area. Our reason for choosing the elderly was that they have already made most of their life choices. James's experiment with heavy drinkers showed that actions speak louder than words. What people decide to do when faced with different options is more revealing than what they say they might do in hypothetical situations. By gathering detailed biographical information about the choices that elderly people made in their lives, reflected in their marriages and other relationships, their professions, and other interests, all against the background of their local social environments, we might bring the influence of their genes into sharper focus. Never mind that I already had too many projects to count. Suddenly, I wanted to have every old person in the city of Binghamton swish with Scope and tell me the story of his or her life.

But would the old people be interested? Would they be threatened by the prospect of learning that their life choices had been influenced in part by their genes? My first opportunity to find out came when I offered to give a lecture at the Lyceum, an association of "adult learners" older than fifty that is affiliated with the national Elderhostel. The Binghamton branch has more than 500 members, and about 75 signed up to hear me give a sales pitch about participating in the "Our Lives, Our Genes" project.

The Lyceum holds its classes in the basement area of the Blessed Sacrament Church, which is just across the Susquehanna River from the university campus. The people who attended ranged from their fifties to their eighties. I was asked to use a microphone and to turn up the volume for the hearing-impaired. I began with the inescapable fact that we are so different from one another. Psychologists have been studying human personalities for many decades, but how would a biologist approach the subject?

Then I posed the question of why species are so different from one another. The answer, crudely speaking, is that they occupy different niches. Many species coexist because there are many different ways to survive and reproduce. Might individual differences within species be explained in the same way as differences between species? The answer is yes. Based on the research of animal behaviorists such as Anne, we know that other creatures in addition to ourselves have distinct personalities and that their differences make a difference for survival and reproduction. I could tell that my audience was charmed by the idea of animal personalities and fascinated by examples of creatures as diverse as bushbabies, crows, sunfish, and water striders.

If individual differences in other species can be explained as different solutions to life's problems, how about our own individual differences? Might there be costs and benefits associated with any particular strategy that balance out in the long run? Do extroverts take life by the horns, sometimes only to get gored? Do shrinking violets play it safe for good reason, at least some of the time? My audience loved thinking about human differences in this way. They were already peppering me with questions without a hint of defensiveness or threat.

Where do our differences come from — our environments, our genes, or both? The answer is both, of course, but the beauty is in the details. To illustrate environmental effects, I talked about how early experiences can cause children to become either securely or insecurely attached to their caregivers and how these attachment styles can have lifelong consequences. To illustrate genetic effects, I asked how many people in the audience had more than one child. Most of them raised their hands. Then I asked whether their own children were different from one another from day one. All of the hands remained raised. I concluded by saying, "You don't need a scientist to tell you that our personalities are influenced in part by our genes."

Then I reviewed some of the research that I have covered in this chapter, emphasizing the importance of context and leading to my key point: a comprehensive study of personality must include information on genes, individual experience, and social context, but very few studies have been this comprehensive. That's where we could contribute. With the "Our Lives, Our Genes" project, we could advance the frontier of knowledge. When I tried to explain why the study focused on the elderly — because they have already made most of their life decisions — I paused because I didn't quite know how to phrase it diplomatically. Fortunately, a wizened old lady in the front row helped me out by shouting, "Because we've been there, done that!" The audience broke into laughter, and I knew that most of them were eager to participate. They saw it as a fascinating learning opportunity, just as I did.

I turned the lecture over to Justin Garcia, an undergraduate student who was part of the new OLOG team. Justin became involved with EvoS as a freshman, and by now he was functioning like a graduate student. He also has an exuberant personality, and I could tell that the audience loved his youthful charm as he told them about neurotransmitters and described the procedure of the study for those who wished to participate. First, they would sign a consent form and provide contact information, then they would swish with Scope ("Your mouths will be minty-fresh!" Justin promised them), and then we would schedule them for a structured interview about their lives that would minimally require about an hour.

The audience clapped boisterously, and most of them were eager to participate. They made their way to the tables in the back to sign up and receive their test tubes of Scope and a short straw. They followed along as Justin instructed them like a bandleader. "Take the Scope into your mouth. Now, swish back and forth for forty seconds. Do not draw it through your teeth, because we don't want food debris. Now, deposit the Scope back into the test tube with the straw. Don't you feel minty-fresh? Go ahead and kiss each other whenever you like!" I'm beaming at the positivity of it all, which is exactly the attitude that I was hoping and striving for.

You might think that I would leave the actual interviews to other members of my team. On the contrary, I make sure to do my share of interviews because it is such a privilege. When an OLOG participant tells me his or her story, I feel that science, literature, and religion have become one. Like Isaac Bashevis Singer's father, the rabbi, a parade of visitors share their lives with me. Their interviews are digitally recorded and form a rich archive of stories, the stuff of literature. The same interviews result in columns of numbers to add to our swelling database, the stuff of science. The point of it all is to understand and improve the human condition, the stuff of religion, literature, and science.

* * *

THE OLOG PROJECT IS STILL in progress as I write these words. The richness of the information that we gather on our participants is beginning to make more sense of the genetic data than other studies with much larger sample sizes but much less contextual information. The Lyceum folks come from all walks of life and joined the Lyceum because of a common love of learning new things, even at an advanced age. In this respect, they are a special breed. As a group, they are exceptionally well adjusted, and not a single person we interviewed committed a criminal offense. Yet they are genetically different from one another, and their genetic differences made a difference in how they navigated their way through the stormy sea of life to reach such a safe harbor.

As the OLOG project expands beyond the initial sample of Lyceum folks, we are chronicling the lives of others who have not fared so well, whether by luck, their own choices, or both. Each person carries genes that have survived in an unbroken chain since a first mutation that occurred in the distant past, which might be 5000, 50,000, 500,000, or 5 million years ago. They also carry cultures from their past that might trace back thousands of years, judging by the crosses and golden domes that grace the churches around our city. Genetic evolution is fast enough and cultural evolution is slow enough for the two to become entwined in a double helix of their own. Both our genes and our cultures have served us well in the past—we're here, after all—but there is no guarantee that they will serve us well in the future. Evolution has no foresight, regardless of whether it is genetic or cultural. It's up to us to become wise managers of evolutionary processes, which requires understanding the complex interactions among our genes, our cultures, and our lives. Not only are we starting to do this with the OLOG project, but we're showing how it can be done in a spirit that a religious person might call joy.

CHAPTER 18

The Natural History of the Afterlife

THE INTERTWINING OF GENETIC and cultural evolution means that we need to think about culture in a new way, not only as scientists but also as concerned citizens. Our conscious intentions are just the tip of the iceberg. Beneath the surface, cultural processes are influencing our behaviors in ways that are no more accessible to our conscious awareness than the mechanisms of gene expression. Only the tools of science can enable us to perceive and reflect on these hidden cultural processes, so that we can act more effectively than we have before.

Of all of the cultural forms that inhabit the human tangled bank, religions captivate and puzzle the scientific imagination the most. For a scientist, the two gold standards for believing something are consistency with natural processes (the laws acting around us, as Darwin put it) and factual evidence. By these standards, most religious beliefs should be quickly abandoned, in the same way that Isaac Bashevis Singer's mother rejected the notion that dead shrieking geese are proof of the creator. Yet religions have not retreated with the advancing tide of science. They remain strong in my city of Binghamton, no less than in the rest of the world. They are also unevenly distributed. Not everyone in Binghamton is religious, and those who are practice their faiths in very different ways. On a planetary scale, western Europe was a boiling cauldron of religious fervor a few centuries ago, but its churches and cathedrals stand largely empty today.

Religion is thriving in America, the Middle East, and most so-called developing parts of the world. Affluence can't be the whole story for the decline of religion, because religiosity remains strong even among many affluent Americans.

Pierre Teilhard de Chardin was ahead of his time when he described religions as life forms that proliferate like species. The gold standards for the survival of any given religion are how it captivates the mind and what it causes the religious believer to do, which in turn depends on the particular environment. Religions are truly like species, with niches that expand and contract over time. As modern-day evolutionists, my colleagues and I are establishing a field of evolutionary religious studies that fulfills Teilhard's prophetic vision. Not only are we explaining the phenomenon of religion in general terms, but we are also in a position to explain the distribution and abundance of religions within a single human population such as Binghamton, New York. The knowledge we are acquiring is of immense practical value in addition to its intellectual import. To the extent that religions contribute to human thriving, understanding them from an evolutionary perspective can help us thrive better.

MY OWN INTEREST IN RELIGION is a natural outgrowth of my interest in the evolution of groups as adaptive units in all species. Religious believers are fond of describing their communities as like bodies and beehives. I'm the guy who studies bodies and beehives, so shouldn't I be adding religious groups to my list? The John Templeton Foundation, which was then just making the scene among philanthropic organizations, announced a program for studying forgiveness from a scientific perspective. It was John Templeton's view that the religious and spiritual traditions of the world contain great wisdom that science can further enlighten. The relationship among religion, spirituality, and science can be positive rather than negative. Like so many virtues, forgiveness is closely associated with religion but is by no means restricted to religion. The Templeton Foundation wanted to know what science had to say about forgiveness and was prepared to pay for it.

As it turned out, science had very little to say about forgiveness up to that point. It was a backwater topic in psychology, along with most other human virtues. For some reason, psychologists focused primarily on what makes us sick rather than what makes us thrive. Today that problem is being remedied by a field called positive psychology, which was initiated as a self-conscious movement in 1998 by Martin Seligman at the University of Pennsylvania. The Templeton Foundation announced its forgiveness program the same year and has been a major supporter of the positive-psychology movement ever since.

I teamed up with anthropologist Christopher Boehm to submit a proposal on forgiveness from an evolutionary perspective. Chris is one of my closest colleagues and has studied everything from the feuding societies of Montenegro to chimpanzee behavior. For the grant, we proposed that he would focus on forgiveness in hunter-gatherer societies, and I would focus on forgiveness in the major religious traditions around the world. We were blessed by the Templeton Foundation and embarked on an adventure that never would have occurred otherwise.

My effort resulted in *Darwin's Cathedral: Evolution, Religion, and the Nature of Society,* published in 2002. The subtitle makes the point that we can't understand religion without understanding the nature of human society in more general terms, which in turn must be grounded in evolutionary theory. The chapter on forgiveness is titled "Forgiveness as a Complex Adaptation." It is predicated on the view that all enduring cultures are complex systems that, like organisms, receive environmental information as input and result in meaningful action as output. Without cultures to organize our experience, none of us could function even for a single day. Émile Durkheim had this insight back in 1912, when he wrote in *The Elementary Forms of Religious Life:* "In all its aspects and at every moment in history, social life is only possible thanks to a vast symbolism."

We rely much more on symbolism than other social creatures do, as Teilhard was among the first to appreciate. Other creatures have culture, as we saw with the crows, and our cultures are thoroughly entwined with our genes, as we saw in chapter 17, but our cultures play such an important role in translating experience into action that they deserve to be called organisms in their own right.

Like a single organism, a culture must be sophisticated in order to survive and reproduce in a challenging world. It's not enough for a biological organism to "do X," as we saw in the parable of the strider. Striders and all other creatures have a large repertoire of behaviors that are appropriately matched to their environmental circumstance, with the matching accomplished by the hammer blows of natural selection. If cultures are like organisms, then they, too, must somehow encode a large repertoire of behaviors that are appropriately matched to environmental circumstance. There must be cultural analogues to the biological concepts of anatomy, physiology, mentality, development, replication, and expression.

Against this background, the idea that forgiveness is a simple behavioral rule in religious cultures, such as the Christian injunction to turn the other cheek, is woefully inadequate. Forgiveness is the failure to retaliate in response to a social transgression. It is an adaptive behavioral response under some circumstances and highly maladaptive under others. The capacity to forgive is part of our innate psychology, although there might also be important individual differences. The expression of the capacity is orchestrated in large part by our cultures. In my chapter on forgiveness in *Darwin's Cathedral*, I showed that the Christian concept of forgiveness is impressively context-sensitive, as one would expect for a cultural organism. Moreover, insofar as religious groups inhabit different environments, their particular rules about the expression of forgiveness must also differ in order to remain adaptive.

Different cultures are like different species. The evolutionary and ecological tool kit that I learned as a biologist could be transferred to the study of culture, explaining both cultural diversity and the details of single cultures. I could explain not only macro differences among major religious traditions but also micro differences, such as among the four gospels of the New Testament. Just as Darwin was bedazzled by his first encounter with the tropical rain forest, I was bedazzled by my first look at religious diversity through the lens of evolutionary theory.

* * *

I WAS NOT THE ONLY EVOLUTIONIST who had become interested in religion. Pascal Boyer's audaciously titled *Religion Explained* was published in 2001, Scott Atran's *In Gods We Trust: The Evolutionary Landscape of Religion* was published in 2002, and Lee Kirkpatrick's *Attachment, Evolution, and the Psychology of Religion* was published in 2004. The number of academic journal articles on religion from an evolutionary perspective was also swelling. Even though we were all evolutionists, we didn't agree on the nature of religion. Pascal, Scott, and Lee saw religion not as a complex adaptive system but as a by-product of traits that evolved for reasons that have nothing to do with religion.

To understand the idea of an evolutionary by-product, consider why moths are attracted to flames. They are adapted to navigate by celestial light sources such as the moon and stars. They fly in a straight line by holding these light sources in a constant orientation, but earthly light sources cause them to spiral inward. The moth's fatal attraction to flame is a by-product of its ability to navigate. Stephen Jay Gould used an architectural metaphor to make the same point. Whenever arches are placed next to each other, they create a triangular space called a spandrel. Arches have a function, but spandrels do not, although they can acquire a secondary function, such as a decorative space. Gould criticized his evolutionist colleagues for failing to appreciate the importance of spandrels and for trying to explain spandrels as adaptations, as if moths evolved to be attracted to flames or buildings were constructed to create spandrels.

Pascal, Scott, and Lee interpreted the elements of religion as by-products without a function. Other evolutionists interpreted religion as adaptive for individuals but not for groups, as adaptive for the cultural trait without necessarily benefiting human individuals or groups, or as adaptive in the past but not the present. All of this squabbling might seem to reflect poorly on evolutionists. If we can't agree with one another, why should anyone else listen to us?

In our defense, scientists are expected to disagree at the beginning of an inquiry; it is only at the end that they are supposed to agree based on a systematic test of their alternative hypotheses. Whenever evolutionists begin studying a biological trait, such as the beaks of finches or the spots on guppies, they must distinguish among a number of reasonable possibilities. The trait might be an adaptation—or not. If it's an adaptation, it might be good for the individual (the parable of the strider) or the group (the parable of the wasp). If it's a human cultural trait, it might spread by virtue of benefiting human individuals, human groups, or only itself as a kind of parasite. If it's not an adaptation, it might have been adaptive in the past but not the present, a by-product of a trait that is currently adaptive, or simply neutral, with no effect on survival and reproduction. These are all reasonable possibilities, and the only way to determine the facts of the matter is by consulting the empirical evidence for each trait on a case-by-case basis. Decades were required for evolutionists to agree that beak finches and guppy spots are adaptations that evolved by natural selection, primarily at the level of individual organisms. The study of religion from an evolutionary perspective is just getting started. As Jack Benny replied when given a choice between his money and his life, don't rush us—we're thinking about it!

I'm optimistic about reaching a consensus on the subject of religion because the major evolutionary hypotheses make testable predictions, and we are overflowing with information about religions around the world and throughout history. There's something you should know about religious scholars. Most of them are not theologians or religious believers. Even when they are, they are expected to check their particular beliefs at the door and study religion as a human construction, like any other cultural system such as a nation or an indigenous culture. Scholars of religion avoid the concept of a God who actively intervenes in the affairs of people as scrupulously as biologists, for the same reasons. In the first place, there's no good evidence for it. In the second place, it's hard to make predictions based on the concept of an intervening God because no one knows the intentions of supernatural agents, even if they exist. For both of these reasons, the vast majority of religious scholars call themselves methodological naturalists. They attempt to explain religion based on the "laws acting around us," as Darwin put it in his tangled-bank passage, including the laws governing human behavior.

The more I became involved in the study of religion, the more I regarded myself in a situation comparable to Darwin's. The natural historians of his day had accumulated an enormous amount of information about plants and animals around the world. Most of it was descriptive, lacking the trappings of modern science, but it was reasonably accurate, because natural historians were careful observers and had strong norms that prohibited making stuff up. Other travelers might invent tall tales about mermaids and dragons, but natural historians held one another accountable to higher standards, and the books and journals of Darwin's day were overflowing with natural-history information. But it wasn't *organized*. Darwin's great achievement was to organize all of that information with his theory of evolution. He approached natural historians with the greatest respect. They had the vital information that he needed, regardless of what they might think of his theory. As one example, Darwin admiringly referred to the great French entomologist Henri Fabre as "that inimitable observer," overlooking the fact that Fabre was a creationist, as were most other naturalists at the time.

My need to consult with scholars of religion was precisely like Darwin's need to consult with natural historians. They had the vital information about religions that I needed, regardless of what they might think about evolutionary theory. By cultivating a respectful relationship, it might be possible to organize the voluminous information on religion in the same way that Darwin organized natural-history information. Religion is not a single trait but a collection of many interacting traits. Each trait needs to be evaluated with respect to the major evolutionary hypotheses, similarly to finch beaks and guppy spots. All of the hypotheses are reasonable, so the challenge is to make a determination on a case-by-case basis. After a large number of cases are examined, we'll be closer to having a fully rounded understanding of religion from an evolutionary perspective, which can be applied on a small spatial scale, such as my city of Binghamton, in addition to around the world and throughout history.

AFTER *DARWIN'S CATHEDRAL* WAS COMPLETED, the Templeton Foundation announced a new funding initiative, "Competitive Dynamics and Cultural Evolution of Religions and God Concepts." Cultural evolution, like forgiveness, was a scientific backwater until the Templeton Foundation got involved; it had a knack for turning backwater subjects into active areas of scientific inquiry. I teamed up with Bill Green, the religion scholar and administrative dynamo who hosted the Evolution Institute's first workshop on childhood education, and we submitted a proposal with two ambitious goals. First, we would attempt to establish a community of evolutionists and religion scholars that would work together to study the nature of religion, including the creation of an Evolutionary Religious Studies Web site (http://evolution.binghamton.edu/religion). Second, we would evaluate beliefs in the afterlife as cultural traits that arose at particular times and places in human history. Why did a given belief in the afterlife arise when and where it did? Why did it spread compared with other beliefs? Why did it eventually become restricted to some human groups and not others? What influence, if any, did it have on human behavior? Was it good for the group, for the individual, or only for itself? Was it useful when it arose, only to become obsolete or a liability in today's world? Was it a detrimental by-product from the start, like a moth to flame? Or was it simply neutral, like *grey* as an alternative spelling of *gray?* All of these are reasonable possibilities that must be evaluated for each afterlife belief, letting the chips fall where they may. When a large sample of beliefs had been examined, we would have a more fully rounded understanding of afterlife beliefs from an evolutionary perspective, providing a model for the study of other elements of religion.

Our proposal wouldn't have had a snowball's chance in hell of being funded by an agency such as the National Science Foundation, but it was exactly what the Templeton Foundation had in mind for its new initiative. The officers of the foundation were deeply committed to religion and spirituality, but they were equally committed to science and unafraid to see where this purely naturalistic examination of afterlife beliefs might lead.

Our first step was to organize a meeting of evolutionists who currently disagreed with one another about the nature of religion. We might favor different hypotheses at the moment, but could we agree on the information that would be required to settle the matter among ourselves? The meeting was an experiment in its own right. Scientists easily lapse into us-versus-them thinking when they disagree with one another, just like anyone else. Present them with a common goal, and they come together as easily as they divided. When our group was presented with the common goal of identifying the information required to settle our differences, we came together as a team and enjoyed one another's company more than when we were bickering over our differences. We collectively wrote a protocol consisting of twenty questions about any given belief in the afterlife that needed to be answered in narrative form, plus seventy-three questions to be answered on a numerical scale from one to ten. The protocol was designed to be used by religious scholars, regardless of what they might know or think about evolution, so that they could provide their detailed knowledge in a way that could be compared across afterlife beliefs. A second meeting brought the evolutionists together with a team of religion scholars organized by Bill so that the collaboration could move forward.

I was so excited to get started that I decided to conduct my own inquiry on the mother of all afterlife beliefs—at least, if you are a Christian—the concept of the resurrection associated with early Christianity. One of my sources was the book *Resurrection: The Power of God for Christians and Jews*, by Kevin J. Madigan and Jon B. Levenson. Madigan is a Christian, Levenson is a Jew, and both are professors at the Harvard Divinity School. They are theologians, which means that they attribute profound meaning to religion, rather than treating it as merely an object of intellectual interest. They are also world-class scholars of religion, which means that they try to explain the history of Judaism and Christianity based on the laws acting around us. They are secure enough in their faiths to be unafraid about where methodological naturalism might lead them.

BELIEF IN HEAVEN AND HELL is part of the beating heart of Christianity, which historically emerged from Judaism, as everyone knows. It is therefore remarkable that belief in heaven and hell was not an important part of Judaism during most of its long history. Many people are amazed to learn this fact—I was—but it is common knowledge among scholars of the period. Every enduring religion must strongly motivate its believers, but Jews were motivated by the prospect of establishing the nation of Israel on earth, not a glorious afterlife. God rewarded and punished his people by granting or denying them abundant progeny, a homeland, and conquest over their neighbors. There was a belief in life after death, but it was conceptualized as Sheol, a gloomy place where everyone goes regardless of whether he or she is good or bad. Here's how Johannes Pederson, a prominent scholar of the Hebrew Bible, quoted by Madigan and Levenson, describes the concept of Sheol:

> Everyone who dies goes to Sheol, just as he, if everything happens in the normal way, is put into the grave.... Firmness, joy, strength, blessing belong to the world of light; slackness, sorrow, exhaustion, curses belong to the realm of the dead.

Similar conceptions of the afterlife were common throughout the ancient world, including the Greek and Roman conception of Hades. These conceptions were not materialistic; there *was* belief in life after death, just nothing to look forward to. Moreover, the ancient Jews didn't spend a lot of time thinking about Sheol, even when they thought about death. As good scholars, Madigan and Levenson report that the word occurs fewer than seventy times in the entire Hebrew Bible, compared with roughly a thousand references to death. According to Philip Johnston, another Judaic scholar:

> "Sheol" never occurs in the many narrative accounts of deaths, whether of patriarchs, kings, prophets, priests, or ordinary people, whether of Israelite or foreigner, of righteous or wicked. Also, "Sheol" is entirely absent from legal material, including the many laws which prescribe capital punishment or proscribe necromancy. This means that "Sheol" is very much a term of personal engagement.

"Personal engagement" means the *circumstances* of one's death. Judging by what they wrote, the ancient Jews contemplated their death with composure as long as they had been successful in life. The horrifying prospect was to die without accomplishing anything. According to Ruth Rosenberg, another scholar of the period, "Whenever death is due to unnatural causes, Sheol is mentioned; whenever death occurs in the course of nature, Sheol does not appear."

These facts were common knowledge among Judaic scholars, but they had not been related to formal theories of religion, as strange as that might seem. Consider one of the most venerable theories of religion as a way to relieve anxiety over death. According to this theory, self-awareness is an adaptation that evolved in our species, providing tremendous advantages but also making us aware of our own death as a negative by-product. Religion arose to allay our fear of death, making it a secondary adaptation. This theory is so old and common that it is often described without using the E-word, but any careful formulation must invoke both genetic and cultural evolution. It is still taken seriously and provides the foundation for a branch of modern psychology called terror-management theory. It's a perfectly reasonable hypothesis worthy of testing. If the primary function of religion is to allay our fear of death, then belief in a comforting afterlife would be a religious universal. But the history of Judaism and many other ancient religions enables us decisively to reject this hypothesis. Few people regard death as an appealing prospect, but they do not need to invent elaborate belief systems to shield themselves from it. Religions might well play an important role in alleviating anxiety but not in such a simple-minded fashion.

If the beliefs in the afterlife that we associate with Christianity weren't always a part of Judaism, when did they arise? According to Madigan and Levenson, the first unmistakable description of the resurrection in the Hebrew Bible is Daniel 12: 1–3:

> At that time, the great prince, Michael, who stands beside the sons of your people, will appear. It will be a time of trouble, the like of which has never been since the nation came into being. At that time, your people will be rescued, all who are found inscribed in the book. Many of those that sleep in the dust of the earth will awake, some to eternal life, others to

> reproaches, to everlasting abhorrence. And the wise will be radiant like the bright expanse of sky, and those who lead the many to righteousness will be like the stars forever and ever.

When was this written, and what was the social context? According to Madigan and Levenson, "One of the most secure and long-lived contributions of modern biblical studies is the dating of Daniel to the time of the Seleucid persecution of 167–164 B.C.E, which makes it one of the latest texts in the Hebrew Bible." In other words, even though the events took place long ago and the documentation is fragmentary, there is sufficient evidence that can be pieced together like a detective story for scholars to reach a consensus on a question such as when a given document was written. The challenge facing the Jews at that time was Antiochus IV, a Hellenistic monarch, and the main problem was not persecution but the fact that many Jews were attracted to Hellenistic culture. The conflict reflected a division among the Jews, not a common threat to all Jews. According to the book of Daniel, eternal life would be granted to Jews who maintained the traditional ways, and all other Jews would suffer everlasting abhorrence. Belief in the resurrection was arguably a key factor that enabled the Maccabees, the traditionalist faction, to resist Antiochus in a victory that is still celebrated in the festival of Hanukkah.

Madigan and Levenson think that belief in the resurrection arose before the Maccabean revolt but in the same social context of groups of Jews that had separated themselves from other Jews and regarded themselves as the chosen people. It was a profoundly collective concept, they write, having little to do with individual accountability or anxiety about death:

> the Jewish expectation of a resurrection of the dead is always and inextricably associated with the restoration of the people of Israel; it is not, in the first instance, focused on individual destiny. The question it answers is not the familiar, self-interested one "Will I have life after death?" but rather a more profound and encompassing one, "Will God honor his promises to his people?"

Once again, there will always be a frontier of knowledge where scholars of religion disagree, but they have already reached a consensus on matters that are profoundly relevant to evolutionary theories of religion. Belief in the end of days, a collective judgment, and eternal life for the righteous have all the earmarks of a group-level adaptation in the context of extreme between-group conflict. Groups whose members adopt such beliefs sincerely enough to act on them will display greater solidarity than other groups. Solidarity does not guarantee success, and history is littered with messianic religious cults that failed despite their solidarity. But the question is not how many times a belief fails but, rather, its likelihood of being represented among the survivors.

THE RELIGION THAT FORMED AROUND Jesus was initially a twig on a branch of this cultural tree that sprouted several centuries previously. The idea of a Messiah who would die and return to signal the end of days was thoroughly familiar. The followers of Jesus thought that he was the Messiah, that he had indeed risen three days after his crucifixion, and that the end time had begun. The Jesus sect would probably have remained a footnote in religious history were it not for a key difference that had nothing to with its otherwise standard (for its time) afterlife beliefs: Gentiles could now become the chosen people. A belief system that previously had been restricted to one ethnic group could now spread throughout the world.

By the time that Paul was working to establish the early Christian church, Jesus had been dead for about twenty years, but the belief that he would return any day remained as strong as ever. It is clear from Paul's letters, which are the oldest documents to be included in the New Testament, that the first Christian communities would fall apart without their belief in the resurrection. Paul insisted that it was not just a story, that it would happen within their lifetimes, and that it would be a bodily, not just a spiritual, resurrection. If it didn't happen, then their beliefs would be a cruel joke: "If the dead are not raised, 'Let us eat and drink, for tomorrow we die' " (1 Corinthians 15: 32).

It might seem baffling that such a belief could persist in the face of contrary evidence. Why don't people stop believing in the resurrection when it doesn't come to pass? This question has a simple and powerful answer from an evolutionary perspective. Beliefs persist primarily on the basis of what they cause people to do. If belief in the resurrection creates communities whose members are highly motivated to cooperate and even sacrifice their lives for one another, then those beliefs will persist and spread compared with beliefs that do not create solidarity. The factual validity of the belief is beside the point.

This is a hard lesson for rationalists to learn. Rationalists, by definition, regard factual validity as a cardinal virtue. It is sacred for them to accept or reject beliefs purely on the basis of factual evidence. But they naively mistake their values for human nature, as if humans are designed to accept beliefs purely on the basis of factual evidence and are malfunctioning when they behave otherwise. Pre-Darwinian rationalists could claim that God made us that way. Some post-Darwinian rationalists have tried to claim that evolution made us that way, but the real implications of evolutionary theory for rationalism are much more complex and interesting. The factual validity of a belief can be tremendously important for survival and reproduction but only in some contexts. Factually invalid beliefs can be highly adaptive in other contexts, as we have seen in the case of belief in the resurrection. Thus, we have evolved the ability to function as rationalists, but we have also evolved the ability to turn our rationalism on and off with great facility, depending on the context. I'll bet that those early Christians, who believed in eternal life to the depth of their souls, also functioned as perfectly good rationalists when it came to aspects of their lives that required knowing the world as it really is.

My own dabbling in religious scholarship convinced me that the Afterlife Project would be a tremendous success. Conceptions of the afterlife originated at a certain time and place. They spread and came to the limits of their distribution for reasons that scholars broadly agree on, even when the events took place thousands of years ago. Scholarly knowledge is like a fossil record of cultural evolution so detailed that it puts the biological fossil record to shame. By collaborating with religious scholars, evolutionists can come to an agreement about which of the major hypotheses best explain the nature of religion. We can trace the phylogenetic tree of Christianity as it grew from a twig on one branch of Judaism to a mighty trunk of its own. Christianity is not a single religion but a phylum of religions. Concepts of the afterlife have diversified within Christianity. Each variation originated at a certain time and place and can be examined from an evolutionary perspective, just like the original concept of the resurrection. As for Christianity, so also for the other great religious traditions as cultural phyla in their own right, along with the thousands of indigenous religions that came before the great religious traditions, many persisting as evolving entities to this day. Cultural evolution is not the same bifurcating process as biological evolution, in which each species is derived from a single ancestral species. For that matter, even biological evolution is not as bifurcating as we used to think, given what we now know about the movement of genes across taxonomic boundaries. A given religion can borrow elements from many other religious and nonreligious traditions, but the fate of each element in competition with other elements is still a matter of cultural evolution.

CAN EVOLUTIONARY THEORY EXPLAIN current afterlife beliefs in addition to those that arose in the distant past? To find out, I decided to ask historian Ronald Numbers about Seventh-day Adventism, which originated in the nineteenth century and is one of the fastest-growing religions worldwide today. Ron should know about Seventh-day Adventism, because he was born within the faith before leaving it to become one of its foremost scholars. Ron knows this religion inside and out.

Seventh-day Adventism rose from the ashes of another religious movement called Millerism in the nineteenth century. William Miller was an affluent farmer, a devout Baptist, and an amateur student of the Bible in the section of New York State called the Burnt Over District for its religious fervor. In the early 1830s, he became convinced through his study of the Bible that Christ was going to return in the very near future. He began to share his views with a few friends and was disappointed by their reaction. As he put it, "To my astonishment, I found very few who listened with any interest. Occasionally, one would see the force of the evidence, but the great majority passed it by as an idle tale."

Miller had not taken leave of his senses. He was a normal human being reasoning on the basis of his senses, using the most trusted source of information as far as his culture was concerned. With persistence, he did end up attracting a large following, thanks largely to the recent advent of the printing industry in America. Newspapers and periodicals shot up like mushrooms to spread the news. The movement remained centered in the Burnt Over District but reached as far as Australia and Hawaii. It's fascinating to think about this historical episode from a cultural evolutionary perspective. Miller's belief was like a mutation that arose by pure happenstance and spread like a virus. Some people were immune, but others were highly vulnerable. The printing industry made the belief far more contagious than it would be otherwise, just like the Internet today or airplanes that quickly spread new flu strains over the globe. The belief did not increase the welfare of human individuals or groups; it was merely well designed to spread itself. It had all the earmarks of a cultural parasite, as proposed by evolutionists such as Richard Dawkins and Daniel Dennett.

At first, Miller and others involved in the movement were not precise about the exact date of Christ's return, only that it would be "about the year 1843." When the year passed, the Millerites fastened on a new date of October 22, 1844, with such fervent hope that the failure of Christ to return on this date became known as the Great Disappointment.

It might seem that the story would end there, and so it did for many Millerites, who abandoned that particular faith and resumed more normal lives guided by different faiths. Others kept the faith, however, and decided that the problem must have been with their calculations. There was nothing unitary or coordinated about the search for new solutions. Different people had different specific visions. Belief in the imminent return of Christ is a powerful source of psychic energy, like gasoline or nuclear power, which can be constructive or destructive or can merely dissipate, depending on how it is harnessed. The Millerite visionaries weren't trying to design sustainable communities on earth, since they truly believed that their life on earth was about to end. Nevertheless, their beliefs *varied* in their sustainability on earth, causing some to wink out of existence and others to endure.

Seventh-day Adventism arose from this raw process of cultural variation and selection, based on the visions of its main prophetess, Ellen G. White (1827–1915), who claimed to have actually visited heaven so that she could describe it firsthand. Adventists believe that Christ can return at any time but that they have much to do to prepare for his arrival. The psychic power of belief in imminent return is harnessed to sustainable practices on earth. Not only do Adventists keep themselves healthy, but there will also be a jewel in their crown in heaven for every person they save. White actually saw the crowns on the shelf. If Christ hasn't returned by now, it might be your fault for not preparing hard enough. That's why Seventh-day Adventism is currently one of the fastest-growing religions on earth and often does a better job creating schools and hospitals in underdeveloped countries (in addition to churches) than many governments and secular nongovernmental organizations.

RON NUMBERS WAS BORN into a family that had been steeped in this tradition for four generations. His ancestors are primarily Scottish and became involved in the movement while Ellen G. White was still alive. Almost all of his male relatives were ministers, and his maternal grandfather was president of the international church. One of his aunts was married to a vice president of the General Conference, another to a prominent revivalist. No one could be born into a fundamentalist religion more thoroughly than Ron, which makes the story of how he became one of the most distinguished scholars viewing it from the outside especially interesting.

Ron was intellectually brilliant. The Adventist elementary schools that he attended were small, like the one-room schoolhouses of pioneer days, enabling Ron to advance at his own pace. By the time he was eleven, he had jumped the equivalent of four grades. His parents put him back three grades so that he could become socially adjusted, and Ron still graduated from high school when he was sixteen. Adventist schools taught subjects such as mathematics and the physical sciences, much like other schools, with the conspicuous exception of fiction and evolution. Literature consisted of the Bible and true stories. Young-earth creationism was taught as if it were a rock-solid fact. When I asked Ron how he could be so brilliant and yet not question something as flimsy as young-earth creationism, he repeated what his father told him: "We study to learn how better to disseminate the Truth that we know. We don't study to question the Truth." Adventists capitalize their most important words and treat *Truth* and *Adventism* as synonyms.

Ron was a straight arrow. It was hard not to be, given his family heritage and rigidly controlled social environment. At the academy, there was vegetarian food, no dancing, no smoking, no drinking, and no jewelry. Today Ron describes it as "no fun," but he was thoroughly accustomed to it at the time and not the kind of person to rebel. At Southern Missionary College (now Southern Adventist University), he studied math and physics but soon discovered that his true love was history. Only when Ron became a graduate student did he leave the Adventist educational system to attend Rice University, Florida State University, and the University of California at Berkeley. Even then, he remained a straight arrow, not only attending the Adventist churches close to him but also teaching Sabbath school and serving as an elder.

Adventism might be growing worldwide, but the congregation was shrinking in Berkeley, and something was also happening inside Ron's head. Adventists learned to spread the Truth, but the whole point of graduate school was to examine received wisdom critically and tear it apart if necessary. Graduate students and their professors did not capitalize the *r* and the *w* in *received wisdom*. Ron liked this game and unsurprisingly became good at it, but it started to eat away at his own faith. The Bible warns its believers about the apple of knowledge for good reason.

One of Ron's epiphanies came when he attended a social gathering of Adventist graduate students and professors. An engineering professor at Stanford had made a study of fossil forests in Yellowstone National Park and concluded that they could not possibly be younger than 30,000 years, even assuming the most rapid possible rates of growth and decomposition. This conclusion, coming from an Adventist professor, was so disturbing to Ron that he sat in his car for hours discussing it with one of his Adventist friends. Ellen White was supposed to be infallible, and she said that the earth was created in six days 6000 years ago. If she was wrong on that point, the authority of all of her claims could be questioned.

Ron was now sliding down a slippery slope but did not take stock of his faith in any comprehensive fashion. He taught at Adventist colleges for five years after obtaining his PhD, although his church attendance became sporadic, and he slowly began to indulge in forbidden pleasures, such as a sip of alcohol now and then. Then he started looking into the history of the Adventist movement. His initial motive was not to tear it apart but to make his lectures more entertaining for his Adventist students, but what he discovered astonished him. White had liberally plagiarized from other sources in her own writing, and the church was complicit in keeping it a secret. She was beloved as God's prophet for the Last Days, centuries ahead of the medical experts, but not a single example of a true prediction, unknown to medical experts at the time, could be documented. The historical evidence left no doubt whatsoever. History is like a crime scene where nobody tries very hard to cover his or her tracks. The smoking guns were everywhere once you knew how to look with critical eyes.

Ron felt obligated to write *Prophetess of Health: A Study of Ellen G. White,* not because he felt vindictive but because he regarded himself as first and foremost a scholar. He was obligated to tell the truth, with a lower-case *t,* about the history of Adventism, come what may. The storm clouds started to gather when copies of the unpublished manuscript fell into the hands of the church fathers. Ron was informed by the dean of the Adventist medical school that he was no longer welcome on the faculty. The dean didn't want to fire him and offered a year of severance pay in return for his resignation, which Ron accepted. The church actually talked Harper & Row into delaying the publication of Ron's book so that they could publish a point-by-point response. A whole page of *Time* magazine was devoted to the controversy. Ron's extended family was devastated. His father became convinced that Ron was possessed by Satan and did not reconcile with him until a year before his own death. Ron's marriage fell apart, and Adventists were warned about him wherever he went. Another historian of Adventism, Jonathan M. Butler, has written an essay about this period of Ron's life, "Historian as Heretic," which is included in the second edition of Ron's landmark work of scholarship.

After Ron was ejected from Adventism like Jonah from the belly of the whale, he lucked into a job at the University of Wisconsin, where he became a distinguished historian of creationism, among other subjects. He's a myth buster wherever he turns his gaze, and his most recent book is *Galileo Goes to Jail and Other Myths about Science and Religion*. When I asked him about his current religious beliefs, he replied that he's "not even a theist." Yet he also displays little bitterness, and when he talks about Adventism as a scholar, it is with much more sympathy and understanding than the typical angry atheist. He appreciates the virtues of Adventism in addition to its many shortcomings, when judged as an enduring cultural system. It's not easy for a culture to function as an organism, and Adventism does a pretty good job from an impartial evolutionary perspective.

During my most recent phone conversation with Ron, I was able to teach him something about Adventism, as strange as that might seem. I quoted the following passage that I had found on www.adventist.org, titled "What Adventists Believe":

> Because you and God are friends, you will spend time together as friends do. Each morning you'll share a hello and a hug and discuss how you can face the day's events together. Throughout the day you'll talk with Him about how you feel. You'll laugh with Him at funny things and ache with Him over sadness and hurts. It's pleasant being God's friend, able to snuggle comfortably into the safety of your relationship. You can always trust Him to treat you well, because He loves you.

Ron's response to this passage was a startled *"Huh?"* This wasn't the Adventism that he knew. He was taught that God is judgmental and Jesus intercedes on our behalf. You didn't snuggle with God. You were constantly on your guard and worked your butt off to earn points so that you could have jewels in your crown and wouldn't roast in hell for an amount of time carefully calibrated to your sins, after which you would turn to dust without any existence at all. Since Ron left Adventism thirty years ago, it has morphed into a different beast to keep up with modern sensibilities. This is true for many religions in America, as religious scholars Christian Smith and Melina Denton have documented in their book *Soul Searching: The Religious and Spiritual Lives of American Teenagers*. The majority of American teens have converged on a single generic conception of God as their snuggle partner and life coach, regardless of the name and past traditions of their denominations. Never underestimate the power of cultural evolution to transform conceptions of God, the afterlife, and every other element of religion. When it comes to altering the sacred story, nothing is sacred. Thank heaven we have good scholars to keep track of it all.

GOD MIGHT BE YOUR SNUGGLE partner in some faiths, but he remains judgmental in others, as we can see by following the fate of Carlton Pearson, a man oozing with charisma who rose to the pinnacle of success as a bishop in the Church of God in Christ, a contemporary Pentecostal denomination. Pearson is already known to the public through his own writing and an episode of the National Public Radio program *This American Life* devoted to him called "Heretic." Even if you are familiar with his story, it becomes especially poignant when seen through the lens of evolutionary theory.

Pearson grew up in a community steeped in Pentecostalism and gained an early reputation for his ability to cast out the devil. He attended Oral Roberts University and became a protégé of Roberts himself, who regarded Pearson as his "black son," capable of reaching an African-American audience. Before long, he was presiding over his own mega-church in Tulsa, Oklahoma, hosting his own television show, and traveling the world holding revivals. He also acquired a scholarly knowledge of religious traditions other than his own. Not only could he speak in tongues, but he could also speak Greek.

One day, Pearson had an epiphany while watching television with his infant daughter on his lap. A news program on an African famine showed footage of dying babies with their swollen bellies. Suddenly, his own belief that these innocent babies would be sucked into hell if they didn't convert to his particular religious faith struck him as monstrous. He was so moved by this revelation that he startled his congregation the following Sunday by announcing that a new gospel had been revealed to him: hell is not a place that we might go after we die but the misery that we create for one another on earth.

It would be hard to imagine a more perfect "natural experiment" for studying the cultural evolution of a religion, comparable to a knockout experiment in biology where a single gene is inactivated in an organism. Everything was the same—the same charismatic leader, the same congregation, the same configuration of beliefs associated with the Church of God in Christ. Only belief in the nature of the afterlife had changed. It had become more benign, still including a glorious heaven but no longer including hell. If the main purpose of religion is to allay one's fear of death, you'd think that Pearson's congregation would like his new gospel even better than the old one.

Alas, Pearson lost his congregation and became branded as a heretic by his church. One person who left Pearson's congregation explained, "We need to find another church that's solid in the word." One of the youth pastors who stayed with Pearson humorously put it this way in the *This American Life* episode: "What's the best way to get the kids' attention? We'll scare 'em! We'll say, do you like to burn?... That's how most of us got saved. We chose because the alternative was scary!...Threat of judgment day sure is easy to pack a church out. That fear factor is definitely effective. If we take away the requirements of coming to church and paying your dues...you can put some guys out of a job!"

Pearson is a man of principle and did not waver in his new faith, despite losing all of the trappings of success. He now preaches his "Gospel of Inclusion" to a much smaller congregation, including people who grew up in the Pentecostal tradition and became victims of intolerance because they didn't fit in, for their sexual orientation or other reasons. He has become affiliated with the United Church of Christ, a denomination that is much more tolerant of diversity in behavior and theology. In an interview, he said, "I've told God that if I'm wrong tell me; even kill me. God has done neither so I'm going all the way."

I have sympathy for both Pearson and the people who left him. We are such a symbolic species that everyone needs a cultural system in order to function adaptively. Pentecostalism is one enduring system in which fear of hell plays a vital role. That doesn't mean that Pentecostalists are always quaking in fear. Most of the time, they feel saved, but fear is necessary to get them into that state of mind. You might love or hate Pentecostalism and other fundamentalist religions, but you've got to admit that they are complex adaptive systems that organize the lives of their members. Against that background, removing a vital element such as belief in hell is like removing a vital organ from an organism or disrupting a wasp colony by preventing the larvae from feeding the adults. The entire system goes limp. It's no wonder that Pearson was branded a heretic and so many of his followers left him to join congregations that were "more solid in the word."

Pearson himself didn't go limp but became more energized by his new belief in the afterlife than ever before. He had become more cosmopolitan and educated than most members of his congregation. The beliefs that successfully organized their lives and had organized his own life in the past no longer squared with his broader vision. A belief that was maladaptive against the background of Pentecostalism was adaptive against another background. Moreover, even though Pearson's revelation was new for him, belief in hell had become attenuated in mainline Protestant religions long ago, and even in Adventism more recently, as we have just seen. Conservative religions are not invariably stronger than liberal religions. Like biological species, they are adapted to different circumstances, and both prevail in their respective niches. A religion such as Adventism grows in underdeveloped countries but shrinks in Berkeley.

* * *

THE AFTERLIFE PROJECT WOULD BE much more systematic than my own personal tour of the tropical rain forest of religious beliefs. Scholars of the same caliber as Bill Green, Ron Numbers, Kevin Madigan, and Jon Levenson would provide the information directly rather than having me interpreting their work. Nevertheless, I already had confidence about the success of the project based on my own inquiry. I was especially excited by the amount of detail that would be forthcoming.

In *Darwin's Cathedral,* I argued that religions are primarily group-level cultural adaptations, not cultural parasites. The Afterlife Project had the resolution to prove me wrong for a specific example such as Millerism, even though I might be right in broad outline. It could document the process of cultural evolution, such as Adventism emerging from Millerism, not just the final product. Finally, it could document the haphazard "Come on down!" side of cultural evolution that I described for biological evolution in the parable of the strider. It was detail that made the study of biological evolution fully rounded, and I could look forward to the same kind of detail for the study of religion.

Unfortunately, this scientific and scholarly enterprise has been disrupted by an assault on religion by two of my evolutionist colleagues, Daniel Dennett (author of *Breaking the Spell: Religion as a Natural Phenomenon*) and Richard Dawkins (author of *The God Delusion*). Dan and Richard are among the world's foremost interpreters of evolutionary theory for the general public. Together with Sam Harris (author of *The End of Faith: Religion, Terror, and the Future of Reason*) and Christopher Hitchens (author of *God Is Not Great: How Religion Poisons Everything*), they are the four horsemen of the "New Atheism" movement. Sam and Christopher are not authorities on evolution but proudly wave the banners of rationalism and science.

The four horsemen want you to know that supernatural agents don't exist. They also want you to know that religion is bad, bad, bad for the human race. Rather than pandering to religious believers, we should get tough and force them to face reality. Finally, they want you to know that atheists are a persecuted minority and should come bursting out of the closet to claim their rights.

I agree with two of these propositions. As an atheist myself, I also believe that supernatural agents don't exist—at least, not the ones that actively intervene in the affairs of people. The vast majority of religion scholars also function as atheists when they adopt methodological naturalism, regardless of their personal beliefs. I also agree that atheists are unfairly stigmatized, especially in America. It's a pluralistic world, and if people of different religious faiths are supposed to get along with one another, then they also need to get along with atheists. An atheist in America should have a fair chance of being elected to a public office, for example, which currently is not the case.

I disagree with the four horsemen that religion is bad, bad, bad for the human race. You might think that I am caricaturing their position, but look at their book titles, which clearly imply that religion is a delusion, a poison, a source of terror, and the enemy of reason, which, of course, is our salvation. Is this beginning to sound a bit like the polarized good-against-evil views of religious fundamentalists?

The portrayal of religion as bad, bad, bad might be factually correct if the elements of religion turned out to be cultural parasites or costly by-products most of the time, but the scholarly study of religion from an evolutionary perspective weighs against that conclusion. Instead, religions emerge much as Émile Durkheim described them: "a unified system of beliefs and practices relative to sacred things...that unite into one single moral community called a Church, all those who adhere to them." Like all organisms, unified human communities can be bad or good, as we have seen with the parables of the strider and the wasp. If we want to avoid the problems associated with religion, we need to understand the nature of religion from a sophisticated evolutionary perspective. What really makes me mad about the New Atheists is that they wrap themselves in the mantle of rationality, science, and evolution, while ignoring what evolution has to say about the nature of religion.

They also ignore the fact that all people require a cultural system of one sort or another in order to function on a daily basis. Just because a cultural system is secular doesn't make it rational. There are countless ways to depart from factual reality without invoking the gods. Every secular belief will be selected primarily on the basis of how it causes people to behave, regardless of its factual validity. Sometimes factual validity will contribute positively to effective action, but sometimes it will get in the way. Thus, people guided by secular cultural systems will toggle between rational and irrational beliefs with great facility, just like people guided by religions. The patriotic histories of nations provide an obvious example: our forefathers who could do no wrong, triumphing over their enemies who deserved everything they got. Who needs the gods when we can so freely distort every other aspect of the world to motivate and justify our actions?

The situation is worse than the New Atheists imagine. They naively believe that by getting rid of religion, we can get rid of irrationality. They also naively believe that factual knowledge invariably leads to more effective action in the world. They don't understand that the relationship between the factual validity of a belief and its effect on human action is more complex and that all enduring cultural systems, whether religious or secular, are shot through with irrational beliefs that persist because they work better for the believer or the community of believers than more rational beliefs. Adaptive irrational beliefs might harm others and even everyone in the long run, but that does not prevent them from evolving on the strength of their local and short-term advantages. Solving these problems will require much more than shaking our fists at religion.

NOW THAT I AM TRYING to make a difference in my city of Binghamton, I find myself working with religious leaders and believers all the time. Every audience I address includes churchgoers, and many of the meetings take place on church premises. We do better than tolerate one another's differences. We have a shared interest in fostering community that makes us genuine allies. Most religious believers are not threatened by science and are among the first to capitalize on technology. When I portray evolutionary science as a valuable tool kit for making the world a better place, they are as eager to use the tools as they are to use the Internet. For my part, I regard them as kindred spirits to the extent that we share the same objectives, without insisting that they embrace my commitment to methodological naturalism. They are welcome to believe whatever they like, as long as it doesn't cause harm to others.

Recently, the Templeton Foundation has blessed me a third time by funding a project with Harvey Whitehouse, a distinguished scholar of religion at Oxford University in England. We will add the study of religion and spirituality to the Binghamton Neighborhood Project as a prototype for establishing similar field sites around the world. We will chart the full panoply of beliefs, religious and secular, like so many species in an ecosystem. By combining this new information with other information in our ever-expanding database, we will be able to understand the causes and consequences of the various beliefs, as practiced in the real world, in more detail than ever before. We won't merely study all of these beliefs; we'll do it in collaboration with the believers. Most of the people I talk with in Binghamton, religious or secular, profess a desire to make their own lives, their city, and the world a better place. We'll provide the scientific tools to listen and reflect on the hidden cultural processes, along with the hidden genetic processes, providing the knowledge required to realize our common desire.

CHAPTER 19

Evonomics

WHEN BERNARD WINOGRAD asked me what evolution might have to say about the world economic crisis, before disappearing into the maelstrom to deal with it himself, he was like Gandalf the Grey in *The Lord of the Rings* trilogy sending the gentle hobbit Frodo on a mission. I knew almost nothing about economics and the world of financial markets that Bernard inhabited. Economic jargon mystified me, an embarrassing confession, since I'm a theoretical biologist fully at home with mathematical and computer-simulation models. Economists were very smart and very powerful, and they spoke a language that I didn't understand. They won Nobel Prizes. Well, not quite *the* Nobel Prize, which was established by the will of Alfred Nobel in 1895, but an add-on that was established in 1968 by the Central Bank of Sweden. Its formal name is the Sveriges Riksbank Prize in Economic Sciences in Memory of Alfred Nobel, but most people call it the Nobel Prize in economics, making it seem that a Nobel laureate in economics has done something comparable to discovering the structure of DNA.

Nevertheless, I had faith that evolution could say something important about the regulatory systems that economists preside over, even if I did not yet know the details. After all, financial markets and other regulatory systems are products of cultural evolution, based on psychological processes that evolved by genetic evolution. That's my area of expertise. If I can't connect the dots between my area of expertise and modern regulatory systems, then something's wrong, and it might be on their end, not mine. I therefore took up Bernard's challenge and prepared to make my way toward Mordor. Little did I know that my adventure would result in insights that would be useful back in my shire of Binghamton, New York.

The Evolution Institute was in its infancy. Jerry Lieberman had managed to procure the funds for our first workshop on childhood education, and we would need to raise additional funds for a second workshop. Bernard might be willing to contribute, since it was his suggestion, but I also had a card of my own to play. The National Science Foundation had created the National Evolutionary Synthesis Center (NESCent), whose purpose was to foster new synthetic research, in part by holding "catalysis meetings" that could include as many as thirty-five people. I wrote a proposal for a catalysis meeting titled "The Nature of Regulation: How Evolutionary Theory Can Inform the Regulation of Large-scale Human Social Interactions," and it was funded. A sizable conference had become a reality without costing Bernard a dime. That was the good news. The bad news was that I now had to deliver the goods. There was no backing out. It was Mordor or bust.

Luckily, not all of the responsibility would be on my shoulders. My job as codirector of the Evolution Institute was to find and assemble the experts who were already studying human regulatory systems from an evolutionary perspective, just as I had done for childhood education. I needed to be the conductor of an orchestra, but at least they would be playing the music.

The study of human genetic and cultural evolution has become a wonderful blend of scientists from all disciplines, including economics. Some of my closest colleagues with economic backgrounds include Herbert Gintis, Samuel Bowles, and Ernst Fehr. They are highly respected among economists and have acquired an advanced knowledge of evolution, publishing widely in the best scientific journals in addition to the best economic journals. I already knew from their work that the main body of modern economics, called neoclassical economics, was being challenged by a new school of thought called experimental and behavioral economics. Proponents of the new school had captured the public imagination with books such as *Nudge: Improving Decisions about Health, Wealth, and Happiness,* by Richard Thaler and Cass Sunstein; *Predictably Irrational: The Hidden Forces That Shape Our Decisions,* by Dan Ariely; and *Animal Spirits: How Human Psychology Drives the Economy, and Why It Matters for Global Capitalism,* by Nobel laureates George Akerlof and Robert Shiller. Sunstein had recently been appointed President Obama's regulatory czar, so the new school of thought was beginning to have an impact on public policy.

According to the new school, neoclassical economic theory is based on a false conception of human nature and must be replaced with a more realistic conception. As Thaler and Sunstein put it in *Nudge,* economic theory and policy needs to be based on "humans," not "econs." I cheered when I first read these lines and looked forward to learning about the new and improved conception of human nature, which presumably would be informed by evolutionary theory. I was disappointed. My colleagues, such as Herb, Sam, and Ernst, confirmed my own impression. They appreciated the relevance of evolution but were a tiny minority among behavioral and experimental economists, who in turn were a tiny minority among neoclassical economists. Modern evolutionary theory was having virtually no impact on currently economic theory and policy.

THE MORE I LEARNED about economics, the more I discovered a landscape that is surpassingly strange. Like the land of Mordor depicted by J. R. R. Tolkien and portrayed so vividly in Peter Jackson's *Lord of the Rings* films, it is dominated by a single theoretical edifice that arose like a volcano early in the twentieth century and still dominates the landscape. The edifice is based on a conception of human nature that is profoundly false, defying the dictates of common sense, before we even get to the more refined dictates of psychology and evolutionary theory. Yet efforts to move the theory in the direction of common sense are stubbornly resisted. There is plenty of dissention among economists, and some of the best are working the hardest for change. The folks who award the Nobel Prize in economics don't like the edifice that much, either, and often add their weight by awarding the prize to the contrarians. Yet even with all that talent and effort and the prestige associated with the Nobel Prize, the edifice remains standing in one spot, like a volcano adding to its own height and spewing out toxic policies. Why does it resist change? One reason is ideological, as we shall see, but another reason involves path dependence. Neoclassical economics provides an outstanding example of the "you can't get there from here" principle in academic cultural evolution. It will never move if we try to change it incrementally. It must be replaced wholesale with a more realistic conception of human nature.

ECONOMISTS WERE MUCH MORE CLOSELY attuned to common sense and evolutionary theory before the volcano erupted. Adam Smith (1723–1790) observed that people following their narrow concerns somehow combine to make the economy work well, as if guided by an invisible hand. Today we use terms such as *emergence* and *self-organization* to describe this phenomenon. It is spectacularly demonstrated by social-insect colonies, as I showed in the parable of the wasp. Honeybees and yellow jackets definitely don't have the welfare of their whole colonies in mind. They're just responding to local environmental cues in ways that make their colonies work well as a whole. They could respond in an infinite variety of ways, most of which would not contribute to the success of their colonies, but natural selection has winnowed the responses that work at the colony level. The same point can be made for the cells in our bodies. They don't have the welfare of our whole bodies in mind; they don't even have a mind. They're just responding to local environmental conditions by switching genes on and off in a way that leads to the welfare of the collective. There is an infinite variety of ways in which cells might respond to local environmental conditions, but natural selection has winnowed the ones that work at the level of the whole organism. When Smith invented the metaphor of the invisible hand, he was saying that human economies are like bodies and beehives in this regard.

Smith did not say that people are entirely self-interested. That would come later. He had a sophisticated and nuanced conception of human nature that he described most fully in *The Theory of Moral*

Sentiments. According to Smith, people have genuine concern for others in addition to themselves. They have a sense of right and wrong that leads to the establishment of norms enforced by punishment. They are interested in their reputation as much as their monetary income, and so on. Shakespeare would have felt comfortable with Smith's conception of human nature. People still responded primarily to their local social environment, so the invisible hand metaphor remained apt, but their preferences couldn't be collapsed into a single generic concept of self-interest.

Charles Darwin and Alfred Russell Wallace were influenced by Smith and other economists, most famously Thomas Malthus (1766–1834), in their formulation of evolutionary theory. Economists, in turn, were influenced by the new theory of evolution. Thorstein Veblen (1857–1929) even wrote a paper in 1898 titled "Why Is Economics Not an Evolutionary Science?" At the dawn of the twentieth century, economics was both closely attuned to commonsense views of human nature and receptive to the new theory of evolution that it had influenced. Then the volcano erupted.

The volcano was the laudable attempt to place economics on a mathematical foundation using physics, not biology, as the inspiration. There was nothing sinister or self-interested about this effort. It made perfect sense at the time. Physicists such as Isaac Newton (1642–1727) had succeeded in describing the laws of physical motion in the precise language of mathematics. Wouldn't it be magnificent to do the same thing for the laws of human society? That became the goal of Léon Walras (1834–1910) and other economists, as recounted by Eric Beinhocker in his excellent book *The Origin of Wealth: Evolution, Complexity, and the Radical Remaking of Economics*. Here is how Eric describes the zeitgeist of the times:

> Following Newton's monumental discoveries in the seventeenth century, a series of scientists and mathematicians, including Leibniz, Lagrange, Euler, and Hamilton, developed a new mathematical language using differential equations to describe a staggeringly broad range of natural phenomena. Problems that had baffled humankind since the ancient

> Greeks, from the motions of planets to the vibrations of violin strings, were suddenly mastered. The success of these theories gave scientists a boundless optimism that they could describe any aspect of nature in their equations. Walras and his compatriots were convinced that if the equations of differential calculus could capture the motions of planets and atoms in the universe, these same mathematical techniques could also capture the motion of human minds in the economy.

It's easy to imagine the allure of this goal. If economics could be placed on a mathematical foundation comparable to physics, it would achieve a status greater than any other human-related discipline such as sociology, anthropology, psychology, or philosophy. Biology held no appeal as a source of insight because it wasn't mathematical. Walras appeared to achieve this goal in his masterwork *Elements of Pure Economics,* published in 1874.

Walras's main challenge was to provide a mathematical demonstration of the invisible hand. If individuals maximizing their local concerns really do self-organize into well-functioning economies, then top-down regulation by government leaders is unnecessary. The government merely needs to stop meddling, and the economy will run itself. Walras succeeded in providing a mathematical proof of this possibility but only by making certain assumptions—lots of them. Anyone who builds theoretical models, as I do, knows how many simplifying assumptions must be made for a model to be mathematically tractable. The world described by mathematics becomes detached from the real world, not because of any ideological bias but just so that you can grind through the equations. It was at this point that economists began to rely on a conception of human nature that defies the dictates of common sense, even before we get to the more refined dictates of psychology and evolutionary theory.

The people who inhabit the economic models, often referred to as *Homo economicus,* are driven purely by self-regarding preferences. Mathematically, this means that they care only about maximizing their own interests, without reference to anyone else's interests. What I want cannot depend on what you want. Exactly what I want can be broadly interpreted, but operationally, it is usually assumed to be my monetary economic interests. The single assumption of self-regarding preferences banishes a large fraction of real-world human preferences from the mathematical kingdom.

Next, people who inhabit the mathematical kingdom are infinitely wise in pursuit of their self-regarding preferences. Lest you think that I am creating a straw man, here is how Thaler and Sunstein, two economic insiders, describe it in *Nudge:* "If you look at economics textbooks, you will learn that *Homo economicus* can think like Albert Einstein, store as much memory as IBM's Big Blue, and exercise the willpower of Mahatma Gandhi. Really."

Once again, these absurd assumptions were driven not by ideological bias but by the tyranny of mathematical tractability. The theory couldn't be pushed in the direction of common sense, because it would become impossible to grind through the equations. Yet the theory commanded enormous prestige, because it represented the ideal of a science of economics comparable to the science of classical physics. Abandoning the ideal would knock economics off its pedestal and back into the ranks of the other human-oriented sciences such as sociology, psychology, political science, and anthropology. Heaven forbid!

AS I DELVED DEEPER into the history of economics, I saw it as a struggle between those who pledged allegiance to the mathematical edifice and those who chafed against its rule, including Thorstein Veblen way back in 1898. Here is a tale of three Nobel laureates to illustrate what I mean.

Nobel laureate number one is Maurice Allais, a French economist who received the prize in 1988 and is most famous for a paradox that he discovered in the 1950s. As an example, imagine that you can choose between getting $1 million with certainty versus $1 million with an 89-percent chance, nothing with a 1-percent chance, and $5 million with a 10-percent chance. If you're like most people, you'll opt for the sure bet. Now you're given a second choice between nothing with an 89-percent chance and $1 million with an 11-percent chance versus nothing with a 90-percent chance and $5 million with a 10-percent chance. If you're like most people, you'll go for the $5 million, even though you have a slightly higher chance of getting nothing. These decisions reflect the fact that when real people make decisions, they are sensitive to both the *average* outcome and *variation around the average*. So are other animals. My evolutionist colleagues have performed dozens of experiments on birds and other creatures showing that they will choose either risky high-payoff options or safe low-payoff options depending on how well fed they are. Their flexible behavior with respect to risk makes perfect sense from an evolutionary perspective.

If people behave the same way, then where's the paradox? The mathematical edifice of economic theory needs the assumption of mean expected utility to remain tractable. For those who have pledged allegiance to the mathematical edifice, risk sensitivity becomes a paradox. Never mind that real people behave that way; it seems like common sense and also makes perfect sense from an evolutionary perspective.

Let's pause to savor the irony of the situation. A purely fictional world defined by mathematical equations acquires so much authority that it becomes the real world for its adherents. Aspects of the real world that cannot be related to the imaginary world are so dumbfounding that they are labeled paradoxes by the faithful. Here is how Allais described his struggle with the faithful in a 1987 essay:

> For nearly forty years the supporters of the neo-Bernoullian [expected utility] formulation have exerted a dogmatic and intolerant, powerful and tyrannical domination over the academic world; only in very recent years has a growing reaction begun to appear. [The Allais paradox] is fundamentally an illustration of the need to take into account not only the mathematical expectation of cardinal utility, but also its distribution as a whole around its average, basic elements characterizing the psychology of risk.

Does this seem a little bit like religion?

Nobel laureate number two is George A. Akerlof, who received the prize in 2001 and was elected president of the American Economic Association in 2007. His presidential address was titled "The Missing Motivation in Macroeconomics." The missing motivation was norms. When I first encountered this paper, I rubbed my eyes in disbelief. *Norms?* Economists have been primary advisors on public policy, and they're only newly considering a little thing called *norms?* What planet do they come from?

Norms don't exist in the mathematical kingdom, because they violate the assumption of self-regarding preferences. When you're *Homo economicus,* your preferences don't depend on anyone else's. Norms are all about the coordination and enforcement of preferences among members of a group. Adding norms would require a massive restructuring of the edifice. Here's Akerlof's own diagnosis of the problem from his presidential address, starting with John Maynard Keynes (1883–1946), a giant figure in the field of economics who made the mistake of pledging his allegiance to common sense rather than the mathematical edifice:

> But a new school of thought, based on classical economics, objected to the casual ways of these folks. New Classical critics of Keynesian economics insisted instead that these relations be derived from fundamentals. They said that macroeconomic relationships should be derived from profit-maximizing firms and from utility-maximizing by consumers with economic arguments in their utility functions. The new methodology... overturned aspects of macroeconomics that Keynesians had previously considered incontestable.

In other words, the mathematical kingdom became the new reality, banishing norms along with risk-sensitive behavior from its conception of human nature. Allais expressed hope in his 1987 essay that the edifice was crumbling, but Akerlof was still standing outside the ramparts in 2007.

Nobel laureate number three is Elinor Ostrom, who was awarded the economics prize in 2009 and was the first woman to receive it. I was fortunate to attend a workshop with Lin, as she prefers to be called, a few months before she received the prize. Already an admirer, I asked her to tell me her story and was delighted to discover that it combines many of the themes that I have been weaving together in this book: how a person who comes from the ranks of everyday life can reach the pinnacle of scientific success, how working in the trenches of "blue-collar" subjects can strike gold for "white-collar" subjects, how evolutionary theory can provide a new foundation for economic theory and policy on a worldwide basis, and how it can be equally relevant for improving the quality of life in my city of Binghamton, one neighborhood at a time.

ELINOR OSTROM'S ANCESTORS WERE PROTESTANT refugees who migrated from Europe to the United States early in the history of our nation on her mother's side and Jews who migrated from eastern Europe in the early 1900s on her father's side. She was born in California in the 1930s, at a time when women were expected to be housewives, secretaries, and schoolteachers. Neither of her parents had a college education, and they did not nurture Lin's academic ambitions. Fortunately, her school provided the support she needed, much like Daphne Fairbairn's. As I recall the many people I have profiled in this book, I am struck by how nurturance must come from somewhere, but exactly where can vary from family, to school, to religion, and even to the kindness of strangers.

Shy and burdened with a mild stutter, Lin gained confidence by participating in the school's debate and speech teams. College was sufficiently affordable in California at the time that she could work her way through without her parents' support. She attended UCLA and became a political-science major after taking a course in American government that captivated her. She had no idea what she might do with a degree in political science but merely enjoyed the subject. She also took some courses in economics, business, and personnel with a livelihood in mind.

Lin married her high school sweetheart and put him through Harvard Law School after they graduated from college. Her first job in Boston was for an electronics company that started its employees out in the mailroom so that they could get to know the whole company. From there, she went to the export department, where she worked in a large clerical room with many other women, overseen by a man sitting at an elevated desk. Lin laughs good-naturedly at the experience and says that it was useful for learning how firms work from the inside.

After one year at the electronics company, Lin boldly talked her way into a job as assistant personnel manager at a large-scale manufacturing company. At first, she didn't get the job, but they reconsidered when she offered to work without charge for a trial period. At that time, there were no Catholics, Jews, or blacks in the entire company and no women in upper positions. When she first arrived, her boss spoke to another staff member over an intercom, who wanted to know if "Lin" was a male or a female. His response upon receiving the answer was "Damn!" Lin laughs at the memory and notes with pride that she managed to hire some people who weren't Protestant males during her two years there.

After her husband graduated from Harvard Law School, they returned to Los Angeles, where he began working for a large corporation. Lin found a job in the personnel department at UCLA and started taking graduate courses part-time. At first, she intended to earn her master's in public administration to increase her job qualifications, but then she switched to more academically oriented courses for the love of it. When she told her husband that she wanted to get her PhD, he found this unacceptable, and they soon divorced. She recalls this without bitterness, adding that she also found the life of a corporate lawyer's wife unappealing. She physically shuddered when she recalled the parties she had to attend, with all of the men in one room and all of the wives in another.

Lin's PhD research was on the decidedly blue-collar subject of water management. Water use in California was unregulated in those days. Anyone could sink a well on their property, but the water they were tapping was a shared resource. When the water table began to decline, the various users realized that water was a limited resource and that something must be done to regulate its use. Lin studied the process whereby the various stakeholders negotiated a workable set of rules without requiring government intervention. It was a success story that she called public entrepreneurship.

Unbeknownst to her, Lin had stumbled on a radically different configuration of ideas from the mathematical empire dominating economic theory. The mathematical empire was founded on the assumption that *self-interest automatically leads to collective well-being.* Lin's work was founded on a stubborn fact of life: *self-interest often leads to the overexploitation of resources and other problems that make life worse for everyone, not better.* When everyone was allowed to suck as much water out of the ground as he or she pleased, there was no invisible hand to rescue the situation.

Yet centralized government intervention was not required to solve the problem. Instead, the local stakeholders got together and negotiated their own agreement, including rules about water use enforced by punishment. Collective well-being was achieved, at least roughly, and there was something emergent and self-organizing about the way it happened—but the process of negotiation bore no resemblance whatsoever to the assumptions of the mathematical empire, which even has trouble accommodating the concept of norms, as we have seen.

Lin married Vincent Ostrom, one of her professors, while she was getting her PhD and followed him when he moved from UCLA to Indiana University. By now, it was the early 1970s, and Lin was in her mid-thirties. No one thought of offering her a job until the department needed someone to teach an early-morning course in American government. Only when she was asked to serve as graduate advisor for the department did she become a full-fledged faculty member. She acted like a mother hen to a very large brood of graduate students as she started to launch her own research career.

Lin used her first opportunity to work intensively with a group of graduate students to focus on the subject of public goods. A public good is anything that can be shared, such as a radio station or a national defense force. The cost of providing a public good might not be shared, however, such as in the case of people who listen to public radio without contributing or who enjoy the security of a national defense force without paying taxes. The temptation to free-ride makes public goods difficult to maintain, as ecologist Garrett Hardin observed in a famous paper, "The Tragedy of the Commons," published in the journal *Science* in 1968. It was still a backwater subject in political science when Lin started working on it, as strange as that might seem today. Many different public goods were studied by people from many disciplines. There was no unified theory or methodology. Lin and her students spent their first semester reading the scattered literature, and she expected them to choose a public good for them to start studying collectively themselves during the next semester. Her only stipulation was that it couldn't involve water, because she had grown tired of that subject from her PhD thesis. Her students chose police departments, a mundane blue-collar subject compared with neoclassical economic theory's lofty goal of creating a physics of social behavior.

Centralization was in vogue among policy makers at the time, with recommendations to consolidate the number of departments from 40,000 to 400 nationwide. Nobody had actually studied police service organizations, however, and reform recommendations were just relying on surface logic. Lin and her students studied the social organization of police departments, using rigorous methods to show that smaller units were often better integrated with the communities they served and more responsive to their needs. They also showed that the optimum scale of police services depended on the particular service; the answer for forensic laboratory services might be different from that for neighborhood patrol services, for example. The best social organization surrounding each service had to be determined on a case-by-case basis, and the people most closely involved were the ones likely to make the best decisions.

Lin's police work continued to explore the new intellectual territory that she had discovered previously with her work on water regulation, which emphasized the importance of decentralization and emergence but in a way that required richly structured interactions at the local level. The emergence emphatically did not result from individuals following only their self-regarding preferences. Sometimes it didn't happen at all. Lin had not yet encountered evolution as an academic subject, but she was still studying the process of cultural evolution in the real world. Many groups of people were negotiating solutions to their problems, some more successfully than others, and the successes were more likely to persist and be emulated than the failures. The variation-and-selection process resulted in outcomes that could be pretty good, if not perfect, and could be improved by studying and assisting the process scientifically. Lin was pioneering the art and science of managing the process of cultural evolution without ever using the E-word.

BY THE 1980S, Lin was gaining a reputation for her work on public goods and was asked by the National Research Council, a branch of the U.S. National Academy of Sciences, to join a committee on common property resources such as irrigation, forestry, and fisheries. Once again, she found the information on the subject in frightful disarray. Many different resources were being studied by people from many disciplines in many parts of the world using many different methods, and nobody was communicating across the boundaries. Lin began the herculean task of imposing order on this information by developing a general framework for analyzing social organizations that could be applied across case studies. At first, she was overwhelmed by the diversity that she encountered. Her database included 100 case studies of farmers' associations in the tiny nation of Nepal alone, each managing its irrigation systems in its own way. Nevertheless, she was able to show that the diverse solutions reflected a smaller number of design principles, which she articulated in her most influential work, *Governing the Commons: The Evolution of Institutions for Collective Action,* published in 1990. By this time, Lin had read a little bit about evolution on her own, but her use of the E-word in the title primarily reflected her own extensive experience studying how groups of people adapt to the challenges of their respective environments in the real world.

Here are the eight ingredients that enable groups to manage their affairs effectively (in my own words), which I like to call Lin Ostrom's recipe for success. She frames the ingredients in terms of managing common resources, but they apply equally well to any situation that requires coordination and cooperation.

LIN OSTROM'S RECIPE FOR SUCCESS

Follow this recipe, and your group will be able to manage its own affairs

1) *Clearly defined boundaries.* Members of the group should know who they are, have a strong sense of group identity, and know the rights and obligations of membership. If they are managing a resource, then the boundaries of the resource should also be clearly identified.
2) *Proportional equivalence between benefits and costs.* Having some members do all the work while others get the benefits is unsustainable over the long term. Everyone must do his or her fair share, and those who go beyond the call of duty must be appropriately recognized. When leaders are accorded special privileges, it should be because they have special responsibilities for which they are held accountable. Unfair inequality poisons collective efforts.
3) *Collective-choice arrangements.* Group members must be able to create their own rules and make their own decisions by consensus. People hate being bossed around but will work hard to do what *we* want, not what *they* want. In addition, the best decisions often require knowledge of local circumstances that *we* have and *they* lack, making consensus decisions doubly important.
4) *Monitoring.* Cooperation must always be *guarded.* Even when most members of a group are well meaning, the temptation to do less than one's share is always present, and a few individuals might try actively to game the system. If lapses and transgressions can't be detected, the group enterprise is unlikely to succeed.
5) *Graduated sanctions.* Friendly, gentle reminders are usually sufficient to keep people in solid citizen mode, but tougher

measures such as punishment and exclusion must be held in reserve.

6) *Fast and fair conflict resolution.* Conflicts are sure to arise and must be resolved quickly in a manner that is regarded as fair by all parties. This typically involves a hearing in which respected members of the group, who can be expected to be impartial, make an equitable decision.
7) *Local autonomy.* When a group is nested within a larger society, such as a farmers' association dealing with the state government or a neighborhood group dealing with a city, the group must be given enough authority to create its own social organization and make its own decisions, as outlined in items (1) to (6) above.
8) *Polycentric governance.* In large societies that consist of many *groups,* relationships among groups must embody the same principles as relationships among individuals within groups.

When I first read about these ingredients, I was struck by their immense practicality. They made as much sense for the neighborhoods of Binghamton as for worldwide regulatory policy.

THE REACTION TO OSTROM as the recipient of the 2009 Nobel Prize in economics (along with Oliver Williamson, who also studies how decisions are made outside markets) spoke volumes about the field of economics. Here is what Steve Levitt, the highly regarded University of Chicago economist best known to the public for his bestselling book *Freakonomics,* wrote on his blog in the *New York Times* on the day after the October 12, 2009, announcement:

> If you had done a poll of academic economists yesterday and asked who Elinor Ostrom was, or what she worked on, I doubt that more than one in five economists could have given you an answer. I personally would have failed the test. I had to look her up on Wikipedia, and even after reading the entry, I have no recollection of ever seeing or hearing her name

> mentioned by an economist. She is a political scientist, both by training and her career — one of the most decorated political scientists around. So the fact I have never heard of her reflects badly on me, and it also highlights just how substantial the boundaries between social science disciplines remain.
>
> So the short answer is that the economics profession is going to hate the prize going to Ostrom even more than Republicans hated the Peace prize going to Obama. Economists want this to be an economists' prize (after all, economists are self-interested). This award demonstrates, in a way that no previous prize has, that the prize is moving toward a Nobel in Social Science, not a Nobel in economics.
>
> I don't mean to imply this is necessarily a bad thing — economists certainly do not have a monopoly on talent within the social sciences — just that it will be unpopular among my peers.

Levitt is refreshing in his honesty. Let's pause to reflect on what he is saying. Economists are proud to be different and don't want to share their status with other branches of the social sciences. They also don't bother to read much outside their own field. They live in a self-contained world and want to keep it that way. Finally, they have adopted the assumptions of their theory as their personal credo. If their theory says that all people are driven entirely by self-interest, then so it must be.

It would be wrong to say that all economists are like this. As we have seen, some of the best and brightest have resisted the mathematical empire and still defend a commonsense view of human nature, from Veblen to Akerlof. Levitt's estimate of one in five is about right, as we can affirm by eavesdropping on the chatter that took place within the field of economics following Lin's award. What better place to listen than the "Economics Job Market Rumors" Web site (www.econjobrumors.com), which is frequented by economics PhD students, postdoctoral students, and young faculty looking for jobs.

In the spirit of *The Lord of the Rings,* I will divide the commentators into the Orcs and the Fellowship of the Ring.

The Orcs

Nobel BULLSHIT!!!! Who the fuck are these idiots? Never heard of them...ever.

What kind of bullshit is this? This year is the worst.

Well, they had to give it to a woman at some point. Why not just throw a dart at a board.

I never saw their work on any reading list during PhD.

A stupid Nobel pick to accompany a stupid job market this year. Our field is falling apart.

Never heard of Ostrom in my life. Lame.

This girl seems to be a political scientist. I don't think she has published original research in any major economics journal.

Multidisciplinary?? Other disciplines are all rubblish [sic]. Why let them conteminate [sic] our purity?

Economics is superior. Don't let political science conteminate [sic] us!

This is the problem with Affirmative Action: last time a woman tried to go to the moon, the Challenger exploded 73 seconds after the launch. Now, this is the end of Economics.

The Fellowship of the Ring

All you guys need to READ MORE. The market rewards multidisciplinary work more and more.

These postings really do show the narrow training of many economists. In fact, economics departments in most universities are highly isolated places in the larger world of social science. To trash a scholar as serious and insightful as Ostrom is shameful.

What if the commons is actually an important field of study, and the fact that most of us never read anything about it during graduate school is something that economic-theory lecturers should take into account when formulating their syllabuses?

Remember: The Orcs outnumber the Fellowship of the Ring five to one and will be heavily represented in the next generation of economists. Be afraid.

IS THERE ANY WAY TO hasten the demise of the mathematical empire or at least force it to gravitate toward common sense? Some clear thinking about the role of mathematical models for the study of any subject will help. Every subject hits a wall of complexity beyond which a comprehensive mathematical description cannot go. The motion of a single celestial body moving through space can be easily described mathematically as a straight line. The motion of two nearby celestial bodies can also be precisely described mathematically as they make their way in a straight line orbiting around each other. The motion of three nearby celestial bodies becomes so complicated that the "three-body problem" pushes analytical mathematics to its limits. In his wonderful book *Chaos: Making a New Science,* James Gleick describes how analytical mathematicians ignored complex interactions because they couldn't cope with them yet still arrogantly looked down on those who used other tools, such as computer-simulation models, better suited for the study of complexity. Eventually, their prejudice was overcome, and now there is a vibrant science of complexity.

There is plenty of mathematics in complexity science, but it plays a different and more humble role from its previous regal status. First, a particular topic is chosen that merits interest. Then the topic is explored with a variety of models, which might be verbal, graphical, mathematical, or computer-driven. Each kind of model has its advantages, and all are simplifications of the real-world topic that is being studied. Models are always in danger of departing from factual reality by making faulty assumptions. The only way to keep them anchored to reality is by testing them against data. Models that don't bring our understanding of the topic closer to reality are rapidly discarded. Every topic becomes the center of a family of models that are only loosely related to one another.

My own field of evolution underwent the same transition. Evolution began to be described mathematically early in the twentieth century. The shuffling of genes by sexual reproduction and the winnowing process of natural selection can be described precisely with equations for one or two loci but rapidly become intractable for three or more loci, just like the three-body problem in physics. Ronald Fisher produced an equation that he called the fundamental theorem of natural selection in 1930, which states that the rate of evolution is proportional to the amount of heritable genetic variation, but so many exceptions to this rule exist that it is vain to call it a theorem, and it certainly wasn't followed by a second and third one. The fact is that the other side of the complexity wall is not like physics during the nineteenth century and never will be. As soon as it becomes more widely known that the mathematical empire of economics is based on a conception of physics that isn't even true for physics, it might lose some of its hypnotic power.

THE WEIRDNESS OF ECONOMICS that I have described so far is based on the impossible ideal of a "physics of social behavior." In addition, writers such as Ayn Rand endowed self-interest with an even more God-like status during the second half of the twentieth century. Rand was not an economist, and professional economists don't necessarily know or care about Rand. Thus, she adds a new layer of weirdness to a story that is already weird enough.

Rand was a Russian immigrant who believed passionately in capitalism, which is understandable for someone who suffered under communist rule. In her novels, such as *Atlas Shrugged,* and essays that more formally developed a philosophy that she called Objectivism, she portrayed the pursuit of self-interest as the highest moral virtue and any restrictions imposed by others as evil, no matter how virtuous they might seem. This was not a license to behave antisocially. Rand was highly moral, but for her, the invisible hand was Truth with a capital T. Once self-interest was properly understood, it always worked out best for everyone in the long run, no matter how much they might protest along the way.

Rand gave self-interest a nobility and infallibility that went way beyond the formal theorizing of professional economists pursuing their own Holy Grail of a "physics of social behavior." Her writing has the same rhetorical structure as a fundamentalist religion. A fundamentalist religion provides very little room for genuine decision making. Everything is portrayed in such polarized terms that the only choice is between good and evil, heaven and hell, riches and ruin. Rationalists hate this feature of fundamentalist religion, because it violates the spirit of genuine inquiry that they hold sacred.

An evolutionist must take a more nuanced stance. When judged by its survival value, a detailed how-to manual for how to live your life, with an answer prepared for every question, might be good or bad. Likewise, the burden of examining every question for yourself might be either good or bad, when judged by the survival value of what you decide. Different costs and benefits cause fundamentalism and rationalism to coexist like different species, each having a distribution and abundance over time and space. Either one can be religious or secular. This is how an evolutionist thinks about cultural diversity as being like biological diversity, which I am applying at the scale of a single city and also worldwide. When I am thinking in this mode, I am no more tempted to judge a fundamentalist or a rationalist than I am to judge a lion, a tree, or a butterfly.

When this way of thinking is applied to Rand's creed of Objectivism, it can be unambiguously classified as fundamentalist—ironic, since it is wrapped in the mantle of rationalism. The followers of Rand were true believers, including Alan Greenspan, one of her disciples, who went on to become chairman of the U.S. Federal Reserve Board for nearly three decades. Greenspan came from the world of professional economics, but it was Rand who turned him into a zealot. As he put it in a 1974 *Newsweek* article: "What she did...was to make me think why capitalism is not only efficient and practical, but also moral."

I'm not the first person to call free-market capitalism a religion. George Soros popularized the term *market fundamentalism* in 1998 in *The Crisis of Global Capitalism,* and the phrase has led a lively existence ever since. Unfortunately, in the polarized world of popular and political discourse, it becomes just another arrow shot ineffectually against an enemy. As an evolutionist, it is important to reflect less judgmentally on what it means for any belief to spread in competition with other beliefs. Evolution has no foresight. Genes and beliefs alike spread on the basis of their local advantages, no matter what the consequences over the long term. Sometimes they spread by virtue of benefiting individuals compared with their immediate neighbors, sometimes by benefiting groups compared with other groups, and so on up a multitiered hierarchy of groups. Adaptations that evolve at one level of the hierarchy are seldom adaptive higher up the scale and more often become part of the problem. Individual selfishness is good for the individual but likely to be bad for the group. Nepotism is good for the family but likely to be bad for a multifamily village. Nationalism is good for the nation but likely to be bad for the community of nations. In this sense, the invisible hand metaphor is profoundly wrong.

Adaptation at a given level of the hierarchy requires a selection process that takes place at the same level. Organisms evolve by between-organism selection, social-insect colonies evolve by between-colony selection, and so on. Higher-level selection structures the lower-level interactions to result in outcomes that are adaptive at the higher level. The final product is emergent, self-organizing, and usually decentralized. The lower-level units obey local rules, do not necessarily have the welfare of the whole system in mind, and don't even require minds (e.g., the cells in our body). In this sense, the invisible hand metaphor is profoundly right. However, higher-level selection is required to discover the local rules that work, which are like needles in a haystack of local rules that don't work. The great error of economic theory is to suppose that people automatically converge on the local rules that work at the collective level merely by following their own self-interest.

HUMAN CULTURAL EVOLUTION is in part a raw process of variation and selection. Life consists of many inadvertent social experiments, and only a few endure. The rules that work are truly unknown to the people who follow them. They weren't consciously invented but merely inherited, like genes. People also consciously invent their own rules, which can be regarded as guided mutations. They are more likely to achieve their objectives than random variation, but they still can easily go wrong based on unforeseen consequences, especially when they interact with rules invented by other people with conflicting interests. Life is complex, and our understanding is severely limited. At the end of the day, we need to try out multiple solutions, designing them as best we can, and select the ones that work based on a careful evaluation of their consequences. We need to manage the process of cultural evolution.

Against this background, we can evaluate belief systems such as Seventh-day Adventism and market fundamentalism as dispassionately as a biologist studying a lion, a tree, or a butterfly. Both spread in a proximate sense because they are psychologically compelling. Believers are convinced that they are right and that they are saving the world for everyone, not just themselves. So did the followers of Jim Jones who drank the poison Kool-Aid and the followers of Osama bin Laden who brought down the World Trade Center. It seems that we can be psychologically compelled to do almost anything. Even among compelling beliefs, some wink out immediately, and others spread based on what they cause people to do.

Seventh-day Adventists are psychologically compelled to work very hard for the coming of the Lord by converting as many people as possible and providing for their health and education. They are highly adaptive at the scale of their own communities and aren't belligerent toward other faiths, but they still can become part of the problem at a larger scale. To my knowledge, Seventh-day Adventists aren't very attentive to worldwide environmental problems, such as global warming, for example.

Market fundamentalists are psychologically compelled to maximize their self-interest, which they conceptualize primarily in monetary terms, secure in their faith that everyone will benefit in the long run. It's easy to see how such a belief can spread on the strength of lower-level advantages, such as individuals gaining over their neighbors or groups gaining over other groups, but it spells nothing but trouble at a larger scale. Evolutionary theory enables us to make this statement at an elementary level. The authority of science can be used to reject definitively the central claim of market fundamentalism, enabling us to proceed with the difficult task of negotiating solutions to our commons problems at all scales.

It is fascinating to read accounts of Greenspan in the wake of the financial collapse, acting as if he is only now regaining his senses after being possessed by a demon. In one article on the Huffington Post on September 28, 2009, he said that the intellectual edifice buttressing radical free-market ideology had "collapsed," leading to a crisis of faith in the ability of banks and other organizations to regulate themselves. When Henry Waxman, a Democratic congressman from California, pointedly asked if his view of the world was not working, Greenspan replied: "Absolutely, precisely."

We can be thankful for people such as Greenspan and Levitt, who can acknowledge when their beliefs might have been wrong. Moreover, rejecting free-market fundamentalism need not entail accepting the most commonly imagined alternatives. Centralization is not necessarily the solution, as Lin discovered in her study of police departments. Many social enterprises are crippled by inappropriate rules imposed by agencies that have no way to assess their consequences. The more I thought about how public and economic policy should be formulated from an evolutionary perspective, combining our best scientific knowledge with managed variation-and-selection processes matched to the scale of the policy issue, the more it seemed like a blend of liberal and conservative principles that might just be appealing to both.

MY EDUCATION IN ECONOMICS TOOK PLACE mostly during 2009, the year of Darwin. All of those trips that seemed like too much proved to be a blessing in disguise, enabling me to speak directly with key people such as Lin Ostrom, rather than merely reading their work. I encountered so many excellent people who could contribute to the dialogue that I started to organize a virtual group even larger than the conference itself. I began to think of the conference as a node in a social process that could begin earlier and continue indefinitely into the future.

After I became comfortable thinking about the mathematical

edifice of neoclassical economics and market fundamentalism, a major piece of the puzzle remained. Experimental and behavioral economists were already attacking the mathematical edifice and claiming to have the answers. They had captured the public imagination with books such as *Nudge* and were beginning to influence public policy. Their battle cry was similar to mine—economic theory and policy need to be based on a realistic conception of human nature—on "humans," not "econs." How did they arrive at their conception of human nature, and did it make sense from an evolutionary perspective?

As I started to plow through the literature on experimental and behavioral economics, I found much to admire, especially the emphasis on performing experiments to anchor theory to reality and the inventory of clever "games" for studying social behavior in carefully controlled settings. Dan O'Brien and I were already drawing on this literature in the games we played with Binghamton High School students. One of the most famous experimental games is called the Ultimatum Game and examines how people behave in negotiations. Suppose that I give you $100 and allow you to offer any amount that you choose to a second person. The second person can accept or reject the offer, and nobody gets anything if the offer is rejected. If people care only about their monetary payoff, then the second person should accept any offer, no matter how small, because anything is better than nothing. Knowing this, the first person should offer a very small amount, such as one dollar, to maximize his or her monetary payoff. Neoclassical economists were dumbfounded when real people didn't behave this way and instead demonstrated a sense of fairness. The average offer was generous, frequently 50 percent, and was likely to be refused if it fell below 30 percent. The second person would rather have nothing than be a victim of unfairness, even when the experiment is carefully designed so that the two people don't know each other's identity and have no possibility of interacting in the future.

The Ultimatum Game speaks volumes about how neoclassical economists have set themselves apart from the rest of the academic world. In the mathematical empire, people have no sense of fairness, along with no risk aversion or norms. Neoclassical economists are so bewitched by their theory that they truly believe that real people don't have a sense of fairness, either. They have already abandoned common sense, and they don't bother to read the social-science literature, so they are comfortable in their hermetically sealed world until some of their own have the bright idea of conducting experiments to check the assumptions of their theory. This is like Ron Numbers ignoring the mountains of evidence for evolution until learning otherwise from a fellow Adventist. Only then—the exact date was 1982 in the case of the Ultimatum Game—did the concept of fairness begin to dawn on them. They couldn't easily incorporate the concept of fairness into their mathematical edifice—that's why it was excluded in the first place—so they just acted alarmed, as if an earthquake had occurred.

On the other hand, there is something really nifty about the Ultimatum Game and the many other games devised by experimental economists. Each one is a tiny microcosm of human social preferences that can be measured precisely under carefully controlled environmental conditions. The conditions can be systematically varied to explore how human social preferences change in response to environmental change. The other branches of the social sciences might be more firmly grounded in common sense and might study fairness in their own ways, but they hadn't converged on a standardized set of tools for measuring social preferences across diverse environmental conditions. Thus, experimental economics games represent an important development for the social sciences as a whole, even if they originated in the hermetically sealed world of neoclassical economics itself.

ECONOMISTS WEREN'T THE ONLY ONES who expected people to behave rationally most of the time. Before the 1970s, many psychologists also expected people to reason on the basis of available information, like natural-born scientists, until psychologists such as Amos Tversky and Daniel Kahneman began to reveal glaring exceptions. For example, in a decision that involves saving the lives of 80 out of 100 people, it makes a huge difference whether the information is described in terms of 80 people who might live or 20 people who might die. In a complex decision such as buying a car, people tend to base their choice on a single factor rather than duly considering multiple factors. Even expertly trained people such as doctors often fail miserably at reasoning about important matters such as whether testing positive for a disease means that you actually have the disease. The discovery of these and other foibles was a humbling demonstration that we are not natural-born rationalists.

The field of behavioral economics was inspired largely by this branch of psychology. In a rare example of reaching outside the discipline, a young economist named Richard Thaler met Tversky and Kahneman at Stanford University. They were in residence as fellows of the Center for Advanced Studies in the Behavioral Sciences (an institution designed to foster interdisciplinarity interactions), and he was a visiting professor at the Stanford branch of the National Bureau of Economic Research. Thaler realized that their work posed severe challenges for the assumptions of neoclassical economics, and the field of behavioral economics was born.

Today the fields of experimental and behavioral economics have largely merged and have demonstrated repeatedly that "econs" and "humans" are different beasts. A theory that informs policy on the basis of a faulty conception of human nature can't be a good thing. Something must be done, but that something depends on providing a more accurate conception of human nature.

It is here that experimental and behavioral economics fails to deliver. Recall the Allais paradox, which can be regarded as an early harbinger of behavioral economics. There is nothing paradoxical about risk aversion from an evolutionary or even a commonsense perspective, as we have seen. It is only paradoxical against the background of neoclassical economic theory. Calling it a "paradox" or an "anomaly" is a sure sign that it is not yet being explained on its own terms. Yet the modern field of experimental and behavioral economics consists primarily of a long list of phenomena treated as anomalies, like tiny satellites orbiting around the mother planet of neoclassical economics independently of one another, unable to escape its gravitational pull. They do not come remotely close to providing an alternative conception of human nature.

I was accustomed to interacting with economists such as Sam Bowles, Herb Gintis, and Ernst Fehr, who are trying to formulate a more coherent alternative conception of human nature and therefore appreciate the relevance of evolutionary theory. They claimed to be a tiny minority among behavioral economists, and I now agreed based on my own reading. To put a number on it, I checked to see how many times the word "evolution" is used in three books on my Kindle: *Nudge* (0), *Predictably Irrational* (0), and *Animal Spirits* (2). These folks are looking for a new conception of human nature, but they don't feel the need to consult evolutionary theory.

I've learned enough to know how behavioral economists are likely to respond to my critique. Why do they need to consult evolution when they can consult psychology? It goes without saying that our psychology is a product of evolution, but why speculate about the distant past when we can directly study the final product?

I have two good answers to this question. First, behavioral economists don't consult psychology in a broad sense but stick primarily to one island of the Ivory Archipelago, the tradition of cognitive heuristics and biases. They need to travel more widely just to visit the other islands associated with psychology, not to speak of other fields devoted to the study of humanity. Second, if you're a biologist, the idea that any species can be studied without reference to evolution is patently absurd. Evolution is consulted in the study of each and every trait in a variety of different ways. The same tool kit is equally essential for understanding our own species. Behavioral economists are at the forefront of challenging economic theory in some respects, but those who think that humans can be studied without reference to evolution are sadly out of date.

I might have been harsh in my assessment of experimental and behavioral economics, but I was not so foolish as to think that my evolutionist brethren or I had all of the answers. Recall from chapter 12 that the field of evolutionary psychology is also provincial and needs to become integrated with other islands of the Ivory Archipelago. An adequate conception of human nature to replace *Homo economicus* is still in the future. It will require an unprecedented degree of integration across the disciplines, using evolutionary theory as the common language. My challenge for the "Nature of Regulation" conference was to set this process in motion.

THE HEADQUARTERS OF NESCent is an attractive brick building in Durham, North Carolina, that was once a cotton mill. Its spacious interior with fifteen-foot-high ceilings, roughhewn wooden posts, and brickwork has been turned into an ideal space for scientific discourse. I arrived feeling strangely calm. The conference was about to take place, but the process had already started with electronic discussions among members of the virtual community, in addition to the conference participants. Other than an introductory talk by myself, there would be no formal presentations. Our job was to use our time together to discuss five major issues that had already been articulated.

First, *rethinking the theory of human regulatory systems from the ground up.* My own inquiry had made it crystal-clear to me that path dependence operates as forcefully for academic cultural evolution as for any other kind of evolution. When it comes to incrementally changing neoclassical economic theory, you can't get there from here. Better to begin from scratch with a conception of human nature based on common sense, all branches of the human sciences, and evolutionary theory.

Second, *learning from other biological systems about the nature of regulation.* Multicellular organisms, social-insect colonies, and the human brain are miracles of self-organized regulation that face the same challenges of cheating and coordination as human groups. We should be consulting nature for ideas about regulation in the same way that we do for ideas about pharmaceuticals.

Third, *reaching a consensus on what constitutes human nature.* Anyone who attempts to reason about human action, formally or informally, employs a set of assumptions about the rules that govern human action. A formal scientific theory must make these assumptions explicit. If every island of the Ivory Archipelago employs a different set of assumptions, then integration needs to take place until a consensus is reached.

Fourth, *appreciating the importance of environmental mismatch.* Some of our regulatory problems reflect the fact that our adaptations to past environments are malfunctioning in our current environments. Neoclassical economics is utterly incapable of dealing with this kind of problem, which can be addressed only from an evolutionary perspective.

Fifth, *taking cultural evolution seriously.* It is common to assume that human nature can be described as a list of culturally universal traits. If one list (such as *Homo economicus*) is wrong, then another list will be right. This assumption ignores the fact that human nature is inherently open-ended, causing people and cultures to evolve different sets of rules based on variation-and-selection processes operating in the recent past. To state the argument in an extreme form, there might be no universal human nature, other than a genetically evolved capacity for open-ended cultural change. If so, then each culture will have a separate nature that must be understood to formulate regulatory policy for that culture and for interactions among cultures. Although this statement is too extreme, it emphasizes the need to study cultural variation in addition to cultural universals from an evolutionary perspective.

Confronting these issues was like staring up at Mount Everest, as I emphasized by actually including an image of Mount Everest in my introductory talk. However, quoting from the NESCent Web site, the purpose of a catalysis meeting is to "increase the scale and ambition of our scientific vision. By allowing interaction beyond the 'usual suspects,' these meetings will facilitate the assembly of networks to collect the primary data needed for synthesis." In other words, the purpose of a catalysis meeting is not to scale Mount Everest but to begin to plan a route up Mount Everest. That was something we could do.

The people sitting in a big square around the room went well beyond the "usual suspects." In stark contrast with the hermetically sealed world of neoclassical economics, they came from all corners of the Ivory Archipelago: animal behavior, anthropology, business, cognitive psychology, economics, ecology, evolutionary psychology, finance, history, law, neurobiology, peace studies, political science, prevention science, social-insect biology, sociology, theoretical biology.

All were highly respected within their disciplines and also saw their subjects as part of one big tangled bank produced by laws acting around us, as Darwin put it so long ago. Evolutionary theory enabled them to communicate across their disciplinary boundaries, not as amateurs but at the highest level of intellectual discourse.

Edward O. Wilson, still active at the age of eighty, planned to be present but had to cancel because of illness. He would still be involved as a member of the Evolution Institute's executive board, and the conference was very much in the spirit of his book *Consilience*, which stresses the unity of knowledge. Jerry Lieberman brought the tradition of humanism and his grassroots experience to the table, in addition to his academic training in political science. Bernard Winograd, who initially posed the question of what evolution might have to say about the world economic crisis, would at last be joining us the next day, adding his considerable direct experience from the world of high finance.

I felt that I was witnessing a historic gathering of academic clans, the First Congress of the Ivory Archipelago, although I didn't actually use the phrase. I also felt that my effort to understand the forbidding world of economics had resulted in insights that would be immensely useful when I returned to the Binghamton Neighborhood Project, like Frodo returning to the shire. Large-scale human society cannot be built directly from individuals. It needs to be multicellular, like a multicellular organism. Individuals need to function within small groups, which in turn need to be organized into larger groups such as cities, which in turn need to be organized into still larger groups such as states, nations, and the global village. Empowering the small groups was just as important as, indeed a prerequisite for, empowering the large ones.

CHAPTER 20

Body and Soul

AT THE BEGINNING OF THIS BOOK, I wrote that there is such a thing as a soul in addition to a body. Bodies and souls can transcend the skins of single individuals, making us part of something larger than ourselves. A city can have a body and a soul, for example. The whole earth can become a single body with a soul, although that would be a tall order.

I like writing things that seem outrageous at first but then make perfect sense after they have been fully explained. Let's see if I can deliver on my claim about bodies and souls.

Bodies are the easy part. Evolutionary science tells us unequivocally that single organisms are extremely cooperative groups—not just metaphorically but literally. You and I are groups of groups of groups, whose lower-level elements lived a more conflictual existence in the past. The lower-level elements became more cooperative by higher-level selection, which winnowed the lower-level interactions, finding the "needles" that cause groups to hang together among the "haystack" that causes them to fall apart.

Organisms are not defined by their physical boundaries. When an organism is riddled with parasites, the parasites do not become part of the organism, even though they are within the same physical boundary. Organisms are defined by their organization. In our nervous system, every neuron is separated from every other neuron by tiny gaps that chemicals must swim across in order for communication to occur. The insects in a social-insect colony are separated by larger gaps that chemicals must float across, along with other modes of communication, but they still make up a single group organism. The specialized cells of our immune systems are like social-insect colonies inside our bodies, physically separated from one another but still intimately communicating. If by "body" we mean "organism," then I have delivered on my claim that bodies can transcend the skins of single individuals, making them part of something larger than themselves. Every social-insect colony is a testimonial to this statement.

It follows that human groups become "bodies" or "organisms" to the extent that they become internally cooperative. Once again, this is not a metaphor but literally as true as it can be. The prospect of being part of something larger than ourselves is fraught with ambivalence. Some people spiritually long for it, while others regard it as a form of bondage and mindlessness. Who wants to be a worker bee or a skin cell?

I believe that the organization of cooperative human groups leaves ample room for the kind of individual freedom and autonomy that we cherish—I certainly do—and that makes the concept of a group organism seem repugnant. Human egalitarianism is and must be guarded. There is always a danger of exploitation from within. People have a horror of placing their fate in the hands of others precisely because it might lead to one form of bondage or another. If by autonomy we mean the capacity to do what we want, in ways that contribute to or at least do not harm society as a whole, and to oppose others who interfere with our worthwhile goals, then autonomy is required for human group organisms to function.

As for mindlessness, an individual mind is designed to receive environmental information as input and results in meaningful action as output. A group organism must also have a mind that functions in the same way. There is more than one way to design a mind at both the individual and group levels. One way is by having an elaborate rule book that is consulted without questioning the rules. Another way is by questioning the rules at each and every opportunity. People who value the questioning mode like to call people who follow the rule-book mode "mindless," but that's silly on their part. Any individual or group that is truly mindless wouldn't survive a day. Moreover, everyone operates in rule-book mode to some degree. Questioning everything would be incapacitating and therefore a form of mindlessness. When we think more clearly about what it means to have a mind at both the individual and group levels, there is ample scope for critical thinking on the part of individuals, now more than ever, in fact.

One thing's for sure: our future is bleak if we don't turn our groups into organisms. Evolutionary science states unequivocally that adaptation at any level of a multitier hierarchy requires selection at the same level and tends to be undermined by selection at lower levels. The invisible hand metaphor is profoundly misleading when it implies that group welfare emerges spontaneously from self-interest. Adaptive groups do self-organize, as we have seen, but only because a process of group-level selection has structured them that way.

When groups aren't organisms, life becomes nasty, brutish, and short for their members. If conflict doesn't get them, then neglect and decay will. When groups do become organisms, life becomes good for their members, but the problems of conflict, neglect, and decay reappear at a higher scale. The only way to solve this problem is by building up the scale of cooperation until the entire planet is a single cooperative group. Otherwise, we will suffer from conflict, neglect, and decay at a massive scale.

Pierre Teilhard de Chardin was prophetic about cultural evolution and the increasing scale of human cooperation, but he got one thing wrong. His concept of the Omega Point treats the emergence of a planetary human organism as inevitable when it is nothing of the sort. Selection and other designing processes become more difficult as we climb each rung of the multitier hierarchy. It's entirely possible for a higher-level organization to collapse, something that has happened many times in the past. We might even drive ourselves extinct while striving to achieve our lower-level goals. If we do, then creatures with our kind of intelligence might never come again. Evolution has no foresight and can easily take us where we don't want to go unless we become wise managers of evolutionary processes.

Evolutionary theory does not provide the comfort of showing that planetary cooperation is inevitable, but it does show that planetary cooperation is possible and provides the conceptual tools for getting the job done.

IT'S FUTILE TO THINK THAT we can solve problems of cooperation at a global scale when they are rampant at a local scale. If healthy organisms require healthy organs and cells, then healthy nations require healthy cities and neighborhoods. One of the best choices I ever made was to adopt my city of Binghamton as my study organism. If I can't put evolutionary theory to work at this scale, then the prospects for group organisms at a higher scale are indeed bleak.

One of the insights that I derive from the concept of a group organism is that organisms are highly complex. That goes without saying for biological organisms, with their anatomy, physiology, nervous systems, and replication machinery. Yet who thinks of a neighborhood or a city that way? When I work with Dan O'Brien and our many city partners to create a single integrated database, I think of it as the city's nervous system. Like any organism, the city needs to receive environmental information as input and process it in a way that leads to effective action as output. If we want to decrease smoking, obesity, and school dropout rates while increasing breastfeeding and nurturance in all of its forms, we need to monitor them, just as our bodies monitor the carbon-dioxide levels in our blood. Then we need to establish a consensus about the need for change so that it is our decision, not someone else's. Finally, we need to establish an incentive system so that the desired behaviors are selected by consequences, following the lead of Tony Biglan and Dennis Embry's successful effort to reduce cigarette sales to minors at a statewide scale. Our minds are already equipped to do these things spontaneously at a small scale when appropriate conditions are met, but tweaking is necessary for the same process to take place at a larger scale or even at a small scale when the appropriate conditions aren't met. I judged Richard Thaler and Cass Sunstein harshly for ignoring evolution in their book *Nudge,* but the concept of tweaking human social dynamics remains a powerful policy tool from an evolutionary perspective.

I'm also inspired by Peter Gray's ideas about self-motivated learning and Lin Ostrom's design principles for effective collective action. People should be in charge of solving their own problems. No one knows the next step better than those taking it, for groups no less than for individuals. The Sudbury Valley School thrives as a community because its members are continuously examining their options as an egalitarian group. What if neighborhoods could become structured that way and their solutions systematically shared so that the best can be chosen? Work would be required to create such an infrastructure, but then it would become an engine for home-grown solutions.

SO MUCH FOR WHAT IT means to have a body. What does it mean to have a soul? I'm fascinated by words for things that seem to have no tangible existence. The fact that we are impelled to use the words suggests that something tangible is at stake, after all.

In a forum on "What Is Religion?" held at the City University of New York on the day after the "Nature of Regulation" conference, I shared the stage with Daniel Dennett and John F. Haught. Before Dan became famous as one of the four horsemen of the current New Atheism movement, he was famous for his book *Darwin's Dangerous Idea*, which explored the profound philosophical implications of evolutionary theory. Before that, he became famous as a philosopher of mind. I still remember being dazzled by his book *Brainstorms*, published way back in 1981, which showed how words for intangibles such as "belief" had essential meaning that could not be entirely reduced to materialistic words such as *brain*. I love talking with Dan even when we disagree, as we do on some, but not all, aspects of the New Atheism movement.

John Haught is also famous as a Roman Catholic theologian who calls himself a theistic evolutionist, which means that he is fully comfortable with the facts of evolution but also remains secure in his religious faith. At the forum, he described the scientific study of religion as like ants crawling around the outside of a goldfish bowl, trying to study what's inside the bowl. Then he explained what it is like to be a goldfish, living on the inside of the bowl as a committed religious believer:

> The feeling of being grasped by, or carried away by what they have taken to be another dimension of reality, other than the profane, other than the mundane, other than the proximate environment that they live in, there's an ultimate environment that's infinite in its scope, that's inexhaustible in its mysteriousness. They've made reference to this dimension primarily through symbolic expression.... Experiences such as these gave people a kind of transparency, a sense of how the divine would influence their lives.... I should add that religion is usually done in community with others.... Above all, I want to emphasize a sense that there is another dimension that is not empirically available but yet is to the religious person even more real than what is empirically available.

I was grateful for John's eloquent account, which struck me as exactly right. I was fascinated that he used the words "proximate" and "ultimate" in the opposite way from how evolutionists use the same two words. For John speaking on behalf of religious believers, the other dimension is more real and vastly more meaningful than the tangible world, which is profane and mundane. The other dimension is the "ultimate" reality, and this world becomes a dreary "proximate environment" by comparison. Even other people become an afterthought.

For an evolutionist, beliefs evolve on the strength of what they cause people to do, just like any other trait. Imagine two religious beliefs. Both are equally gripping in the way that John describes, but one causes people to commit suicide to hasten their passage to the afterlife, while the other causes them to work industriously for themselves and one another. Evolution does its thing, and we are left with only one religious belief, which exists by virtue of its effect on survival and reproduction. That is the ultimate explanation for the belief, as evolutionists use the term. The proximate explanation is its gripping psychological effect, which governs the behavior of the believer in a mechanistic sense.

Evolutionary theory's proximate/ultimate distinction goes a long way toward explaining why the ants have such a hard time communicating with the goldfish. When enduring religions are studied from an evolutionary perspective, their secular utility is impossible to deny. Yet religious believers are not necessarily driven by utilitarian impulses in the psychological sense. Falling under the spell of a perfect God who commands us to help our fellow man can work better as a proximate mechanism than directly wanting to help our fellow man. How else can we explain John's mention of real human communities as an afterthought in his discourse on what it's like to live inside the fishbowl?

Actually, the situation is more complicated than this. Some religious believers think like ants, and some nonbelievers think like goldfish. The whole concept of the fishbowl needs to be dissociated from the concept of religion. Let me explain with two examples.

Meet Myles Horton, one of the great social activists of the early twentieth century, who was born into a Calvinist community in Appalachia. His mother was a pillar of her church, the kind of person everyone in the community relied on. Here's a story that Horton recounts in his autobiography, *The Long Haul:*

> One day I went to my mother and said, "I don't know, this predestination doesn't make any sense to me, I don't believe any of this. I guess I shouldn't be in this church." Mom laughed and said, "Don't bother about that, that's not important, that's just preacher's talk. The only thing that's important is that you've got to love your neighbor." She didn't say "love God," she said "Love your neighbor, that's all it's all about." ... It was a good non-doctrinaire background, and it gave me a sense of what was right and what was wrong.

Horton's mother knew that the true meaning of religion (its ultimate explanation, in evolutionary terms) was "Love your neighbor" and that the theology was "just preacher's talk" for cajoling people into doing the right thing (the proximate mechanism, in evolutionary terms). To her son, she could spell it out and even laugh about it.

She was looking at her religion like an ant, yet she still remained a fully committed member of her religious community.

As my second example, meet Nathaniel Branden, a young man who became Ayn Rand's disciple and lover before eventually leaving the Objectivism movement. Here is how he describes the thrill that he felt when the movement was young in his memoir, *Judgment Day:*

> This is how we were back then, Ayn and I and all of us—detached from the world—intoxicated by the sensation of flying through the sky in a vision of life that made ordinary existence unendurably dull.

The Ayn Rand movement was stridently atheistic, but it existed inside a fishbowl as surely as any religion. Its members thought like goldfish.

If believers and atheists can be both inside and outside fishbowls, then what do we mean by a fishbowl? A fishbowl is any system of beliefs that is detached from factual reality but nevertheless powerfully motivates the believer. People inside fishbowls come alive psychologically, bursting with energy to do what needs to be done. Exactly what that is depends on the details of the belief system. It can be almost anything, from drinking poison, to flying a plane into the World Trade Center, to building schools and hospitals in undeveloped countries, to trying to rid the world of religion, to establishing capitalism as the new world order. Whatever it is, people inside fishbowls become true believers who will not be deterred as long as they remain gripped by their beliefs.

People outside fishbowls are trying to operate in methodological naturalism mode. We live in a material world governed by laws acting around us. It's up to us to manage our affairs. If we don't, we are doomed to a life that is nasty, brutish, and short. Perhaps we will even cause ourselves to wink out of existence as a species. The only hope for managing our affairs is to see the world clearly as it really is and then to act compassionately on the basis of our knowledge. That is the mantra of people who live outside fishbowls.

It might seem that life outside the fishbowl is unbearably dull, compared with the enchantment that can be found only inside the fishbowl. Once again, the situation is more complicated than this. A recent memoir titled *Closing Time* by social commentator Joe Queenan, who is of Irish descent, describes his life growing up poor in Philadelphia. The church was a huge part of his culture and his life, including attending Catholic schools, serving as an altar boy, contemplating the priesthood, and briefly attending a seminary, but there was no enchantment anywhere, for Queenan or any of his associates, judging from his memoir. Countless other accounts testify to the dull, bureaucratic, ham-fisted, conformity-inducing side of religion, not to speak of the deception, backbiting, social climbing, and exploitation that can take place under its cloak. How many people have you met who describe themselves as "recovering Catholics"? These dreary aspects of life take place inside the fishbowl, in addition to the kind of enchantment that John Haught describes.

Then there are people like Carl Sagan, who celebrate life on the outside of the fishbowl with such giddiness that I want to avert my eyes. Myles Horton wasn't giddy, but he had as much grit and determination as any religious believer in his fight for social justice. "Love your neighbor" was good enough for him, without the need for preacher's talk. I'm also filled with zeal, as you might have noticed. Zeal and slackness, like religion and atheism, exist both inside and outside the fishbowl.

When we examine how words such as "soul" and "spirit" are used inside the many fishbowls associated with religion, we find a diverse bestiary, as the Afterlife Project is in the process of documenting. What are the behaviors that these beliefs have evolved to motivate as the "needles," compared with the "haystack" of beliefs that didn't evolve? And why do people outside fishbowls use words such as "soul" and "spirit" almost as much as those on the inside? I have not yet made a formal study, as I have for words such as "altruism" and "selfish," but I think I know what the answer will be to a first approximation. People are spiritual, or have soul, to the extent that they are searching for meaning in a way that will lead to a better life for all, not just themselves. We say that people have "heart" in much the same way.

ONE BOOK THAT DEVELOPS THIS thesis is *Spiritual Evolution: How We Are Wired for Faith, Hope, and Love,* by George Vaillant, whose earlier books include *Adaptation to Life* and *The Natural History of Alcoholism.* George is a kindred spirit and a soul mate, as you can tell from his book titles. He is in his seventies, making him my spiritual father of sorts. He also has a connection to my real father. In 1940, my father had just entered Harvard University and became involved in a longitudinal study that was to last his entire life. The purpose of the Grant study, as it came to be known, was to follow a cohort of normal, healthy men throughout their lives. It was a radical idea at the time, compared with the more typical short-term studies focusing on people with problems. George is trained as a psychiatrist and directed the Grant study from the early 1960s until just a few years ago. He knew my father intimately, interviewing him repeatedly and gathering extensive information, just as I was starting to do for the participants of our OLOG study. He even knew about me and what my father thought about me from the time I was born.

The first time I met George was at a conference a few years before I started the Binghamton Neighborhood Project. I had just finished giving my talk and was approached by an elderly man, still in robust health, who was beaming at me like a proud father. That's because he takes a fatherly interest in all of the children of the men in the Grant study. He knew that my father was proud of me and was now sharing the same pride on meeting me for the first time. My father had passed away after a long period of Alzheimer's disease, and I had to fight back tears when George introduced himself, having heard my father talk about the Grant study throughout my own life.

George's life perfectly illustrates how science, literature, and religion can become entwined in the pursuit of spiritual goals. Most of his ancestors came from the British Isles in the seventeenth and early eighteenth centuries, but the family name came from a man who emigrated from France in 1848 to escape the socialistic and liberal revolts taking place at the time. George's father was an archeologist working in Mexico at a time when archeologists really were a bit like Indiana Jones. He wrote a classic book on the Aztec civilization and mingled with the elites of Mexican intellectual society, such as the artist Diego Rivera. By the time George was born, his father had settled down as a museum curator, first at the American Museum of Natural History and then at the University of Pennsylvania.

George was equally drawn toward science and the humanities and ended up majoring in literature and history at Harvard University. He attended medical school to become a psychiatrist, which is "what English majors who go to medical school do," as he put it. His intention was to establish a practice, with his name on a brass plate. Training in Freudian psychotherapy was required and appealed to George's literary sensibilities, but even then, he could see that it was not even remotely scientific and was more like attending divinity school. He was more strongly drawn to biological approaches that were being pioneered at the time.

George stumbled upon a career in academic psychiatry when he started to work with old medical records of people with drug addictions or severe disorders such as schizophrenia. Doctors didn't know how to cure these afflictions, but when George tracked down the patients many years later, he discovered that some had recovered on their own. He began to think of old records as like trees that had been planted earlier by others and now were ready to be harvested. The fact that some patients recovered without the help of doctors meant that doctors had much to learn from carefully studying the lives of their patients. George was a good writer, and longitudinal studies provided plenty of material that stimulated a lot of interest. Becoming an academic psychiatrist was like living the life of Riley for George, or "the leisure of the theoried class," as he likes to put it.

George discovered another example of patients curing themselves when he became director of an alcoholism clinic and started to attend meetings of Alcoholics Anonymous (AA) as an observer. Psychiatric science had a miserable track record curing alcoholism, but AA was doing something right, and spirituality seemed to be a key feature. Here is a passage that I found after only a few seconds of surfing the Internet (www.barefootsworld.net/aaspiritual.html). The capitalizations and emphasis are part of the passage and have not been added by me:

Why Alcoholics Anonymous Is Spiritual

One of the many terms for Spiritual Practice is ***"avodah,"*** a word from hebrew. It is also a synonym for ***"work" or "discipline."***

Spirituality is a Discipline.

When people say to me, "I'm a spiritual person," they often mean that they treasure some vague feeling of connection with God, nature, or humanity, that is most often divorced from any behavioral obligation.

The disembodied spirituality so often spoken about by those who do not practice any spiritual discipline rarely obligates them to anything, and often excuses the grossest behavior. We have witnessed this many times, in ourselves and in others.

Spirituality is not a feeling, nor is it vague. Spirituality is a conscious practice of living out the highest ethical ideals in the concreteness of our everyday life, and it is that continued practice that brings the awakening of our own spirit into a conscious contact with our Higher Power.

The Program of Alcoholics Anonymous is not a religion, it is a Spiritual Discipline. The conscious practice of the principles of the 12 Steps and their virtues of ***Honesty, Hope, Faith, Courage, Integrity, Willingness, Humility, Brotherly Love, Justice, Perseverance, Prayer and Meditation, and Service to One Another,*** in all our daily affairs is a Spiritual Discipline requiring rigorous honesty and perseverance, and a responsibility to our fellows, to our Higher Power, each as we understand or don't understand it, and to ourselves. The various 12 Step Programs are a mode of living out our daily lives ***Sober,*** one day at a time, under the rigor of ***Spiritual Discipline,*** which may or may not be addressed by any particular religion to which a person adheres.

I could not ask for a more perfect demonstration of the points that I am trying to make in this chapter, right down to the capitalized words in italics. Spirituality might seem intangible, but, properly defined, it is concrete and has a potent effect on our behavior. It is not to be confused with religion. It is a form of discipline that results in virtuous behavior. Calling it a discipline means that it exercises a kind of authority over us. We obey its dictates even when it's not easy over the short run. The capitalized words in italics help to convey authority. These are commandments. The phrase "Higher Power" is even used, even while stressing that spirituality is not a religion. Finally, those who prefer to think of spirituality in the context of religion are welcome to do so. Ants and goldfish are both welcome to join AA, but everyone must become spiritually disciplined.

George was profoundly influenced by AA, so much so that he almost wished he were an alcoholic so he could experience its power firsthand. He did the next best thing by becoming one of the few trustees of AA without a personal history of alcoholism. He also made the kind of spirituality embodied by AA the agenda for his scientific research, culminating in his most recent book, *Spiritual Evolution*.

NOW THAT WE HAVE CLARIFIED what words such as "spirit," "soul," and "heart" mean, we are in a position to say when bodies possess them. It is precisely when they have the virtues listed in the above-cited passage. Groups can possess these virtues, in addition to individuals, so I have delivered on my claim that a neighborhood, a city, or even the entire planet can have a soul in addition to a body.

Next, we need to ask how a body can acquire a soul if it doesn't already have one. Goodwill is not good enough. In addition to wanting the best for everyone, we also must know how to provide it. Goodwill must be supplemented with know-how. Know-how requires thinking like ants, outside the goldfish bowl. This has always been the case, but it is true now more than ever. If we don't use the tools of science to solve the problems of modern existence, *fuhgeddaboudit*, as the Brooklynites like to say.

I therefore arrive at the satisfying conclusion that methodological naturalism is required in order to have a soul. If we don't make an earnest attempt to understand the world the way it really is and to base our actions on our factual knowledge, then we cannot be virtuous at any scale, and we can expect our lives to be nasty, brutish, and short.

The only way to get everyone to behave in a certain way is to establish it as a norm. I therefore propose the following norm, which should be followed by all people who wish to be virtuous and which I will call the First Ant Commandment in honor of John Haught's colorful imagery:

The First Ant Commandment

To defy the authority of empirical evidence is to disqualify oneself as someone worthy of critical engagement in a dialogue.

I thank His Holiness the Dalai Lama for the wording of this commandment, which comes from his book appropriately titled *The Universe in a Single Atom: The Convergence of Science and Spirituality*. A second commandment is needed to keep the first one on the right track:

The Second Ant Commandment

If you're undermining the commons, then you're degrading your soul.

Beneath my playful language is a serious intent. Adopting norms such as the First and Second Ant Commandments requires both a personal commitment and a social commitment to make others conform. When a sufficient number of people uphold a given norm, it becomes solidly established and easy to maintain. The incentives reward conformance most of the time, transgressions are easily detected and punished, most transgressors are contrite rather than defiant, and so on. If we could measure the First Ant Commandment as an established norm, it would have a distribution and abundance in everyday life. We might document that it is solidly established among scientists and serious scholars but sadly lacking among politicians and people engaged in public discourse. We might discover that among politicians, the norm is more solidly established in Denmark than in America. If we could add the dimension of time to our measurements, we would see the boundaries of the norm shrinking or expanding. We might document that the norm was better established among American public figures in the past than in the present. We might discover that the norm is currently eroding even among scientists, especially when science becomes dominated by profit motives. These are just guesses, since nobody has done a study like this, to my knowledge, although it is technically feasible. The point is that we are capable of influencing the distribution and abundance of norms. Establishing the First Ant Commandment more widely is a hopeful and feasible cause, not a hopeless cause. It all depends on the personal and social commitments that we decide to make. If we don't establish the First Ant Commandment more widely, then we won't have the know-how to solve our problems, no matter how much goodwill we possess.

The Second Ant Commandment requires a bit of exegesis. The idea that self-interest invariably promotes the common good is profoundly wrong, but a more subtle formulation of the invisible hand metaphor is valid, as we have seen. Once the local rules that do promote the common good are winnowed from those that don't—the needles from the haystack—then the common good indeed emerges from individuals who do not necessarily have the common good in mind. When we are attempting actively to make decisions, however, our decisions become the winnowing process. If we are not making decisions on behalf of the common good, then we will be generating conflict, neglect, and decay at some level of the multitier hierarchy. Our souls will become degraded.

EARLIER, I SAID THAT ONE way to design a mind is by having an elaborate rule book, and another way is by questioning the rules. Rule questioners like to call rule followers mindless, which is unfair, but neither one is *mindful* as the word is used in Buddhist thought and mindfulness-based therapies such as ACT. Mindfulness can become a key for acquiring a soul, in the sense of helping us thrive at all scales of a multitier hierarchy. ACT involves managing the variation-and-selection process. Behavioral flexibility is increased by accepting and distancing oneself from one's problems. Selection is based on an explicit consideration of values and goals. If a city can have a body, a mind, and a soul, then it can undergo therapy. The problems that we must accept and distance ourselves from include not only our individual experiences but also our genes and cultures, which we have brought with us from all corners of the globe. They worked well enough in the past to survive up to now, but how well they work in the present and future is anyone's guess. No matter — they are like so many people on a bus whom we need to manage the best we can as we drive to the destination of our choice.

CHAPTER 21

City on a Hill

IN HIS SERMON ON THE MOUNT, Jesus is reputed to have said, "You are the light of the world. A city that is set on a hill cannot be hidden" (Matthew 5: 14). Ever since, the image of a city on a hill has been used to inspire people to greatness. John Calvin's city of Geneva was called the city on a hill in the 1500s because the version of Protestantism that he helped to implement made it work so well. The Puritan John Winthrop used the same image in 1630 on the ship *Arbella* before it arrived in Massachusetts to inspire the first American colonists to greatness:

> For we must consider that we shall be as a city upon a hill. The eyes of all people are upon us. So that if we shall deal falsely with our God in this work we have undertaken... we shall be made a story and a by-word throughout the world. We shall open the mouths of enemies to speak evil of the ways of God.... We shall shame the faces of many of God's worthy servants, and cause their prayers to be turned into curses upon us till we be consumed out of the good land whither we are going.

Fast-forwarding to the twentieth century, President John F. Kennedy used the same image at the dawn of his administration:

> I have been guided by the standard John Winthrop set before his shipmates on the flagship *Arbella* three hundred and thirty-one years ago, as they, too, faced the task of building a new government on a perilous frontier. "We must always consider," he said, "that we shall be as a city upon a hill—the eyes of all people are upon us."... Courage—judgment—integrity—dedication—these are the historic qualities of the Bay Colony and the Bay State—the qualities which this state has consistently sent to this chamber on Beacon Hill here in Boston and to Capitol Hill back in Washington.

Ronald Reagan had very different political views from Kennedy's, but he repeatedly used the same image, most poignantly in his farewell speech to the nation:

> The past few days when I've been at that window upstairs, I've thought a bit of the "shining city upon a hill." The phrase comes from John Winthrop, who wrote it to describe the America he imagined. What he imagined was important because he was an early Pilgrim, an early freedom man. He journeyed here on what today we'd call a little wooden boat; and like the other Pilgrims, he was looking for a home that would be free. I've spoken of the shining city all my political life, but I don't know if I ever quite communicated what I saw when I said it. But in my mind it was a tall proud city built on rocks stronger than oceans, wind-swept, God-blessed, and teeming with people of all kinds living in harmony and peace, a city with free ports that hummed with commerce and creativity, and if there had to be city walls, the walls had doors and the doors were open to anyone with the will and the heart to get here. That's how I saw it and see it still.

I'm not here to give either a sermon or a political speech, you'll be happy to know. For me, the phrase "city on a hill" has a more concrete meaning. My efforts to understand and improve the human condition in Binghamton result in GIS maps with hills and valleys of well-being. If I manage to succeed with the help of my many associates, then the valleys will start to rise until eventually, the entire city will be sitting on a hill of well-being. It won't be an inspirational metaphor. It will be a fact.

THE BINGHAMTON NEIGHBORHOOD PROJECT IS entering its fifth year as I conclude this book. In our third year, we administered the Developmental Assets Profile, the survey that started it all, again to the public-school students as the start of a longitudinal database. The plan is to do this in perpetuity, so that we can chart the movement of the GIS hills and valleys over time, just as geologists chart the movement of the physical earth.

We had not started any valley-lifting projects at that point, so we did not expect to see any movement on the basis of our efforts, but we did discover a very encouraging result. Some of the students were too young to take the survey in 2006 and were taking it for the first time in 2009, but hundreds took the survey both times, enabling us to measure how they changed over the three-year period. Of these, about 20 percent moved their residential locations within the city. Some moved from hills to valleys, and others moved from valleys to hills. Even though we have not yet changed the topography of the city, we can use this "natural experiment" to see what happens to individuals when the topography changes for them by virtue of moving locations. The answer is that they respond quite quickly to their new environment, which can be good or bad depending on whether they have moved up or down the hills of well-being.

This beautiful result addresses the fundamental issue of how fast people adapt to environmental change. The theoretical possibilities range from very slow to very fast, as we have seen. At the slow end, change might never occur if the "you can't get there from here" principle of path dependence operates. Or generations might be required, as with genetic evolution and some kinds of cultural evolution. Or events that take place early in life might stay with us for the rest of our lives, as we saw with the inspirational first-grade teacher Miss A and the potent effects of the Good Behavior Game during the first and second grades. Or we can be like thirsty plants that quickly respond to changes in our growth conditions. All of these are legitimate possibilities, and they probably all take place to some degree. Our result provides convincing support for the fast end of the continuum. When teens in the city of Binghamton change their environment by moving, they respond as individuals within a period of three years. If we can change their environment for the better without requiring them to move, we will have the satisfaction of seeing them improve quickly, like watering thirsty plants.

All creatures have evolved to listen and reflect in ways that lead to sustaining action. In less poetic terms, they receive and process information in ways that enable them to survive and reproduce. A social group is not different from an individual organism in this regard, as we saw with the parable of the wasp and even our own immune systems, which function like social-insect colonies inside our bodies. Small human groups function spontaneously like organisms, at least under the right conditions, but larger groups require additional cultural evolution before they can function like organisms. Efficient business organizations closely monitor and respond to relevant aspects of their environment. They qualify as organisms, although their self-serving objectives often put them on a collision course with other organisms. Cities do not have the infrastructure to monitor and process information in the same way as a biological organism or an efficient corporation. They cannot function as an organism for the simple reason that they don't have a nervous system.

But Binghamton is on its way to being the first city with a nervous system. Imagine a city in which all of the information gathered by its various sectors — health, social services, environment, crime, school — feeds into a single, integrated database that is updated at frequent intervals and is spatially resolved to the level of neighborhoods. Imagine also that the information is vigilantly protected so that it cannot be used against the interests of individuals. The rules that govern scientific research on humans are more respectful of privacy and human rights than any other system of rules governing human conduct. Stripped of personal identity, the information can be used to monitor and improve the quality of life in ways that everyone can agree on, such as prenatal and early-childhood care, high school graduation rates, stress-related diseases, happiness, and longevity. All decisions are likely to go wrong when they are based on bad information or imposed by some on others without their consent. Our only hope of making wise decisions is on the basis of good information and being agreed on by consensus. Any city can decide to create a nervous system for itself, but we're at the forefront, thanks to the imaginative leap of regarding a city as an organism.

An organism the size of a city must be multicellular. The cells are small groups of people with the authority to manage their own affairs in ways that contribute to, or at least do not interfere with, the larger common good. People come alive in small groups working together to solve problems of common interest. That is where they feel safe, known and liked as individuals, and respected for their contributions to the group. It is the ancestral human social environment, and we will never feel fully at home when we depart from it.

We also must re-create the ancestral natural environment in order to feel fully at home in our cities. Barren urban landscapes poison the body and the soul and always will. If hospital patients recover faster and feel less pain in rooms with plants, think of how we can improve the quality of life for everyone by welcoming nature back into our cities. You'd think that science wouldn't be necessary to reach this conclusion—why don't we just do what is so manifestly natural and good for us? For better or for worse, we are such a cultural species that our theories trump our instincts all the time. If an idea seems to make sense, we act on it, like those fools who are granted wishes in folk tales. We're even worse than those fools, because we never come to our senses. We just continue acting on our wishes, heedless of the unforeseen consequences wreaking havoc all around us. The only way out of this maze of unforeseen consequences is with the help of a better theory and the back-and-forth process of prediction and test that anchors theory to reality. After five years of listening and reflecting on my city of Binghamton from an evolutionary perspective, I feel in a strong position to advise about how to raise its valleys into hills.

* * *

I THOUGHT THAT MY PLATE was full when I started the Binghamton Neighborhood Project, but I'll never regret my decision to create the Evolution Institute with Jerry Lieberman. Human life is indeed a multitier hierarchy, and it is important to work at all levels simultaneously, from the entire planet to the smallest neighborhood. The EI is a David organization compared with the Goliaths of other policy-making organizations, but do you remember how that contest turned out?

My encounter with economics has convinced me that the bigger they are, the harder they fall. When smart people are guided by the wrong theory, they go a very long way in the wrong direction. The inability of neoclassical economics to gravitate toward common sense, not to mention the more refined dictates of psychology and evolutionary theory, provides an outstanding example of path dependence in cultural evolution. Seeing clearly that "you can't get there from here" provides justification for a paradigmatic change in economic theory and policy.

Our partnership with NESCent to establish the evolutionary paradigm is continuing at a very fast pace by cultural evolutionary standards. The University of Chicago Press will be publishing a volume based on the "Nature of Regulation" conference titled *Evonomics: Evolutionary Theory as a New Foundation for Economics and Public Policy.* Geoffrey Hodgson, one of the world's foremost scholars of economics in relation to evolutionary theory, is coediting the volume with me, enabling us to make an authoritative case for the paradoxical fact that modern economic theory is almost completely uninformed by evolutionary theory, even though economics and evolutionary theory have been entwined throughout their histories. The University of Chicago Press leaped at the opportunity to publish the volume, which is highly significant, given the role that the University of Chicago has played in the history of economics.

The virtual community that I started to organize for the "Nature of Regulation" conference has grown, enabling us to speak with a single voice. We recently submitted a white paper to the National Science Foundation, "The Relevance of Evolutionary Science for Economic

Theory and Policy," which included more than sixty signatories reflecting all disciplines. In our ongoing meetings at NESCent headquarters, we have decided to adopt a case-study approach for demonstrating the advantages of the evolutionary perspective for real-world policy issues such as obesity, private property, and environmental policy. Jerry Lieberman and I are expanding our portfolio of focal topics to include intergroup conflict, the quality of life, the nature of discourse, the importance of play, and the ethical dimensions of public policy. We are forming partnerships with think tanks and other policy-making organizations around the world.

Prevention science and evolutionary science are becoming a single island thanks to the efforts of my two amigos, Tony Biglan and Dennis Embry. Tony has organized a consortium called the Promise Neighborhood Research Consortium (PNRC), which will make science-based solutions available to communities nationwide through its Web site and consulting relationships. I am honored to sit at the same table with Tony, Dennis, Steve Hayes, and their colleagues as a member of the PNRC and to see prevention science increasingly framed in terms of evolutionary theory. Ever the entrepreneur, Dennis has a grand scheme to market scientifically validated techniques for changing behavioral practices on a Web site similar to Amazon.com. Each technique will be available for the price of a music CD, the success of each technique will be monitored with a rating system, and controlled experiments can even be performed on the best combinations of techniques by offering them in various combinations with discounted pricing. If Dennis has his way, better behaviors will spread with the rapidity of hit songs.

Even for a methodological naturalist such as myself, committed to upholding the Two Ant Commandments, catalysis is a kind of miracle. Without the presence of a catalyst, some chemical reactions proceed slowly or not at all. Add even a tiny amount of the catalyst, and the rate of change becomes 10, 100, 1000 times faster. We needn't rely on supernatural intervention for this kind of miracle. We can accomplish transformational change by acting as catalysts. Perhaps the rate of change is already starting to quicken.

* * *

ISAAC BASHEVIS SINGER'S FATHER USED the great religious tradition of Judaism to advise people on their everyday problems, and one of my greatest pleasures in writing this book has been to combine the sublime and the mundane. In some respects, I'm like a plumber, here to fix our collective clogged drain with my evolutionary tool kit. But I'm also reflecting on the most profound issues that have ever been pondered by religious sages, philosophers, and storytellers throughout the ages.

Knowledge is sacred, or at least it should be, but it is created by everyday people committed to holding one another accountable for the factual statements that they make about the world. Anyone can join this faith, regardless of their day jobs, expanding the norms associated with science into other arenas that are currently sorely lacking in accountability.

The distinction between basic and applied research, which is reflected in the status hierarchy of academic disciplines from white-collar to blue-collar, can be erased. For evolutionary science, the best basic research is on people from all walks of life as they go about their daily lives, which is also the best applied research. Understanding and improving the human condition go hand-in-hand.

After five years of listening and reflecting on my city of Binghamton, we are beginning to try our hand at valley raising. One of our most ambitious projects is called the Design Your Own Park competition. Like every city during these hard economic times, Binghamton is struggling to make ends meet, but it is wealthy in one respect: it has lots of vacant lots. Each one is potentially a valuable resource for the surrounding neighborhood to create the park of its dreams. Not just a children's park but a park for everyone in the neighborhood. Not somebody else's idea of a park but *their* idea of a park, based on consulting everyone in the neighborhood, from the youngest to the oldest.

The DYOP competition is designed to make this happen throughout the city. Each group that enters the competition is assigned a facilitator, who might be me or a bright, eager student from the EvoS program. The facilitation process and judging criteria for the competition are designed to provide the ingredients required for groups to manage their commons effectively, as identified by Lin Ostrom. The group must be well identified and must include the full participation of the neighborhood. There must be proof that all ages have been consulted. There must be something resembling a contract or a constitution ensuring that the costs and benefits are fairly distributed, so that all of the work doesn't fall on the shoulders of a few beleaguered volunteers. Decision making must be by consensus. There must be a way to monitor participation, graduated sanctions to prevent slacking, and a fast, fair system for resolving conflicts. Groups that don't include these elements in their plans won't win the competition. In short, the DYOP competition first provides a valuable common resource for a neighborhood in the form of a park and then provides the ingredients required to achieve any collective goal in the form of the facilitation process and judging criteria. Brilliant, as Watson said to Holmes.

Once a favorable social environment for collective action has been created, groups are free to design their parks however they like, just as students at the Sudbury Valley School are free to structure their learning however they like. They must submit an economy version of their plan, in addition to a deluxe version, to remain within reasonable budget constraints. The more they can rely on their own artistic talent, construction skills, labor, and material, the better. Finally, they must include a plan for maintaining the park over the long term at minimal expense to the city. Local maintenance is an asset for the neighborhood, as it is to the city, because it provides a continuing context for interactions among neighbors after the park has been built.

The city of Binghamton is an enthusiastic partner for the DYOP competition. So is our local branch of the United Way, which will help us raise money for developing the parks. Landscape architects involved with the project will turn the ideas generated by the neighborhood groups into concrete designs. A workshop that Peter Gray and I organized brought world experts to Binghamton to discuss the interconnected subjects of play, public spaces, and empowering neighborhoods to manage their own affairs. I was proud to see members of the neighborhood groups at the workshop, speaking and listening, alongside community leaders and my colleagues from the academic world.

The competition is designed to be friendly, so that every park that exceeds a certain score is scheduled for implementation as soon as funds can be raised for them. The competition will be repeated, so that every neighborhood can potentially have its own park. Even better, the neighborhoods that create and maintain the parks will be well adapted to solve other collective-action problems. Already, we are discussing neighborhood homework clubs and addressing the problems of health care, job training, and crime prevention, and more.

Each neighborhood that enters the competition is surveyed to assess current strengths and needs, along with a comparison neighborhood matched for demographic variables that has not (yet) entered the competition. These surveys will provide baseline information to compare with future surveys. If entering the DYOP improves the quality of life in a neighborhood, we'll be able to show it with a high degree of certainty. The citywide maps that we create from giving the Developmental Assets Profile to the public-school students will provide a second estimate. The neighborhoods surrounding the parks will literally rise into hills on the GIS maps.

The DYOP project is so appealing that it can be described without using the words "science" or "evolution," but it is richly informed by both. This is not something that I attempt to hide. On the contrary, in addition to writing the guidelines for the competition, Peter and I have written a companion document, "The Science Behind the Scenes of the Design Your Own Park Competition." In plain language that anyone can understand, we invite people backstage to view the scientific principles at work. The importance of allowing groups to manage their own affairs. Harnessing the motivating power of between-group competition while avoiding its destructive potential. Using variation and selection to discover best practices. The importance of beautiful natural surroundings. The importance of unstructured play in mixed-age groups for children. The importance of safety, relaxation, and playfulness in adults. The importance of social control and how it can emerge spontaneously. The scientific evidence in support of these principles is fascinating in its own right and adds weight to what also seems like common sense when viewed from the right perspective.

As I was working with the city to create the DYOP competition, the mayor's office told me about a group on the north side of Binghamton that had been trying to improve an existing neighborhood park for several years to no avail, despite good intentions on all sides. Could I facilitate this group in addition to organizing the citywide competition? I leaped at the opportunity, because it meant that I could start working with a neighborhood right away, as a forerunner for what I hoped would take place throughout the city.

That is how I found myself in the middle of a celebration on upper Murray Street, not long after the Saint Patrick's Day parade. The people who had been trying to improve their existing park were titans of civic virtue, similar to Peggy Wozniak and Doug Stento. They didn't have leadership positions within the city, but they had the same unflagging interest in making their world a better place. At first, they couldn't believe that something might actually happen, given their frustration during the past three years. Once they became convinced, their energy knew no bounds. They decided to hold a brainstorming party at the park. They distributed flyers for a radius of three blocks on each side. They cleaned up the park in preparation. Food was served. A DJ was on hand to play music. There were games for the kids, even a magician and a balloon artist. Dancing spontaneously broke out. Someone stationed at the entrance to the park obtained contact information, while my posse of graduate students and EvoS Irregulars and I circulated among the crowd with audio recorders, asking people what they would like to see in a neighborhood park for all ages.

Talk about acting as a catalyst! I had done almost nothing except turn some kind of mysterious key that caused the neighborhood to spring to life. The atmosphere was as festive as that of the Saint Patrick's Day parade, which brought the city out to celebrate the end of winter. I could almost feel one of the valleys of my GIS map rising beneath my feet. I had been living in Binghamton for more than twenty years, but never before had I felt so much at home.

Acknowledgments

When you have become part of something larger than yourself, there are a lot of people to thank. Each of the programs that I manage—EvoS for higher education, the Evolution Institute for public policy, the Binghamton Neighborhood Project for community-based research, and the Evolutionary Religion Studies Web site for the study of religion—involves working with large networks of people. In all cases, the spirit of cooperation and the excitement of discovery turn our work into pleasure.

My agent, Michelle Tessler (a graduate of Binghamton University); my editor, John Parsley (a Binghamton native); and the staff at Little, Brown have made it possible for me to write a book for a general audience without compromising substance. It would be hard to imagine a better publishing experience. My two sisters, Becka and Lisa, who share my novelistic roots, helped me at numerous junctures. The book's final shape owes much to Becka's fierce advocacy for certain chapters.

I owe a special debt of gratitude to the citizens of Binghamton, New York, who have treated this egghead professor and former local slacker with such hospitality—even though my business card states my profession as "evolutionist."

Finally, if I were forced to choose one person for thanks, it would be my wife, Anne, who is facing me across the dining-room table at this moment, working on her laptop as I work on mine. Our love for each other is founded on our love for our work. Everyone should be so lucky.

Notes

The Neighborhood Project was written for the full spectrum of readers, from those who are newly encountering the evolutionary paradigm to seasoned professionals. These notes provide a portal for advancing your knowledge, including references to the work that I cite and suggestions for further reading.

The programs that I describe in this book—EvoS, the Binghamton Neighborhood Project, the Evolution Institute, and Evolutionary Religious Studies—all have Web sites that provide abundant material for further inquiry. Just type their names into a search engine such as Google, and you'll get there. My previous book, *Evolution for Everyone: How Darwin's Theory Can Change the Way We Think about Our Lives,* provides a basic tutorial that is also written for the full spectrum of readers.

If you are a scientist or a scholar residing on one of the many islands of the Ivory Archipelago, you can start adopting the evolutionary paradigm surprisingly easily. In a survey that I gave to my colleagues at the forefront of studying evolution in relation to human affairs, I discovered that most of them received little to no formal training in evolution and picked up their knowledge on their own. You can join their ranks.

For students at any stage of development, the evolutionary paradigm can provide a way to listen and reflect on course material, even when teachers aren't using the E-word. Like Omar Eldakar, you might find your grades and general thinking skills improving, a hypothesis that we are currently in the process of testing with the many students who have graduated from Binghamton University's EvoS program.

If you are an adult whose day job is not even remotely scientific, I hope you take comfort in the fact that scientists are also everyday people who come from all walks of life and that science itself is an everyday process of people holding one another accountable for their factual statements about the world. Becoming knowledgeable about evolution as a citizen and expanding the norms associated with science and scholarship into other walks of life are essential steps toward wisely managing the cultural evolutionary process.

Rather than providing specific links to Web sites, which often change, I have provided enough information to reach the sites through a search engine such as Google. Most of the scientists I profile have Web sites that enable you to learn more about their work, including pdf files of their articles. Add a keyword such as "evolution" or the institutional affiliation along with the name, and you'll quickly find them. A section of Google called Google Scholar provides another portal to the scientific and scholarly literature. Never before has so much information become available to so many people—waiting to be organized by evolutionary theory.

Introduction: The Listener

John Allman's book *Evolving Brains* (1999) provides a wonderful comparison of brains across species, showing how they are adapted to receive and process information in different ways. Terrence Deacon's book *The Symbolic Species* (1998) shows how the human brain has evolved a unique capacity for symbolic thought.

Chapter 1: Evolution, Cities, and the World

I tell more of my own story in chapter 35 of my previous book *Evolution for Everyone* (Wilson 2007). Stephen Jay Gould's provocative thought experiment about replaying the tape of life is in his book *Wonderful Life* (1989). The unpredictability of anyone's life, or the history of life in general, is based on a property of complex systems called sensitive dependence on initial conditions, which is described for a general audience by James Gleick's *Chaos* (1987), in addition to numerous other sources. For a good history of the sociobiology debate, see Ullica Segerstrale's *Defenders of the Truth* (2000). My work on group selection is described in my book with Elliott Sober, *Unto Others* (1998), academic articles that can be found on my Web site, and a series of posts on my Evolution for Everyone blog titled "Truth and Reconciliation for Group Selection." Two recent books by authors other than myself (Borrello 2010, Harman 2010) cover the fascinating history of the group-selection controversy.

Chapter 2: My City

The first history of Binghamton was written in 1840 by John B. Wilkinson (1840/1967) and published with an update by Tom Cawley and illustrations by Johnny Hart, the great cartoonist who lived in our area. Visit the Web site of the Broome County Historical Society for a selection of books on the Binghamton area, including a lovely coffee-table book by Gerald R. Smith, *Partners All: A History of Broome County, New York* (2006). If you are able to visit Binghamton, I recommend our excellent Roberson Museum, which features exhibits on Binghamton past and present, in addition to its general science and arts exhibits. One of the exhibits forms the basis of my section on the colonial days of Binghamton. Oren Lyons is a traditional faith keeper of the Turtle Clan and a member of the current day Haudenosaunee, in addition to being professor of American studies at SUNY Buffalo. The Benjamin Franklin quote is cited and elaborated in Johansen (1982). George Washington's chilling directive to Sullivan is preserved in the writings of George Washington from the original sources, available at the Electronic Text Center at the University of Virginia library. The poignant letter from a soldier carrying out the operation, delivered to his sweetheart upon his death, is located in the Tioga Point Museum. Mary Jemison's bitter recollection is cited on p. 59 of Seaver (2008). The retrospective passages by Sherman and Flick are included in a Web site called "Sullivan/Clinton Campaign Then and Now," which keeps the memory of this ethnic-cleansing event in our own history alive. Cross (2006) provides a good history of religious fervor and innovation in New York during the 1800s.

Chapter 3: The Parable of the Strider

Typing "water strider" into Google Scholar will provide access to the same literature that I consulted for this chapter. An excellent review is "Walking on Water: Biolocomotion at the Interface" (Bush and Hu 2006). Goodwyn et al. (2008) and Feng et al. (2007) studied the super-hydrophobic properties of strider feet. Stim Wilcox's foundational research on the use of surface waves as a mode of communication made the cover of *Science* (Wilcox 1979), in addition to many other articles. In his retirement, he has returned to his childhood love of making weapons and has written a beautiful book on how to make bows from a single piece of wood (use "Wilcox Bows" as a search term). Visit Daphne Fairbairn's Web site for a complete list of her publications, including her heroic study with Richard Preziosi (Fairbairn and Preziosi 1994; Preziosi and Fairbairn 1996, 2000).

Chapter 4: The Parable of the Wasp

There's lots to read about social-insect colonies as super-organisms, including Bert Holldobler and Ed Wilson's *The Super-organism* (2009), Tom Seeley's *Honeybee Democracy* (2010) and *The Wisdom of the Hive* (1995), in addition to Jim Hunt's *The Evolution of Social Wasps* (2007). Ed Wilson has even written a novel in which an ant colony is one of the main characters (E. O. Wilson 2010). Niko Tinbergen's clever experiment with wasps is described in his classic book *Curious Naturalists* (Tinbergen 1959/1984). Tinbergen (1963) also wrote a paper titled "On Aims and Methods of Ethology," which remains essential reading and is widely taught as "Tinbergen's Four Questions." Lynn Margulis's (1970) groundbreaking work on nucleated cells as bacterial super-organisms was generalized by Maynard Smith and Szathmáry (1995, 1998) to include other major evolutionary transitions. For a more recent update, see Massimo Pigliucci and Gerd Muller's edited volume *Evolution: The Extended Synthesis* (2010).

Chapter 5: The Maps

Visit the Web site of Search Institute to see how it has been helping to improve schools and communities for more than fifty years. I am grateful to Peter Benson and Art Sesma for helping me get started on my own community-based research and allowing me to use DAP. Wilson, O'Brien, and Sesma (2009) is the first scientific publication based on our maps. To see how carefully research on humans is regulated, visit the National Institute of Health's Web site on bioethics resources or the World Health Organization's site on ethical standards and procedures with research on human beings.

Chapter 6: Quantifying Halloween

For more on Machiavellianism from an evolutionary perspective, see Wilson, Near, and Miller (1996 and 1998). A biography by Blass (2009) chronicles the life of Stanley Milgram, one of the fathers of social psychology. Type "Lost letter method" into Google Scholar to peruse the many studies based on his ingenious technique. William Poundstone's *Prisoner's Dilemma* (1992) provides a good introduction to game theory and experimental games. Our use of experimental games to study variation among neighborhoods was inspired in part by similar research at a worldwide scale (Henrich et al. 2004).

Chapter 7: We Are Now Entering the Noosphere

The Jesuit and the Skull (Amir 2007) is an excellent biography of Teilhard and his times. The "Homo Symbolicus" workshop is described on the John Templeton Foundation Web site and will soon be published as an edited volume. Visit Christopher Henshilwood's Web site or use "Blombos Cave" as a search term to keep up with the latest findings on this remarkable site. The Vatican conference can be found on the Internet by using "Biological Evolution: Facts and Theories" as a search term and will also result in an edited volume. Important recent books on human cultural evolution based on our capacity for symbolic thought include Deacon (1998), Jablonka and Lamb (2006), Richerson and Boyd (2005), and Tomasello (2009). The two passages from Julian Huxley are from *Essays of a Humanist* (1964) and *Man in the Modern World* (1947).

Chapter 8: The Parable of the Immune System

I highly recommend Lauren Sompayrac's *How the Immune System Works* (2008) as a basic tutorial on the immune system and Steve Frank's *Immunology and Evolution of Infectious Disease* (2002) as a book-length treatment of immunology from an evolutionary perspective. Use "hygiene hypothesis" as a search term to learn the latest news on the mismatch between our immune systems and our current environments. I explore the comparison between the immune system and our capacity for behavioral and cultural change in D. S. Wilson (2010).

Chapter 9: The Reflection

Scientific articles based on the BNP as of this writing include Wilson and O'Brien (2009); Wilson, O'Brien, and Sesma (2009); and O'Brien and Wilson (2010). Visit the BNP Web site to keep up with our continuing progress. An earlier article that I wrote with psychologist Mihalyi Csikszentmihalyi, "Health and the Ecology of Altruism," explores the same themes using a nationwide database of American youth compiled by Mihalyi and his colleagues (Wilson and Csikszentmihalyi 2007).

Chapter 10: Street-Smart

I highly recommend Dick Nisbett's *Intelligence and How to Get It* (2009) as a book that affirms the capacity for change, as long as one knows what to do. The wonderful story of Miss A is described there on pp. 62–63 and is based on a study by Pederson, Faucher, and Eaton (1978). Charles Murray's quote is from a 2007 *Wall Street Journal* article, "Intelligence in the Classroom," July 10, 2007. Mueller and Dweck (1998) conducted the experiment on praising children for their intelligence versus their hard work, and Heine et al. (2001) compared the consequences of success and failure in children of Asian and European descent. Omar Eldakar continues to thrive, and you can access his wide-ranging publications through his Web site.

Chapter 11: The Humanist and the CEO

The Humanists of Florida Association, American Humanist Association, Discovery Institute, and National Center for Science Education all have Web sites that can be accessed by using their names as search terms.

Chapter 12: The Lost Island of Prevention Science

The Ivory Archipelago is also a major theme of my previous book *Evolution for Everyone*. I hope that this chapter helps the field of prevention science to become better known throughout the Ivory Archipelago. To learn more, visit the Web site of the Society for Prevention Research and the Promise Neighborhood Research Consortium, which I also describe in the final chapter. Watson's famous boast was published in his book *Behaviorism* (1930). Rebecca Lemov's *World as Laboratory: Experiments with Mice, Mazes, and Men* (2005) provides an excellent history of behaviorism. Paul Naour (2009) uses a taped conversation between B. F. Skinner and E. O. Wilson to explore the relationship between radical behaviorism and sociobiology. Tony Biglan (2003) relates Skinner's "selection by consequences" principle to prevention science as a unifying principle. *The Adapted Mind* (Barkow, Cosmides, and Tooby 1992) remains one of the best ways to understand the field of evolutionary psychology at its inception. David Buller's *Adapting Minds* (2005) slightly alters the title to make some of the same points that I do about the human capacity for rapid open-ended change. Writing in the 1960s, John Bowlby was a pioneer in the field of evolutionary psychology, and his work on attachment theory (e.g., 1969, 1990) is still central to the modern field of evolutionary developmental psychology (e.g., Ellis and Bjorklund 2005, Bjorklund 2007). Frank Sulloway's biography of Sigmund Freud (1992) evaluates his work against the background of modern evolutionary theory. For the scientific assessment of therapeutic methods, consult Ackerman and Hilsenroth (2003), Barlow (2007), Herbert and Forman (2011), Kazdin (2007), Lambert (2003), and Norcross et al. (2006). The heroic research of the Oregon Youth Study is described in Capaldi et al. (2001) and Dishion et al. (1997, 2004, 2005). Dishion and Andrews (1995) demonstrated the counterproductive effects of aggregating high-risk youth in intervention programs. To learn more about Tony's work, use his name and "ORI" as search terms. To learn more about Dennis and his unique brand of scientific entrepreneurship, use his name and "Paxis Institute" as search terms.

Chapter 13: The Lecture That Failed

Use "Steve Hayes," "ACT," and "RFT" as search terms to learn more about Steve, acceptance and commitment therapy as a therapeutic method, and relational-frame theory as the scientific foundation for ACT. I especially recommend Steve's self-help book *Get out of Your Mind and into Your Life: The New Acceptance and Commitment Therapy* (Hayes and Smith 2005) and his scientific review article that relates ACT to other therapeutic methods and RFT (Hayes 2004). The Good Behavior Game is marketed by Dennis Embry's Paxis Institute and described in an excellent review article by Dennis (Embry 2002). A special issue of the journal *Drug and Alcohol Dependence* (volume 14, 2008) was devoted to assessing the lasting effects of the GBG played in the Baltimore City Schools during first and second grades.

The amazing "reward and reminder" intervention to reduce cigarette sales to minors at a statewide scale is described in Embry et al. (2010).

Chapter 14: Learning from Mother Nature about Teaching Our Children

Peter Gray's *Psychology* is currently in its sixth edition (2010) and remains one of the best introductions to the broad field from an evolutionary perspective. Peter has also become a popular blogger for *Psychology Today.* His academic articles on the Sudbury Valley School include Gray and Chanoff (1986), Gray and Feldman (2004), and Gray (2007). Visit the Sudbury Valley School's Web site to learn more about this remarkable approach to education, including a listing of other schools inspired by the original one. The *Psychology Today* article on the Sudbury Valley School was published in the May 2006 issue and is available on its Web site. Some good books on childhood in traditional and especially hunter-gatherer societies include Lancy (2008), Lancy et al. (2010), and Konner (2010). The humorist Lenore Skenazy has written a wonderful book on the importance of giving our kids freedom to learn, *Free Range Kids: Giving Our Children the Freedom We Had without Going Nuts with Worry* (2009). Bruce Ellis's research on dandelion and orchid children is summarized in Ellis and Boyce (2008). To learn more about direct instruction, visit the Web site of the Association for Direct Instruction.

Chapter 15: The World with Us

Rasmussen (2007) provides a recent survey of sensory-deprivation research. Ed Wilson, who has a knack for launching entire disciplines, wrote *Biophilia* in 1984. Recent books and articles in the genre include Kellert et al. (2008) and Grinde and Patil (2009). John Maerz is one of the most prolific young scientists studying invasive species, as you can see by visiting his Web site. The impact of invasive earthworms is documented by Migge-Kleian et al. (2006). Blossey and Notzold (1995), Blumenthal (2006), Joshi and Vrieling (2005), and Wolfe et al. (2004) document rapid evolution in invasive plant species. Phillips et al. (2006) provide a brief report on rapid evolution in the cane toad, and you can find more at Richard Shine's Web site. The devastating effects of deer, a native species that has become overabundant, are reviewed in Cote et al. (2004). Allendorf et al. (2008) review the genetic effects of harvesting wild animal populations. Von der Lippe and Kowarik (2008) demonstrated that the city of Berlin is a net exporter of seeds. The literature on the health benefits of breastfeeding is voluminous; one large randomized control trial demonstrates substantial benefits for cognitive development (Kramer et al. 2008). Park and Mattson (2009) demonstrated the powerful effects of plants on hospital recovery rates, and Hietanen et al. (2007) used affective priming to demonstrate the emotional effects of viewing vegetation, building on the line of inquiry initiated by Ulrich et al. (1991). Visit the Web site of the John Bowne High School to learn more about its remarkable agricultural program, and use "BioBlitz" as a search term to learn about—and become involved in—this great movement.

Chapter 16: The Parable of the Crow

The Golden Guide series that inspired Kevin McGowan as a child is being updated for the next generation, thanks to St. Martin's Press. If you resonate to Kevin's story, then think about joining your local chapter of the Audubon Society or another nature-oriented club. Better yet, take the Cornell Lab of Ornithology's home-study course in bird biology taught by Kevin. Some of the scientific publications from the Crow Research Group include McGowan (2001), Clark et al. (2006), Townsend et al. (2009a, 2009b), and Becky Heiss's comparison of growth rates in urban and rural crows (Heiss et al. 2009). Use "New Caledonian crow" as a search term to learn more about this fascinating species, including video clips in addition to the scientific literature. John Marzluff has written a book on crows and ravens (Marzluff and Angell 2005), in addition to his more recent clever experiment involving the use of masks (Marzluff et al. 2010). Nihei and Higuchi (2001) document the crows in Tokyo that have learned to use automobiles as nutcrackers.

Chapter 17: Our Lives, Our Genes

Two books on rapid genetic evolution in humans are *The 10,000 Year Explosion: How Civilization Accelerated Human Evolution* (Cochran and Harpending 2009) and *Before the Dawn: Recovering the Lost History of Our Ancestors* (Wade 2006). The distinction between wanting and liking is made by Arias-Carrión and Pöppel (2007) and Pecina et al. (2003). Dan Eisenberg et al. (2008) examine the consequences of the DRD4 polymorphism in nomadic and settled populations of the Ariaal people of Kenya. The three meta-analyses that examine the relationship between the DRD4 polymorphism and ADHD, personality, and addiction, respectively, are Faraone et al. (2001), Munafo et al. (2008), and McGeary (2009). James MacKillop's careful study of alcohol craving is described in MacKillop et al. (2007). Amato et al.'s display of scientific brawn (2009) demonstrates that disruptive selection is operating more forcefully for psychological traits than for any other category of traits.

Chapter 18: The Natural History of the Afterlife

The Evolutionary Religious Studies Web site is intended to promote evolutionary science as the framework of choice for the study of religion. Visit the site of the John Templeton Foundation to see how they are attempting to address "the big questions." For more on positive psychology, see Seligman (2002) and Frederickson (2009). For more on forgiveness from an evolutionary perspective, see McCullough et al. (2000) and McCullough (2008). Gould and Lewontin (1979) used the architectural example of spandrels to illustrate the possibility that biological traits can be by-products rather than adaptations. The passages from Pederson (1991, p. 461), Johnston (2002, p. 72), and Rosenberg (1981, p. 85) can be found on pages 42, 70, and 71 of Madigan and Levenson (2008). Carlton Pearson chronicles his own odyssey in *The Gospel of Inclusion: Reaching beyond Religious Fundamentalism to the True Love of God and Self* (2008) and *God Is Not a Christian, Nor a Jew, Muslim, Hindu... God Dwells with Us, in Us, around Us, as Us* (2010). I critique the New Atheism movement in a series of posts on my Evolution for Everyone blog titled "Atheism as a Stealth Religion."

Chapter 19: Evonomics

I elaborate on the theme of this chapter in a series of posts on my Evolution for Everyone blog titled "Economics and Evolution as Different Paradigms." I recommend Eric Beinhocker's *The Origin of Wealth: Evolution, Complexity, and the Radical Remaking of Economics* (2006) as an excellent short introduction to neoclassical economic theory and why it needs to be radically remade. For a more extensive treatment of the entwined histories of economic and evolutionary thought, see books by Geoffrey M. Hodgson, including his most recent (with Thorbjorn Knudsen) *Darwin's Conjecture: The Search for General Principles of Social and Economic Evolution* (2010). The small group of economists who had already adopted an evolutionary perspective, with whom I initially consulted to organize my conference, include Sam Bowles, Herbert Gintis, and Ernst Fehr. They are now part of a disciplinary melting pot that includes psychologists (such as Gerd Gigerenzer), anthropologists (such as Joseph Henrich), and evolutionists (such as Peter Richerson and Robert Boyd). Visit their Web sites, or consult the selection of books and articles listed in the References section below. My example of the Allais paradox is taken from the Wikipedia entry on the term. His bitter comment about the economic establishment is in Allais (1987, p. 8). George Akerlof's comment is made in Akerlof (2007, pp. 5–6). To learn more about Elinor Ostrom, read her classic *Governing the Commons: The Evolution of Institutions for Collective Action* (1990) and her more recent works that you can find on her Web site. The term "tragedy of the commons" was coined by ecologist Garrett Hardin in a famous article published in the journal *Science* in 1968. I write more about Ayn Rand as a fundamentalist in a chapter of *Evolution for Everyone* titled "Ayn Rand: Religious Zealot." The ultimatum game was invented by Guth, Schmittberger, and Schwarze (1982). Visit the Web site of the National Evolutionary Synthesis Center (NESCent) to learn about their many other synthetic projects. A volume based on the "Nature of Regulation" conference, coedited with Geoffrey Hodgson and titled *Evonomics: Evolutionary Theory as a New Foundation for Economics and Public Policy,* will be published by the University of Chicago Press.

Chapter 20: Body and Soul

The "What Is Religion?" forum was organized by the City University of New York Graduate School as part of its "Great Issues" series and can be viewed on the Internet by using "Great Issues Forum: What Is Religion" as a search term. I'm grateful to John Haught for his vivid image of scientific ants studying religious goldfish. His most recent book is *Making Sense of Evolution: Darwin, God, and the Drama of Life* (2010). The passage from Myles Horton's autobiography *The Long Haul* (1990) is on p. 7. The passage from Nathaniel Branden's memoir *Judgment Day* is on p. 18. The elegant passage from His Holiness the Dalai Lama, which I have enshrined as the First Ant Commandment, is taken from p. 76 of his book *The Universe in a Single Atom: The Convergence of Science and Spirituality.*

Chapter 21: City on a Hill

John Winthrop's famous "city on a hill" passage is from his sermon "A Model of Christian Charity." Keep up with our progress at the Web sites that I manage.

References

Ackerman, S. J., and Hilsenroth, M. J. (2003). A review of therapist characteristics and techniques positively impacting the therapeutic alliance. *Clinical Psychology Review* 23: 1–33.

Akerlof, G. A. (2007). The missing motivation in macroeconomics. *American Economics Review* 97: 3–36.

Akerlof, G. A., and Shiller, R. J. (2009). *Animal Spirits: How Human Psychology Drives the Economy, and Why It Matters for Global Capitalism*. Princeton, N.J.: Princeton University Press.

Allais, M. (1987). *The Allais Paradox*. In J. Eatwell, M. Milgate, and P. Newman (eds.), *The New Palgrave: Utility and Probability*. New York: Macmillan, pp. 3–9.

Allendorf, F. W., England, P. R., Luikart, G., Ritchie, P. A., and Ryman, N. (2008). Genetic effects of harvest on wild populations. *Trends in Ecology and Evolution* 23: 327–337.

Allman, J. M. (1999). *Evolving Brains*. New York: Scientific American Library.

Amato, R., Pinelli, M., Monticelli, A., Marino, D., Miele, G., and Cocozza, S. (2009). Genome-wide scan for signatures of human population differentiation and their relationship with natural selection, functional pathways and diseases. *PLoS One* 4: e7927.

Amir, A. (2007). *The Jesuit and the Skull: Teilhard de Chardin, Evolution, and the Search of Peking Man*. New York: Riverhead (Penguin).

Arias-Carrión, O., and Pöppel, E. (2007). Dopamine, learning and reward-seeking behavior. *Acta Neurobiologiae Experimentalis* 67: 481–488.

References

Ariely, D. (2008). *Predictably Irrational: The Hidden Forces That Shape Our Decisions.* New York: HarperCollins.

Atran, S. (2002). *In Gods We Trust: The Evolutionary Landscape of Religion.* Oxford: Oxford University Press.

Barkow, J. H., Cosmides, L., and Tooby, J. (eds.) (1992). *The Adapted Mind: Evolutionary Psychology and the Generation of Culture.* Oxford: Oxford University Press.

Barlow, D. H. (2007). *Clinical Handbook of Psychological Disorders,* 4th ed. New York: Guilford.

Beinhocker, E. D. (2006). *The Origin of Wealth: Evolution, Complexity, and the Radical Remaking of Economics.* Cambridge, Mass.: Harvard Business School Press.

Bergson, H. (1911). *Creative Evolution.* New York: Holt.

Biglan, A. (2003). Selection by consequences: One unifying principle for a transdisciplinary field of prevention. *Prevention Science* 4: 213–232.

Bjorklund, D. F. (2004). *Children's Thinking: Cognitive Development and Individual Differences.* Florence, Ky.: Wadsworth.

———(2007). *Why Youth Is Not Wasted on the Young: Immaturity in Human Development.* New York: Wiley-Blackwell.

Bjorklund, D. F., and Pellegrini, A. D. (2001). *The Origins of Human Nature: Evolutionary Developmental Psychology.* Washington, D.C.: American Psychological Association.

Blass, T. (2009). *The Man Who Shocked the World: The Life and Legacy of Stanley Milgram.* New York: Basic Books.

Blossey, B., and Notzold, R. (1995). Evolution of increased competitive ability in invasive nonindigenous plants: A hypothesis. *Journal of Ecology* 83: 887–889.

Blumenthal, D. M. (2006). Interactions between resource availability and enemy release in plant invasion. *Ecology Letters* 9: 887–895.

Boehm, C. (1999). *Hierarchy in the Forest: Egalitarianism and the Evolution of Human Altruism.* Cambridge, Mass: Harvard University Press.

Borrello, M. (2010). *Evolutionary Restraints: The Contentious History of Group Selection.* Chicago: University of Chicago Press.

Bowlby, J. (1969). *Attachment. Attachment and Loss,* vol. 1. New York: Basic Books.

———(1990). *A Secure Base: Parent-Child Attachment and Healthy Human Development.* New York: Basic Books.

Bowles, S. (2003). *Microeconomics: Behavior, Institutions, and Evolution*. Princeton, N.J.: Princeton University Press.

Boyd, R., Gintis, H., and Bowles, S. (2010). Coordinated punishment of defectors sustains cooperation and can proliferate when rare. *Science* 328: 617–620.

Boyer, P. (2001). *Religion Explained*. New York: Basic Books.

Branden, N. (1989). *Judgment Day*. Boston: Houghton Mifflin.

Bruner, J. S. (1973). *Beyond the Information Given: Studies in the Psychology of Knowing*. New York: Norton.

Buller, D. J. (2005). *Adapting Minds: Evolutionary Psychology and the Persistent Quest for Human Nature*. Cambridge, Mass.: MIT Press.

Bush, J. W. M., and Hu, D. L. (2006). Walking on water: Biolocomotion at the interface. *Annual Review of Fluid Mechanics* 38: 339–369.

Capaldi, D. M., Dishion, T. J., Stoolmiller, M., and Yoerger, K. (2001). Aggression toward female partners by at-risk young men: The contribution of male adolescent friendships. *Developmental Psychology* 37: 61–73.

Clark, A. B., Robinson, D. A. J., and McGowan, K. J. (2006). Effects of West Nile Virus mortality on social structure of an American crow (Corvus brachyrhynchos) population in upstate New York. *Ornithological Monographs* 60: 1–14.

Clutton-Brock, T. (2007). *Meerkat Manor: Flower of the Kalahari*. London: Phoenix.

Cochran, G., and Harpending, H. (2009). *The 10,000 Year Explosion: How Civilization Accelerated Human Evolution*. New York: Basic Books.

Cote, S. D., Rooney, T. P., Tremblay, J.-P., Dussault, C., and Waller, D. M. (2004). Ecological impacts on deer overabundance. *Annual Review of Ecology and Systematics* 35: 113–147.

Cross, W. R. (2006). *The Burned-over District: The Social and Intellectual History of Enthusiastic Religion in Western New York, 1800–1850*. Ithaca, N.Y.: Cornell University Press.

Dalai Lama, H. H. (2005). *The Universe in a Single Atom: The Convergence of Science and Spirituality*. New York: Morgan Road Books.

Darwin, C. (1859). *The Origin of Species*. London: John Murray.

Dawkins, R. (2006). *The God Delusion*. Boston: Houghton Mifflin.

Deacon, T. W. (1998). *The Symbolic Species*. New York: Norton.

Dennett, D. C. (1981). *Brainstorms*. Cambridge, Mass.: MIT Press.

———(1995). *Darwin's Dangerous Idea*. New York: Simon & Schuster.

———(2006). *Breaking the Spell: Religion as a Natural Phenomenon.* New York: Viking.

Depew, D. J., and Weber, B. H. (1995). *Darwinism Evolving.* Cambridge, Mass.: MIT Press.

Diamond, J. (1997). *Guns, Germs, and Steel.* New York: Norton.

Dishion, T. J., and Andrews, D. W. (1995). Preventing escalation in problem behaviors with high-risk young adolescents: Immediate and 1-year outcomes. *Journal of Consulting and Clinical Psychology* 63: 538–548.

Dishion, T. J., Eddy, J. M., Haas, E., Li, F., and Spracklen, K. (1997). Friendships and violent behavior during adolescence. *Social Development* 6: 207-223.

Dishion, T. J., Nelson, S. E., Winter, C. E., and Bullock, B. M. (2004). Adolescent friendship as a dynamic system: Entropy and deviance in the etiology and course of male antisocial behavior. *Journal of Abnormal Child Psychology* 32: 651–663.

Dishion, T. J., Nelson, S. E., and Yasui, M. (2005). Predicting early adolescent gang involvement from middle school adaptation. *Journal of Clinical Child and Adolescent Psychology* 34: 62–73.

Durkheim, É. (1912/1995). *The Elementary Forms of Religious Life,* trans. K. E. Fields. New York: Free Press.

Eisenberg, D. T. A., Campbell, B., Gray, P. B., and Sorenson, M. D. (2008). Dopamine receptor genetic polymorphisms and body composition in undernourished pastoralists: An exploration of nutrition indices among nomadic and recently settled Ariaal men of northern Kenya. *BMC Evolutionary Biology,* 8.

Ellis, B. J., and Bjorklund, D. F. (eds.) (2005). *Origins of the Social Mind: Evolutionary Psychology and Child Development.* New York: Guilford.

Ellis, B. J., and Boyce, W. T. (2008). Biological sensitivity to context. *Current Directions in Psychological Science* 17: 183–187.

Embry, D. D. (2002). The good behavior game: A best practice candidate as a universal behavioral vaccine. *Clinical Child and Family Psychology Review* 5: 273–297.

Embry, D. D. (2004). Community-based prevention using simple, low-cost, evidence-based kernels and behavior vaccines. *Journal of Community Psychology* 32: 575–591.

Embry, D. D., and Biglan, A. (2008). Evidence-based kernels: Fundamental units of behavioral influence. *Clinical Child and Family Psychology Review* 11: 75–113.

Embry, D. D., Biglan, A., Galloway, D., McDaniels, R., Nunez, N., Dahl, M. J., et al. (2010). Reward and reminder visits to reduce tobacco sales to young people: A multiple-baseline across two states. Ms.

Fairbairn, D. J., and Preziosi, R. F. (1994). Sexual selection and the evolution of allometry for sexual size dimorphism in the water strider, *Aquarius remigis*. *American Naturalist* 144: 101–118.

Faraone, S. V., Doyle, A. E., Mick, E., and Biederman, J. (2001). Meta-analysis of the association between the 7-repeat allele of the dopamine D4 receptor gene and attention deficit disorder. *American Journal of Psychiatry* 158: 1052–1057.

Fehr, E., and Fischbacher, U. (2003). The nature of human altruism. *Nature* 425: 785–791.

———(2005). The economics of strong reciprocity. In H. Gintis, S. Bowles, R. Boyd, and E. Fehr (eds.), *Moral Sentiments and Material Interests: The Foundations of Cooperation in Economic Life*. Cambridge, Mass.: MIT Press.

Feng, X. Q., Gao, X. F., Wu, Z. N., Jiang, L., and Zheng, Q. S. (2007). Superior water repellency of water strider legs with hierarchical structures: Experiments and analysis. *Langmuir* 23: 4892–4896.

Frank, S. A. (2002). *Immunology and Evolution of Infections Disease*. Princeton, N.J.: Princeton University Press.

Frederickson, B. (2009). *Positivity: Groundbreaking Research Reveals How to Embrace the Hidden Strength of Positive Emotions, Overcome Negativity, and Thrive*. New York: Crown Archetype.

Galatzer-Levy, R., Bachrach, H. A. S., and Waldron, S. W. J. (2000). *Does Psychoanalysis Work?* New Haven, Conn.: Yale University Press.

Geary, D. C. (1996). *Children's Mathematical Development: Research and Practical Applications*. Washington, D.C.: American Psychological Association.

———(2004). *The Origin of Mind: Evolution of Brain, Cognition, and General Intelligence*. Washington, D.C.: American Psychological Association.

———(2009). *Male, Female: The Evolution of Human Sex Differences*, 2nd ed. Washington, D.C.: American Psychological Association.

Gibbons, E. (1962/2005). *Stalking the Wild Asparagus*. Chambersburg, Pa.: Alan C. Hood.

Gigerenzer, G. (2007). *Gut Feelings: The Intelligence of the Unconscious*. New York: Viking.

Gigerenzer, G., Todd, P. M., and Group, A. R. (eds.) (2000). *Simple Heuristics That Make Us Smart*. Oxford: Oxford University Press.

Gintis, H. (2007). A framework for the integration of the behavioral sciences. *Behavioral and Brain Sciences* 30: 1–61.

———(2009). *Game Theory Evolving,* 2nd ed. Princeton, N.J.: Princeton University Press.

Gintis, H., Bowles, S., Boyd, R., and Fehr, E. (eds.) (2005). *Moral Sentiments and Material Interests: The Foundations of Cooperation in Economic Life.* Cambridge, Mass.: MIT Press.

Gleick, J. (1987). *Chaos: Making a New Science.* New York: Penguin.

Goodwyn, P. P., De Souza, E., Fujisaki, K., and Gorb, S. (2008). Moulding technique demonstrates the contribution of surface geometry to the super-hydrophobic properties of the surface of a water strider. *Acta Biomaterialia* 4: 766–770.

Gould, S. J. (1989). *Wonderful Life: The Burgess Shale and the Nature of History.* New York: Norton.

Gould, S. J., and Lewontin, R. C. (1979). The spandrels of San Marco and the panglossian paradigm: A critique of the adaptationist program. *Proceedings of the Royal Society of London* B205: 581–598.

Gray, P. (2007). The educative functions of free play and exploration. *Eye on Psi Chi* (Fall): 18–21.

Gray, P. (2010). *Psychology,* 6th ed. New York: Worth.

Gray, P., and Chanoff, D. (1986). Democratic schooling: What happens to young people who have charge of their own education? *American Journal of Education* 94: 182–213.

Gray, P., and Feldman, J. (2004). Playing in the zone of proximal development: Qualities of self-directed age mixing between adolescents and young children at a democratic school. *American Journal of Education* 110: 108–145.

Grinde, B., and Patil, G. G. (2009). Biophilia: Does visual contact with nature impact on health and wellbeing? *International Journal of Environmental Research and Public Health* 6: 2332–2343.

Guth, W., Schmittberger, R., and Schwarze, B. (1982). An experimental analysis of ultimatum bargaining. *Journal of Economic Behavior and Organization* 3: 367–388.

Hardin, G. (1968). The tragedy of the commons. *Science* 162: 1243–1248.

Harman, O. (2010). *The Price of Altruism: George Price and the Search for the Origins of Kindness.* New York: Norton.

Harris, S. (2004). *The End of Faith: Religion, Terror, and the Future of Reason.* New York: Norton.

Haught, J. (2010). *Making Sense of Evolution: Darwin, God, and the Drama of Life.* Louisville, Ky.: Westminster/John Knox.

Hayes, S. C. (2004). Acceptance and commitment therapy, relational frame theory and the third wave of behavioral and cognitive therapies. *Behavior Therapy* 35: 639–665.

Hayes, S. C., Barnes-Holmes, D., and Roche, B. (eds.) (2001). *Relational Frame Theory: A Post-Skinnerian Account of Human Language and Cognition.* New York: Springer.

Hayes, S. C., and Smith, S. (2005). *Get out of Your Mind and into Your Life: The New Acceptance and Commitment Therapy.* Oakland Calif.: New Harbinger.

Herbert, J. D., and Forman, E. M. (2011). *Acceptance and Mindfulness in Cognitive Behavior Therapy: Understanding and Applying the New Therapies.* Hoboken, N.J.: Wiley.

Heine, S. J., Kitayama, S., Lehman, D. R., Takata, T., Ide, E., and Leung, C. (2001). Divergent consequences of success and failure in Japan and North America: An investigation of self-improving motivation. *Journal of Personality and Social Psychology* 81: 599–615.

Heiss, R. S., Clark, A. B., and McGowan, K. J. (2009). Growth and nutritional state of American crow nestlings vary between urban and rural habitats. *Ecological Applications* 19: 829–839.

Henrich, J., Boyd, R., Bowles, S., Camerer, C., Fehr, E., and Gintis, H. (2004). *Foundations of Human Sociality: Economic Experiments and Ethnographic Evidence from Fifteen Small-scale Societies.* Oxford: Oxford University Press.

Henrich, J., and Henrich, N. (2007). *Why Humans Cooperate: A Cultural and Evolutionary Explanation.* Oxford: Oxford University Press.

Herbert, J. D., and Forman, E. M. (2011). *Acceptance and Mindfulness in Cognitive Behavior Therapy: Understanding and Applying the New Therapies.* Hoboken, N.J.: Wiley.

Herrnstein, R., and Murray, C. (1994). *The Bell Curve: Intelligence and Class Structure in American Life.* New York: Free Press.

Hietanen, J. K., Klemettila, T., Kettunen, J. E., and Korpela, K. M. (2007). What is a nice smile like that doing in a place like this? Automatic affective responses to environments influence the recognition of facial expressions. *Psychological Research—Psychologische Forschung* 71 (5): 539–552.

Hitchens, C. (2007). *God Is Not Great: How Religion Poisons Everything.* New York: Twelve Books, Hachette Book Group.

Hodgson, G. M., and Knudsen, T. (2010). *Darwin's Conjecture: The Search for General Principles of Social and Economic Evolution.* Chicago: University of Chicago Press.

Holldobler, B., and Wilson, E. O. (1990). *The Ants.* Cambridge, Mass.: Belknap Press.

———(2009). *The Super-organism: The Beauty, Elegance, and Strangeness of Insect Societies.* New York: Norton.

Horton, M. (1990). *The Long Haul.* New York: Doubleday.

Hunt, J. H. (2007). *The Evolution of Social Wasps.* Oxford: Oxford University Press.

Huxley, J. S. (1927/1979). *Religion without Revelation.* Santa Barbara, Calif.: Greenwood.

———(1942/2010). *Evolution: The Modern Synthesis,* ed. M. Pigliucci and G. Muller. Cambridge, Mass.: MIT Press.

———(1943). *Evolutionary Ethics.* Oxford: Oxford University Press.

———(1947). *Man in the Modern World.* London: Chatto and Windus.

———(1964). *Essays of a Humanist.* New York: Harper & Row.

Jablonka, E., and Lamb, M. (2006). *Evolution in Four Dimensions: Genetic, Epigenetic, Behavioral, and Symbolic Variation in the History of Life.* Cambridge, Mass.: MIT Press.

Johansen, B. C. (1982). *Forgotten Founders: How the American Indian Helped Shape Democracy.* Cambridge, Mass.: Harvard Common Press.

Johnston, P. S. (2002). *Shades of Sheol: Death and Afterlife in the Old Testament.* Downer's Grove, Ill.: IVP Academic.

Joshi, J., and Vrieling, K. (2005). The enemy release and EICA hypothesis revisited: Incorporating the fundamental difference between specialist and generalist herbivores. *Ecology Letters* 8: 704–714.

Kazdin, A. E. (2007). Mediators and mechanisms of change in psychotherapy research. *Annual Review of Clinical Psychology* 3: 1–27.

Kellert, S. R., Heerwagen, J., and Mador, M. (2008). *Biophilic Design: The Theory, Science and Practice of Bringing Buildings to Life.* New York: Wiley.

Kirkpatrick, L. A. (2004). *Attachment, Evolution, and the Psychology of Religion.* New York: Guilford.

Konner, M. (2010). *The Evolution of Childhood: Relationships, Emotion, Mind.* Cambridge, Mass.: Belknap.

Kramer, M. S., Aboud, F., Mironova, E., et al. (2008). Breastfeeding and child cognitive development: New evidence from a large randomized trial. *Archives of General Psychiatry* 65: 578–584.

Lambert, M. J. (2003). *Bergin and Garfield's Handbook of Psychotherapy and Behavior Change,* 5th ed. New York: Wiley.

Lancy, D. F. (2008). *The Anthropology of Childhood: Cherubs, Chattel, and Changelings.* Cambridge: Cambridge University Press.

Lancy, D. F., Bock, J., and Gaskins, S. (eds.) (2010). *The Anthropology of Learning in Childhood.* Lanham, Md.: AltaMira.

Lemov, R. (2005). *World as Laboratory: Experiments with Mice, Mazes, and Men.* New York: Hill and Wang.

Levitt, S. D., and Dubner, S. J. (2006). *Freakonomics: A Rogue Economist Explores the Hidden Side of Everything.* New York: William Morrow.

MacKillop, J., Menges, D. P., McGeary, J. E., and Lisman, S. A. (2007). Effects of craving and DRD4 VNTR genotype on the relative value of alcohol: An initial human laboratory study. *Behavioral and Brain Functions* 3: 11.

Madigan, K. J., and Levenson, J. D. (2008). *Resurrection: The Power of God for Christians and Jews.* New Haven, Conn.: Yale University Press.

Malthus, T. (1798/2008). *An Essay on the Principle of Population.* Oxford: Oxford University Press.

Mandeville, B. (1714/1998). *The Fable of the Bees: or Private Vices, Publick Benefits.* Indianapolis: Liberty Press.

Margulis, L. (1970). *Origin of Eukaryotic Cells.* New Haven, Conn.: Yale University Press.

Marzluff, J. M., and Angell, T. (2005). *In the Company of Crows and Ravens.* New Haven, Conn.: Yale University Press.

Marzluff, J. M., Walls, J., Cornell, H. N., Withey, J. C., and Craig, D. P. (2010). Lasting recognition of threatening people by wild American crows. *Animal Behavior* 79: 699–707.

Maynard Smith, J., and Szathmáry, E. (1995). *The Major Transitions in Evolution.* New York: W. H. Freeman.

———(1998). *The Origins of Life: From the Birth of Life to the Origin of Language.* Oxford: Oxford University Press.

McCullough, M. (2008). *Beyond Revenge: The Evolution of the Forgiveness Instinct.* New York: Jossey-Bass.

McCullough, M. E., Pargament, K. I., and Thoresen, C. E. (eds.) (2000). *Forgiveness: Theory, Research and Practice.* New York: Guilford.

McGeary, J. (2009). The DRD4 exon 3 VNTR polymorphism and addiction-related phenotypes: A review. *Pharmacology Biochemistry and Behavior* 93 (3): 222–229.

McGowan, K. J. (2001). Demographic and behavioral comparisons of suburban and rural crows. In J. M. Marzluff, R. Bowman, and R. Donelly (eds.), *Avian Ecology and Conservation in an Urbanizing World*. Norwell, Mass.: Kluwer Academic.

Migge-Kleian, S., McLean, M. A., Maerz, J. C., and Heneghan, L. (2006). The influence of invasive earthworms on indigenous fauna in ecosystems previously uninhabited by earthworms. *Biological Invasions* 8: 1275–1285.

Mueller, C. W., and Dweck, C. S. (1998). Praise for intelligence can undermine children's motivation and performance. *Journal of Personality and Social Psychology* 75: 33–52.

Munafo, M. R., Yalcin, B., Willis-Owen, S. A., and Flint, J. (2008). Association of the dopamine D4 receptor (DRD4) gene and approach-related personality traits: Meta-analysis and new data. *Biological Psychiatry* 63 (2): 197–206.

Murray, C. (2007). Intelligence in the classroom. *Wall Street Journal*, July 10, 2007.

Naour, P. (2009). *E. O. Wilson and B. F. Skinner: A Dialogue between Sociobiology and Radical Behaviorism*. New York: Springer.

Nihei, Y., and Higuchi, H. (2001). When and where did crows learn to use automobiles as nutcrackers? *Tohoku Psychologica Folia* 60: 93–97.

Nisbett, R. E. (2009). *Intelligence and How to Get It*. New York: Norton.

Norcross, J. C., Beutler, L. E., and Levant, R. F. (eds.) (2006). Evidence-based Practices in Mental Health: Debate and Dialogue on the Fundamental Questions. Washington, D.C.: American Psychological Association.

Numbers, R. L. (1976). *Prophetess of Health: A Study of Ellen G. White*. New York: Harper & Row.

Numbers, R. L. (2009). *Galileo Goes to Jail and Other Myths about Science and Religion*. Cambridge, Mass.: Harvard University Press.

O'Brien, D. T., and Wilson, D. S. (2010). Community perception: The ability to assess the safety of unfamiliar neighborhoods and respond adaptively. *Journal of Personality and Social Psychology* 100 (4): 606–620.

Ostrom, E. (1990). *Governing the Commons: The Evolution of Institutions for Collective Action*. Cambridge: Cambridge University Press.

Park, S. H., and Mattson, R. H. (2009). Therapeutic influences of plants in hospital rooms on surgical recovery. *Hortscience* 44 (1): 102–105.

Pearson, C. (2008). *The Gospel of Inclusion: Reaching beyond Religious Fundamentalism to the True Love of God and Self.* New York: Atria.

———(2010). *God Is Not a Christian, Nor a Jew, Muslim, Hindu... God Dwells with Us, in Us, around Us, as Us.* New York: Atria.

Pecina, S., Cagniard, B., Berridge, K., Aldridge, J., and Xhuang, X. (2003). Hyperdopaminergic mutant mice have higher "wanting" but not "liking" for sweet rewards. *Journal of Neuroscience* 23: 9395–9402.

Pederson, E., Faucher, T. A., and Eaton, W. W. (1978). A new perspective on the effects of first-grade teachers on children's subsequent adult status. *Harvard Educational Review* 48: 1–31.

Pederson, J. (1991). *Israel: Its Life and Culture.* Atlanta: Scholars.

Phillips, B. L., Brown, G. P., Webb, J. K., and Shine, R. (2006). Invasion and the evolution of speed in toads. *Nature* 439: 803.

Pigliucci, M., and Muller, G. (eds.) (2010). *Evolution: The Extended Synthesis.* Cambridge, Mass.: MIT Press.

Pinker, S. (1997). *How the Mind Works.* New York: Norton.

———(2002). *The Blank Slate: The Modern Denial of Human Nature.* New York: Viking.

Poundstone, W. (1992). *Prisoner's Dilemma.* New York: Doubleday.

Preziosi, R. F., and Fairbairn, D. J. (1996). Sexual size dimorphism and selection in the wild in the waterstrider *Aquarius remigis*: Body size, components of body size, and mating success. *Journal of Evolutionary Biology* 9: 317–336.

———(2000). Lifetime selection on adult body size and components of body size in a water strider: Opposing selection and maintenance of sexual size dimorphism. *Evolution* 54: 558–566.

Queenan, J. (2009). *Closing Time: A Memoir.* New York: Viking.

Rasmussen, J. (ed.) (2007). *Man in Isolation and Confinement.* Piscataway, N.J.: Aldine.

Richerson, P. J., and Boyd, R. (2005). *Not by Genes Alone: How Culture Transformed Human Evolution.* Chicago: University of Chicago Press.

Rosenberg, R. (1981). *The Concept of Sheol within the Context of Ancient Near Eastern Beliefs.* Cambridge, Mass.: Harvard University Press.

Seaver, J. E. (2008). *A Narrative of the Life of Mrs. Mary Jemison.* San Diego, Calif.: ICON Group.

Seeley, T. D. (1995). *The Wisdom of the Hive.* Cambridge, Mass.: Harvard University Press.

———(2010). *Honeybee Democracy.* Princeton, N.J.: Princeton University Press.

Segerstrale, U. (2000). *Defenders of the Truth: The Battle for Science in the Sociobiology Debate and Beyond.* Oxford: Oxford University Press.

Seligman, M. (1995). The effectiveness of psychotherapy. *American Psychologist* 50: 965–974.

———(2002). *Authentic Happiness: Using the New Positive Psychology to Realize Your Potential for Lasting Fulfillment.* New York: Free Press.

Singer, I. B. (1991). *In My Father's Court.* New York: Farrar, Straus and Giroux.

Skenazy, L. (2009). *Free Range Kids: Giving Our Children the Freedom We Had without Going Nuts with Worry.* New York: Jossey-Bass.

Skinner, B. F. (1948/2005). *Walden Two.* Indianapolis: Hackett.

———(1971/2002). *Beyond Freedom and Dignity.* Indianapolis: Hackett

———(1981). Selection by consequences. *Science* 213: 501-504.

Smith, A. (1776/2010). *The Wealth of Nations.* Hollywood, Fla.: Simon & Brown.

———(1759/1982). *The Theory of Moral Sentiments.* Indianapolis: Liberty Fund.

Smith, C., and Denton, M. L. (2005). *Soul Searching: The Religious and Spiritual Lives of American Teenagers.* New York: Oxford University Press.

Smith, G. R. (2006). *Partners All: A History of Broome County, New York.* Virginia Beach, Va.: Donning.

Sober, E., and Wilson, D. S. (1998). *Unto Others: The Evolution and Psychology of Unselfish Behavior.* Cambridge, Mass.: Harvard University Press.

Sompayrac, L. M. (2008). *How the Immune System Works,* 3rd ed. Hoboken, N.J.: Wiley, Blackwell.

Soros, G. (1998). *The Crisis of Global Capitalism: Open Society Endangered.* New York: PublicAffairs.

Sulloway, F. (1992). *Freud, Biologist of the Mind: Beyond the Psychoanalytic Legend.* Cambridge, Mass.: Harvard University Press.

Teilhard de Chardin, P. (1959). *The Phenomenon of Man.* New York: Collins.

Terkel, S. (1970). *Hard Times: An Oral History of the Great Depression.* New York: Pantheon.

———(1974). *Working: People Talk about What They Do All Day and How They Feel about What They Do.* New York: Pantheon/Random House.

Thaler, R. H., and Sunstein, C. R. (2008). *Nudge: Improving Decisions about Health, Wealth, and Happiness.* New Haven, Conn.: Yale University Press.

Tinbergen, N. (1959/1984). *Curious Naturalists,* rev. ed.. Amherst, Mass.: University of Massachusetts Press.

———(1963). On aims and methods of ethology. *Zeitschrift für Tierpsychologie* 20: 410–433.

Tomasello, M. (1999). *The Cultural Origins of Human Cognition.* Cambridge, Mass.: Harvard University Press.

———(2009). *Why We Cooperate.* Boston: MIT Press.

———(2010). *Origins of Human Communication.* Cambridge, Mass.: MIT Press.

Townsend, A. K., Clark, A. B., McGowan, K. J., and Lovette, I. (2009a). Reproductive partitioning and a test of the assumptions of reproductive skew models in the cooperatively breeding American crow (Corvus brachyrhynchos). *Animal Behavior*: 503–512.

———(2009b). Disease-mediated inbreeding depression in a large, open population of cooperative crows. *Proceedings of the Royal Society Biological Sciences* 276: 2057–2064.

Ulrich, R. S., Simons, R. F., Losito, B. D., Fiorito, E., Miles, M. A., and Zelson, M. (1991). Stress recovery during exposure to natural and urban environments. *Journal of Environmental Psychology* 11: 201–230.

Vaillant, G. E. (1977). *Adaptation to Life.* New York: Little, Brown.

———(1983). *The Natural History of Alcoholism.* Cambridge, Mass.: Harvard University Press.

———(2008). *Spiritual Evolution: How We Are Wired for Faith, Hope and Love.* New York: Harmony.

Veblen, T. (1898). Why is economics not an evolutionary science? *Quarterly Journal of Economics* 12: 373–397.

Von der Lippe, M., and Kowarik, I. (2008). Do cities export biodiversity? Traffic as a dispersal vector across urban-rural gradients. *Diversity and Distributions* 14: 18–25.

Wade, N. (2006). *Before the Dawn: Recovering the Lost History of Our Ancestors.* New York: Penguin.

Walras, L. (1874/2003). *Elements of Pure Economics.* London: Routledge.

Watson, J. B. (1930). *Behaviorism.* Chicago: University of Chicago Press.

Weiner, J. (1994). *The Beak of the Finch: A Story of Evolution in Our Time.* New York: Knopf.

Weisman, A. (2007). *The World without Us.* New York: Little, Brown.

Whewell, W. (1833). *Astronomy and General Physics Considered with Reference to Natural Theology (Bridgewater Treatise).* London: Pickering.

Wilcox, R. S. (1979). Sex discrimination in *Gerris remigis*: Role of a surface wave signal. *Science* 206: 1325–1327.

Wilkinson, J. B. (1840/1967). *J. B. Wilkinson's The Annals of Binghamton of 1840 with an Appraisal, 1840–1967, by Tom Cawley.* Binghamton, N.Y.: Broome County Historical Society and Old Onaquaga Historical Society.

Wilson, D. S. (2002). *Darwin's Cathedral: Evolution, Religion, and the Nature of Society.* Chicago: University of Chicago Press.

———(2007). *Evolution for Everyone: How Darwin's Theory Can Change the Way We Think about Our Lives.* New York: Delacorte.

———(2010). Learning from the Immune System about Evolutionary Psychology. In A. Andrews and J. Carroll (eds.), *Evolutionary Review.* Albany: SUNY Press.

Wilson, D. S., and Csikszentmihalyi, M. (2007). Health and the ecology of altruism. In S. G. Post (ed.), *The Science of Altruism and Health,* pp. 314–331. Oxford: Oxford University Press.

Wilson, D. S., Near, D., and Miller, R. R. (1996). Machiavellianism: A synthesis of the evolutionary and psychological literatures. *Psychological Bulletin* 199: 285–299.

———(1998). Individual differences in Machiavellianism as a mix of cooperative and exploitative strategies. *Evolution and Human Behavior* 19: 203–212.

Wilson, D. S., and O'Brien, D. T. (2010). Evolutionary theory and cooperation in everyday life. In S. Levin (ed.), *Games, Groups, and the Global Good,* pp. 155–168. Berlin: Springer.

Wilson, D. S., O'Brien, D. T., and Sesma, A. (2009). Human prosociality from an evolutionary perspective: Variation and correlations on a city-wide Scale. *Evolution and Human Behavior* 30: 190–200.

Wilson, D. S., and Wilson, E. O. (2007). Rethinking the theoretical foundation of sociobiology. *Quarterly Review of Biology* 82: 327–348.

———(2008). Evolution "for the good of the group." *American Scientist* 96: 380–389.

Wilson, E. O. (1975). *Sociobiology: The New Synthesis.* Cambridge, Mass.: Harvard University Press.

———(1984). *Biophilia.* Cambridge, Mass.: Harvard University Press.

———(1998). *Consilience: The Unity of Knowledge.* New York: Knopf.

———(2010). *Anthill: A Novel.* New York: Norton.

Wilson, S. (1955). *The Man in the Gray Flannel Suit.* New York: Simon & Schuster.

———(1958). *A Summer Place.* New York: Simon & Schuster.

Winthrop, J. (1630/2009). *A Model of Christian Charity.* New York: Evergreen Review.

Wolfe, L. M., Elzinga, J. A., and Biere, A. (2004). Increased susceptibility to enemies following introduction in the invasive plant Silene latifolia. *Ecology Letters* 7: 813–820.

Index

Index

Index

About the Author

David Sloan Wilson is SUNY Distinguished Professor of Biology and Anthropology at Binghamton University. He is widely known for his fundamental contributions to evolutionary science and for explaining evolution to the general public. His books include *Evolution for Everyone: How Darwin's Theory Can Change the Way We Think About Our Lives*; *Darwin's Cathedral: Evolution, Religion, and the Nature of Society*; and *Unto Others: The Evolution and Psychology of Unselfish Behavior* (with Elliott Sober). In addition to his own research, Wilson manages programs that expand the scope of evolutionary science in higher education, public policy, community-based research, and the study of religion.